RENEWABLE-RESOURCE MATERIALS

New Polymer Sources

POLYMER SCIENCE AND TECHNOLOGY

Editorial Board:

William J. Bailey, *University of Maryland, College Park, Maryland*

J. P. Berry, *Rubber and Plastics Research Association of Great Britain, Shawbury, Shrewsbury, England*

A. T. DiBenedetto, *The University of Connecticut, Storrs, Connecticut*

C. A. J. Hoeve, *Texas A & M University, College Station, Texas*

Yoichi Ishida, *Osaka University, Toyonaka, Osaka, Japan*

Fran E. Karasz, *University of Massachusetts, Amherst, Massachusetts*

Oslas Solomon, *Franklin Institute, Philadelphia, Pennsylvania*

Recent volumes in the series:

Volume 20　POLYMER ALLOYS III: Blends, Blocks, Grafts, and Interpenetrating Networks
Edited by Daniel Klempner and Kurt C. Frisch

Volume 21　MODIFICATION OF POLYMERS
Edited by Charles E. Carraher, Jr., and James A. Moore

Volume 22　STRUCTURE PROPERTY RELATIONSHIPS OF POLYMERIC SOLIDS
Edited by Anne Hiltner

Volume 23　POLYMERS IN MEDICINE: Biomedical and Pharmacological Applications
Edited by Emo Chiellini and Paolo Giusti

Volume 24　CROWN ETHERS AND PHASE TRANSFER CATALYSIS IN POLYMER SCIENCE
Edited by Lon J. Mathias and Charles E. Carraher, Jr.

Volume 25　NEW MONOMERS AND POLYMERS
Edited by Bill M. Culbertson and Charles U. Pittman, Jr.

Volume 26　POLYMER ADDITIVES
Edited by Jiri E. Kresta

Volume 27　MOLECULAR CHARACTERIZATION OF COMPOSITE INTERFACES
Edited by Hatsuo Ishida and Ganesh Kumar

Volume 28　POLYMERIC LIQUID CRYSTALS
Edited by Alexandre Blumstein

Volume 29　ADHESIVE CHEMISTRY
Edited by Lieng-Huang Lee

Volume 30　MICRODOMAINS IN POLYMER SOLUTIONS
Edited by Paul Dubin

Volume 31　ADVANCES IN POLYMER SYNTHESIS
Edited by Bill M. Culbertson and James E. McGrath

Volume 32　POLYMERIC MATERIALS IN MEDICATION
Edited by Charles G. Gebelein and Charles E. Carraher, Jr.

Volume 33　RENEWABLE-RESOURCE MATERIALS: New Polymer Sources
Edited by Charles E. Carraher, Jr., and L. H. Sperling

A Continuation Order Plan is available for this series. A continuation order will bring delivery of each new volume immediately upon publication. Volumes are billed only upon actual shipment. For further information please contact the publisher.

RENEWABLE-RESOURCE MATERIALS

New Polymer Sources

Edited by

Charles E. Carraher, Jr.

Florida Atlantic University
Boca Raton, Florida

and

L. H. Sperling

Lehigh University
Bethlehem, Pennsylvania

PLENUM PRESS • NEW YORK AND LONDON

Library of Congress Cataloging in Publication Data

International Symposium on Polymeric Renewable Resource Materials (2nd: 1985:
Miami Beach, Fla.)
   Renewable-resource materials.

   (Polymer science and technology; v. 33)
   "Proceedings of the Second International Symposium on Polymeric Renewable
Resource Materials, sponsored by the Division of Polymeric Materials, held April
28–May 1, 1985, in Miami Beach, Florida"—T.p. verso
   Symposium held at the 189th meeting of the American Chemical Society, Apr.
28–May 3, 1985, Miami Beach, Fla.
   Bibliography: p.
   Includes index.
   1. Polymers and polymerization—Congresses. 2. Natural products—Congresses.
I. Carraher, Charles E. II. Sperling, L. H. (Leslie Howard), 1932–   . III. American
Chemical Society. Division of Polymeric Materials: Science and Engineering. IV.
American Chemical Society. Meeting (189th: 1985: Miami Beach, Fla.). V. Title. VI.
Series.
QD380.I594 1985   668.9   86-8153
ISBN 0-306-42271-9

Proceedings of the Second International Symposium on Polymeric
Renewable Resource Materials, sponsored by the Division of Polymeric
Materials, held April 28–May 1, 1985, in Miami Beach, Florida

© 1986 Plenum Press, New York
A Division of Plenum Publishing Corporation
233 Spring Street, New York, N.Y. 10013

All rights reserved

No part of this book may be reproduced, stored in a retrieval system, or transmitted
in any form or by any means, electronic, mechanical, photocopying, microfilming,
recording, or otherwise, without written permission from the Publisher

Printed in the United States of America

> I will plant in the wilderness the cedar
> the acacia-tree
> and the myrtle
> and the oil-tree;
> I will set in the desert the cypress,
> the plane-tree
> and the larch together;
> That they may see, and know
> and consider
> and understand together,
> That the hand of the Lord hath done this,...

Isaiah, 41:19 and 20 (first portion)

The need to improve our utilization of the Earth's natural resources is every one's business, from every country. This book presents papers from all parts of the world on the subject of making new or improved polymers from renewable resources, be they plastics, elastomers, fibers, coatings, or adhesives. In important ways, this book constitutes part II of an edited work published by Plenum Press in 1983, "Polymer Applications of Renewable-Resource Materials." To that extent, about half of the authors are the same. However, their papers present an update of their research three years later. The other half of the authors are entirely new. Both of these books grew out of symposia sponsored by the Polymeric Materials: Science and Engineering Division of the American Chemical Society. The papers for the present book are based loosely on a symposium held at the Miami Beach meeting in April, 1985.

Unfortunately, interest in polymers from renewable resources fluctuates with the price and availability of petroleum oil. At the time of writing this preface, the price is low, and appears to be headed lower still. While this is a good thing for the economics of the world, it tends to hide the fact that the long-term shortage of oil is very real, and sooner or later will come back to haunt all of us. However, research on all subjects, including that of natural products, yields best results when developed sysematically, with regular support. Let us hope that both low oil prices and high interest in renewable resources may both be true in the future! In any case, all of us must be prepared for state of the art utilization of our natural resources, and the time to begin is now.

The present book is divided into six sections. First, there is a review paper that describes the state of the art in a number of areas of polymers from renewable resources. This chapter tends

to emphasize those aspects of the field not covered by the
original papers that follow. The original papers are grouped into
sections on saccharides and polysaccharides, graft copolymers from
polysaccharides, oils and triglyceride oils, proteins and leather,
and rubber, lignin, and tannin. While each of the above general
categories is well represented in the world of commerce, many
natural products exist for which little or no commercial products
exist. One such group of materials are spider webs. For this
reason, the editors included a special section in their review
paper on this topic in the hope that it may arouse some interest
in the mind of some enterprising soul, and to encourage us to
revisit some of nature's solutions with the intent of "borrowing"
important but yet unused concepts.

The editors wish to take this opportunity to thank all of the
authors for their splendid contributions. Since secretaries from
the four corners of the globe contributed to the production of
this volume, it is difficult to single out one individual to
thank. However, they are remembered here.

Charles E. Carraher, Jr.
L. H. Sperling

January, 1986

SECTION I - REVIEW

MODERN POLYMERS FROM NATURAL PRODUCTS

L. H. Sperling* and C. E. Carraher, Jr.**

*Polymer Science and Engineering Program
Department of Chemical Engineering
and Materials Research Center #32
Lehigh University, Bethlehem, PA   18015

**Florida Atlantic University
Boca Raton, Florida   33431

INTRODUCTION

Modern man was not "born yesterday". At the time man evolved, the
only materials about him were inorganics such as rocks, water, and air,
and what are called today the "renewable resources" or "natural products".
Amazingly, most of these materials are in service still today. These
include wool, cotton, animal glue, and natural rubber. This last was used
by the American Indians long before Columbus. Other natural products
served as important monomers, such as linseed oil; and an early
crosslinker for animal skins was tannin, forming leather. More recently,
natural product polymers were chemically modified, as in the
esterification of cellulose.

A surprising number of 20th century polymers, however, also have
their origins in natural products. Sebacic acid is commercially derived
from castor oil; it forms the "10" component of nylon 6,10. Alkyd paints
are based significantly on triglyceride oils. The point is that even
today, natural products are in wide use. While petrochemicals have
certainly gotten the lion's share of publicity, renewable resources have
continued to be grown, harvested, and used.

Three years ago, the subject of renewable resources was reviewed by
the authors (1), to which the reader is referred for many of the basic
aspects of renewable resources. The present review will update reference
(1) with emphasis on subject matter not extensively covered earlier.
Also, since many of the chapters that follow also touch on the topics to
be reviewed, some effort was made to avoid repetition.

The emphasis of the first review centered largely on the chemical
nature of components derived from natural sources. The present review
continues this but also describes the use of raw, chemical mixtures that
can give products with unique combinations of properties arising from the
product's structural nature.

Before proceeding with specific subject matter, other reviews must be mentioned (2-16). While most of these references are in the chemical literature and are well-known to polymer scientists, reference (6) is in the biological literature. However, it contains a wealth of information about renewable resources, especially their mechanical behavior.

FUNCTIONAL GROUPS

An important theme for natural products chemistry is the general similarity between the kind of reactions possible with synthetic feedstock and natural feedstock. For smaller natural chemicals, this similarity is especially close while for biomacromolecules this similarity may be moderated by intra- and intermolecular effects and steric factors.

Most of the common functional groups employed by the synthetic chemists are present in natural products. Table I contains a listing of a number of these functional groups and an example or two illustrating natural products possessing these functional groups.

Nature also offers a number of ring systems that can be exploited in ring-opening polymerizations (Table II).

TREES

The ultimate source of all life and all natural products is the sun. Major beneficiaries of the sun and the associated conversion of carbon dioxide and water into saccharides (photosynthesis) are trees. Trees are woody, perennial plants that contain a stem that remains from year to year. As a tree grows from a seed, it developes separate, but interrelated parts-roots, stem and crown. The major photosynthesis centers are the leaves. The stem or trunk contains, from outside to center, the bark, wood and pith, wood being the major constitutent. The root beneath the earth's surface, which acts as an anchor and draws water and mineral mutrients from surrounding soil for transport to the remainder of the tree.

About one-third of the earth's area, almost 10 billion acres, are occupied by trees. Research in forestry has allowed the development of trees that grow five to ten feet yearly and which can be harvested for lumber and related products within 10 to 20 years. The total wood reserve is about four trillion cubic feet with only about 1% harvested yearly.

Major commercial uses of cut trees are as lumber, pulpwood, veneer logs, posts, pilings, and chemical wood (including particle board).

Wood consists largely of cellulose and lignin, both polymers. The wood cells are arranged longitudinally in the stem, shaped as elongated tubes, actually hollow fibers, giving a tough, flexible but light-weight fiber. Cells vary in exact structure and form depending mainly on the designated function of the particular cell (Figure 1).

Cells are of two major kinds. Food storage cells, parenchyma cells, are short lived, remain alive for about one year. Prosenchyma cells act as suppor and conduction and usually lose their protoplasm in the year they are formed. The major portion of wood consists of these prosenchyma and parenchyma cells.

Table I. Functional Groups Present In Nature.

Name	Structure
Vanillin	4-hydroxy-3-methoxybenzaldehyde structure
Sucrose	glucose-fructose disaccharide structure
Muscone	macrocyclic ketone structure
n-Butyl mercaptan	$CH_3CH_2CH_2CH_2SH$
Poison Ivy Urushiol	catechol with $(CH_2)_7-CH=CH-CH_2-CH=CH-(CH_2)_2-CH_3$ side chain
Diallyl disulfide	$CH_2=CH-CH_2-S-S-CH_2-CH=CH_2$
3,5-Dibromotyrosine	dibromo-hydroxyphenyl amino acid structure
p-Methylnitrosoaminobenzaldehyde	methylnitrosoamino benzaldehyde structure
Chloramphenicol (Chloromycetin)	NO_2-phenyl-$CH(OH)-CH(CH_2OH)-NH-C(=O)-CHCl_2$ structure

Table I. Functional Groups Present In Nature (cont.)

Name	Structure
Norcardamine	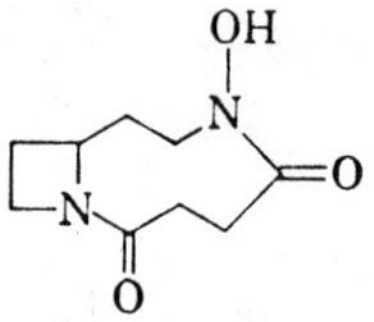
Firefly Luciferin	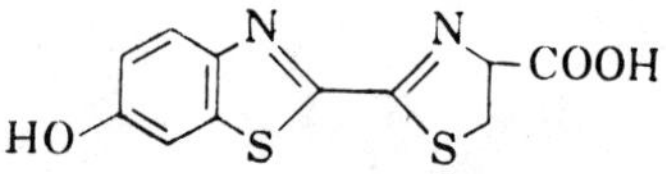
Vitamin K_2	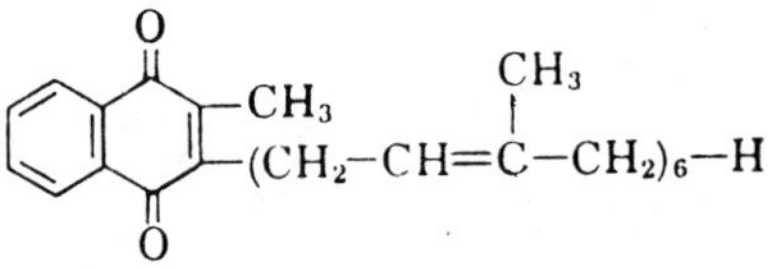
Junipal	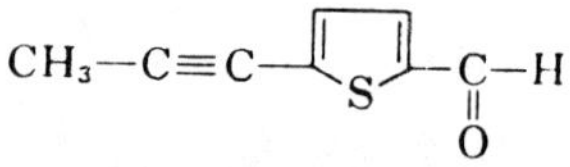
Matricaria ester	$CH_3—CH=CH—(C\equiv C)_2—CH=CH—\underset{\underset{O}{\Vert}}{C}—OCH_3$
Reserpine	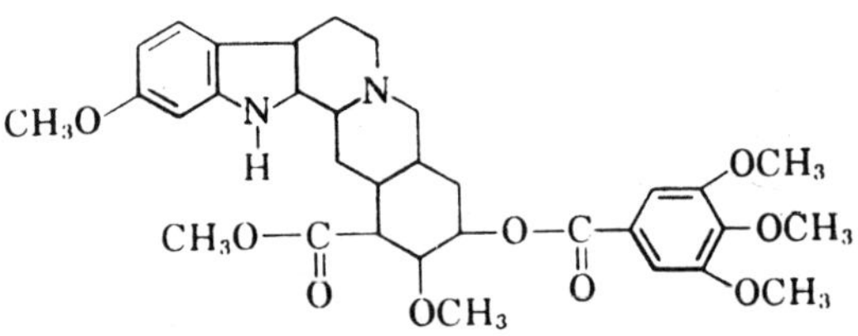

Table II. Selected Small and Strained Heteroatomic Natural Occuring Rings.

Name	Structure
Linaloöl epoxide	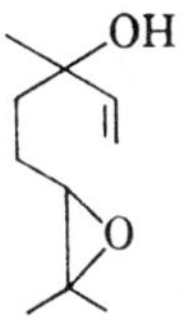
Scopolamine	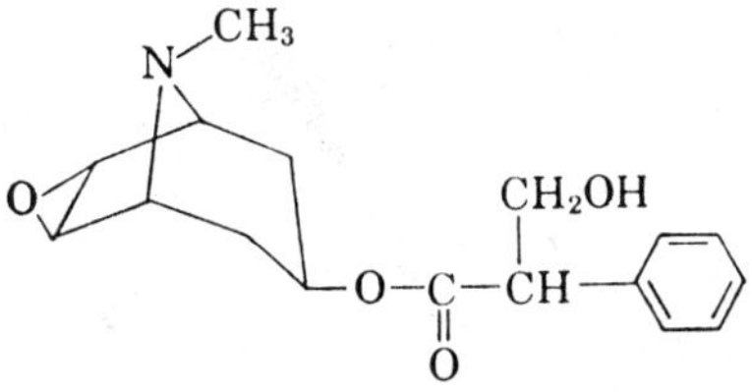
Azetidine-2-carboxylic acid	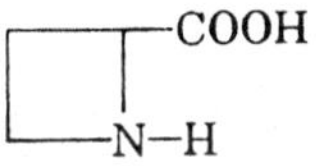
Penicillin R=benzyl, *p*-hydroxybenzyl, *n*-amyl, 1-pent-2-enyl	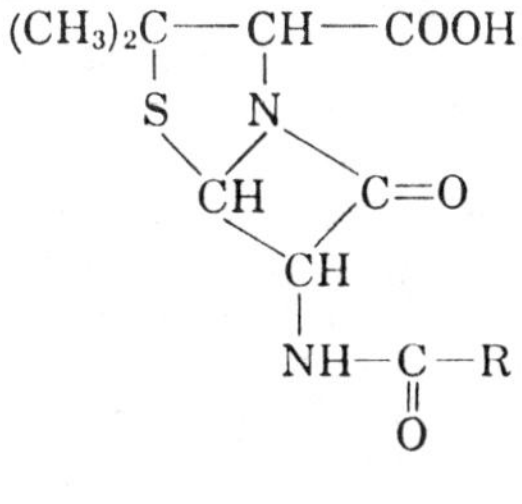
Picrotoxin n	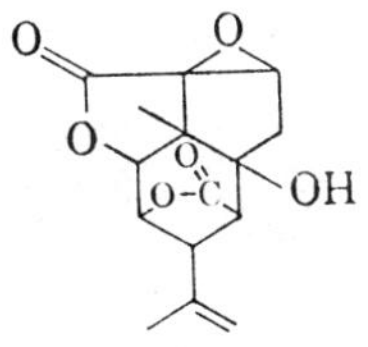

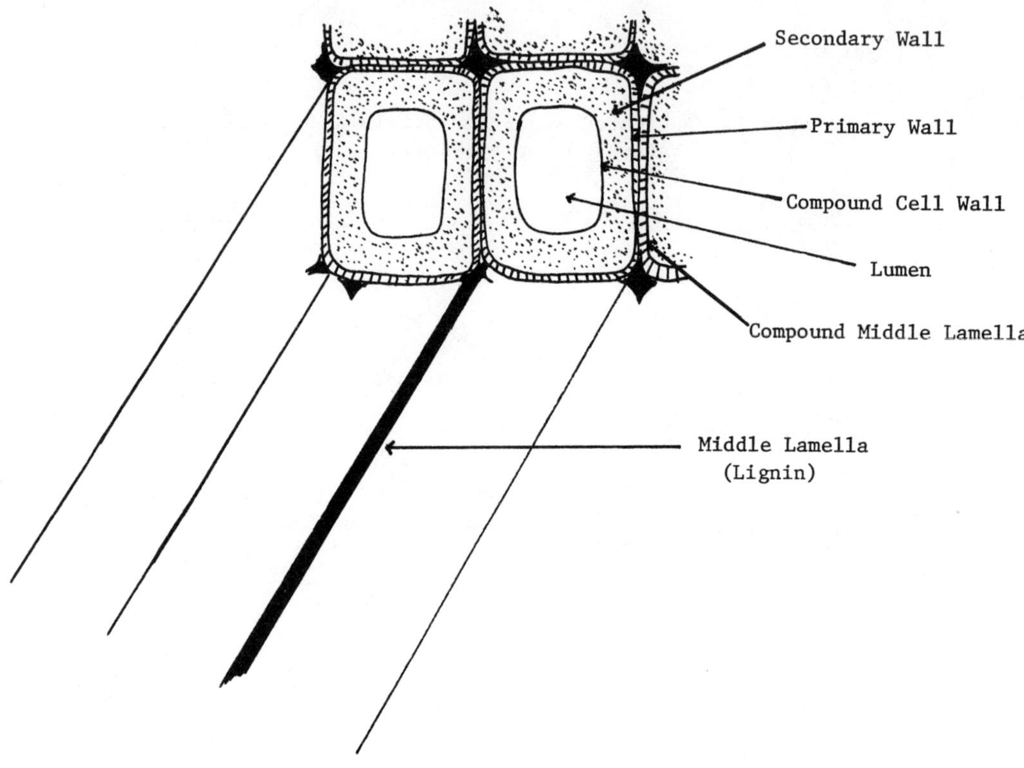

Figure 1. Cross-section of wood, illustrating
cellular structure.

Chemically, trees are truly composites containing a variety of
chemicals, but are largely macromolecular (excepting water). The general
chemical nature of wood is described in Table III. The cell walls are
composed of mainly lignin and polysaccharides and in turn the
polysaccharies are mainly cellulose (glucose basic unit) and a variety of
hemicelluloses derived from pentoses (xylose and arabinose) and hexoses
(glucose, mannose and galactose). Acetic acid, uronic acids, and
methoxyuronic acids also are derived from the hydrolysis of the
hemicelluloses. The extent of each component varies as to the tree age,
particular climatical history, and location and type of wood cell.
Table IV contains a brief listing of typical values for selected trees.
Thus, trees are a ready source of removable natural products, both in
itself and in the nature of its components. Much research is continuing
in taking advantrage of this bountiful, complex natural resource. Wood
has been modified utilizing free radical, redox, ionic and radiation
methods introducing a wide variety of monomeric and polymeric materials.
Composites, grafts, IPN's, etc. have been formed. Still much remains to
be done to take advantage of secondary, tertary and quartinary structures
offered by wood.

Just as the chemical composition of wood is quite variable, so also
are its physical properties. Even so, general values can be given and
found in Table V. Wood is relatively light, varying from about 0.3 g/cc
for western red cedar to 0.75 g/cc for osage orange for common USA trees.
Because of its highly porous nature, wood possesses a large surface area
for chemical modification. Surface areas may reach to 100 to 300 m^2/gram.

Table III. Chemical Composition of Wood by Type of Material.

             Saccharides -- cellulose
                            hemicellulose
                            pectins
                            starch
                            arabinoglactans

         Acids   -- fatty acids

         Alcohols   -- eliphatic alcohols
                            sterols

         Proteins

         Phenols   -- lignin
                            phlobaphenes
                            tannins

         Terpenes   -- terpene
                            terpenoids
                            resin acids

         Inorganic salts and oxides

         Alkaloids

This large surface area also allows for ready swelling by acid and base
solutions allowing the breakage of hydrogen bonding, reshaping, and
subsequent neutralization and reformation of hydrogen bonds locking in the
new structure. It must be noted that most of these physical properties vary
with the direction with which the value is obtained, i.e., along or against
the grain, etc.

Shrinkage is a problem for wood products but this can be largely
controlled through surface treatment. Dry wood is an excellent electrical
and thermal insulator with an extremely low coefficient of linear expansion.
Wood also has a high tensile strength comparable to cotton fibers and
greater than many metals. Thus, wood itself offers a number of properties
that may be usefully incorporated in a modified product.

Plain old wood constitutes the basic material for building homes,
making furniture, and even garden stakes, etc. According to Jeronimidis
(17), wood has a cellular composite structure with four levels of
organization, molecular, fibrillar, cellular, and macroscopic. Wood
contains about 40-50% cellulose by dry weight. The primary organization of
the cellulose is in the form of microfibrils. These are wound around the
cells, or tracheids, that make up the larger part of wood in a complex
helical fashion, see Figure 1 (18). The S_2 wall makes up to 80% of the
total cell wall area. Because of its low microfibrillar angle, it is the
major load bearing component in wood.

Jeronimidis (17) points out that cellulose in wood has a very high
theoretical modulus, 250 GPa. Because of amorphous material, lignin, and
pores, the actual modulus of wood is nearer 10 GPa, see Figure 2 (17).
However, this value remains higher than that of polystyrene at room
temperature, 3 GPa, pores and all. However, wood can be improved upon
significantly. Wood impregnated with poly(methyl methacrylate) has
sufficient environmental resistance to be used for knife handles. Such
compositions survive many years of daily washing in hot, soapy water!

Table IV. Average Saccharide Composition of Selected
Woods (percentage composition–dry weight).

	Glucan	Mannan	Galactan	Xylan	Arabinan	Uronic Anhyride	Acetyl
Sugar Maple	52	2	.1	15	.8	4	3
White Elm	53	2	.9	11	.6	4	4
White Pine	45	11	2	6	1	4	1
Douglas Fir	44	11	5	3	3	3	.8
Eastern Hemlock	45	11	1	4	.6	3	2
White Spruce	47	12	1	7	2	4	1
Loblolly Pine	45	11	2	7	2	4	1
White Birch	45	2	.6	25	.5	5	4
Beech	48	2	.8	16	.4	3	3
Trembling Aspen	57	2	1	18	.5	5	4

Table V. Typical Physical Values for Wood.

Bulk Resistivity	3×10^{17} to 3×10^{18} ohm-cm (dry)
	10^9 to 10^{10} ohm-cm (10% moisture)
Dielectric constant	1.5 to 3.0
Power-Loss factors	2.5×10^{-2} to 8.5×10^{-2}
Specific heat	0.3 cal/g °C (dry)
Thermal conductivity	2×10^{-4} to 4×10^{-4} (cal/sec cm^2)
	(°C/cm)
Modulus of elasticity (1000 psi)	1,200 to 2,000
Modulus of rupture (psi)	7,000 to 16,000
Shear parallel to grain-maximum shearing strength (psi)	700 to 2,400
Fiber stress proportional limit (psi)	5,200 to 9,500
Tension perpendicular to grain – maximum crushing strength (psi)	180 to 800
Compression parallel to grain – maximum crushing strength (psi)	4,500 to 8,500
Compression perpendicular to grain – fiber stress at proportional limit (psi)	380 to 2,100
Compression parallel to grain – fiber stress at proportional limit (psi)	3,600 to 6,200
Tensile strength (fibers)	3,300 to 10,000 kg/cm^2

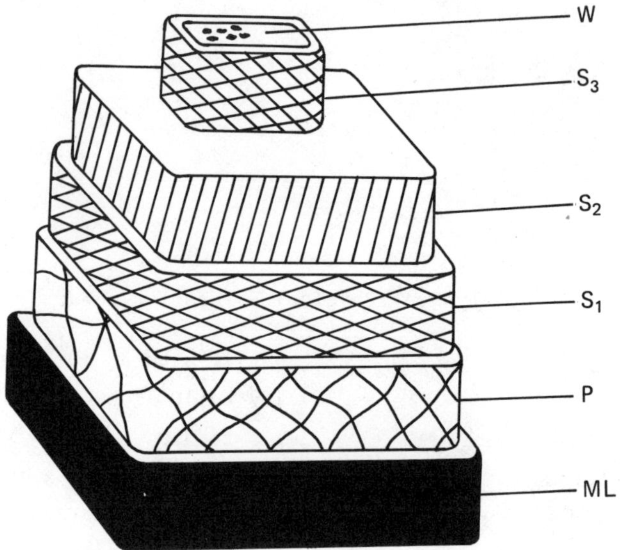

Figure 2. Schematic representation of the tracheid cell structure and surrounding lignin. ML, lignin; P, primary wall; S_1, S_2, and S_3, cell wall layers of cellulose. Lines indicate orientation directions (18).

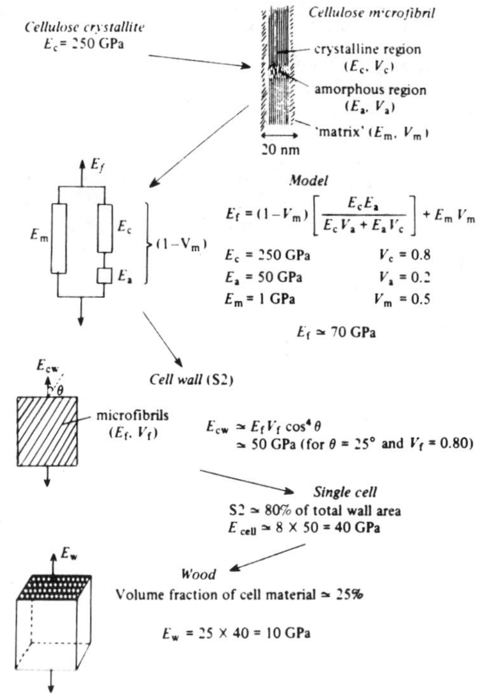

Figure 3. Modulus of cellulose and wood (17).

POLYSACCHARIDES

The most important renewable resource polymers are the polysaccharides. The most widely used are cellulose and starch. However, other
important polysaccharides include amylopectin, glycogen, chitin, carragee-
nan, pectic acid, dextran, and agar. Important chemical derivatives of
cellulose include the nitrate, the acetate and higher esters, and ethers.

Cellulose

Cellulose occurs throughout the plant world. However, much of the
industrially utilized cellulose is derived from wood, recovered by pulping.
After pulping, much wood cellulose goes into the manufacture of paper.
Newspapers, books, and magazines are all based on a composite of cellulose,
filler, and binder. Here, the polymer additive to paper should not be
underestimated. Long ago, paper was "sized" with starch or animal glue to
fill up the pores. Without such additives, blotting paper is formed, which
cannot be written on. Today, synthetic polymers, often in latex form, are
used to fill up the pores, which improves the writing quality of the paper.

Another major source of cellulose is cotton, which is used widely for
clothing. While cotton makes excellent clothing in the pure form, a section
of this review (below) will be devoted to fiber blends, which often employ
cotton as one of the components.

Cellulose can be reacted with carbon disulfide and dissolved in sodium
hydroxide solutions producing a solution known as viscose. On regeneration
in an acid bath, products known as rayon or cellophane result, depending on
whether a fiber or a film is made. Interestly, because cellulose has three
hydroxyl groups on each glucoside residue, a variety of products can be made
from each chemical derivative. For example, amorphous cellulose diacetate
makes a clear plastic, while crystalline diacetate makes a clear plastic,
while cellulose triacetate makes quality fibers (known under trade names
such as "Arnel" in the U.S.).

Since cellulose, in its various forms, constitutes about half of all of
the polymer consumed industrially in the world, its importance is obvious.
While important research on methods of improving cellulose are continuing
(several teams are reporting their recent research in this book), the
quantity of research being carried out remains far less than the importance
of the product. All too often, cellulose is treated as "arrived", with
little attention paid to methods of improving the product.

Starch

While cellulose is indigestible by humans, only forming roughage,
starch forms a major part of the human diet. To grow renewable resources to
eat remains a noble profession! However, an increasing number of
applications of starch and starch derivatives are being found. Some of the
derivatives of starch are shown in Table VI. Item No. 4 is of special
interest, because sodium carboxymethyl cellulose, CMC, is also used for
soluble laundry detergent bags. The idea is to have a film-forming polymer
that rapidly dissolves in water. The polymer should be part of the
formalation serving to prevent oils from redepositing on the surface of the
fabrics being cleaned; here, ionic or electrostatic forces are apparently
important.

Table VI. Derivatives of Starch

No.	Description of Starch or Starch Derivative	Real or Potentional Uses	Reference(s)
1	Glycol glucosides	Urethane foams, surfactants, alkyd resins	1-5
2	Combination with poly(ethylene-co-acrylic acid) and polyethylene	Biodegradable materials as mulch film	6
3	Substitute for carbon black	Processing, production rubber	7
4	Substitute for poly(vinyl alcohol)	Water soluble laundry detergent bags	8
5	Starch encasement of rubber droplets (powdered rubber) in latex	Assist in manufacture of rubber	9
6	Insoluble starch xanthates	Removal of heavy metal contaminants from industrial process water	10
7	Graft starch products (including "Super Slurper")	Absorption of body fluids; establishing trees, industrial thickening agent; controlling forest fires; removal of water in fuel alcohol mixtures	11-13

References

1. F. H. Otey, B. Zagoren and C. Mehtretter, *Ind. Eng. Chem.*, *Prod. Res. Dev.* 2:256 (1963).
2. R. Leitheser, C. Impola, R. Reid, and F. H. Otey, *Ind. Eng. Chem.*, *Prod. Res. Dev.*, 5:276 (1966).
3. W. McKillip, J. Kellen, C. Impola, R. Buckney and F. H. Otey, *J. Paint Techn.* 42:312 (1970).
4. F. H. Otey, R. Westhoff and C. Mehltretter, *J. Cell. Plast.*, 8:156 (1972).
5. P. E. Throckmorton, R. Egan, D. Aelony, C. Mulberry and F. H. Otey, *J. Am. Oil Chem. Soc.*, 51:486 (1974).
6. F. H. Otey, R. Westhoff and W. Doane, *Ind. Eng. Chem.*, *Prod. Res. Dev.*, 19(4):592 (1980).
7. R. Buchanan, W. Kwolek, H. Katz and C. R. Russell, *Staerke*, 23(10): 350 (1971).
8. F. Otey, A. Mack, C. Mehltretter and C. R. Russell, *Ind. Eng. Chem.*, *Prod. Res. Dev.*, 13:90 (1974).
9. T., Abbott, W. M. Doane and C. R. Russell, *Rubber Age*, 195(8):43 (1973).
10. R. Wing, L. Navickis, B. Lasberg and W. Rayford, EPA Final Report EPA-600/2-78-085 (1978).
11. M. Weaver, E. B. Bagley, G. F. Fanta, and W. M. Doane, *Appl. Polym. Symp.*, 25:97 (1974).
12. E. B. Bagley, G. F. Fanta, R. C. Burr, W. M. Doane and C. R. Russell, *Polym. Eng. Sci.* 17(5):311 (1977).
13. C. L. Swanson, G. F. Fanta, R. G. Fecht and R. C. Burr, "Polymer Applications of Renewable-Resource Materials", C. E. Carraher and L. H. Sperling, Eds., Chapt. 5, Plenum, NY, 1983.

TRIGLYCERIDE OILS AS MONOMERS

Fats and oils constitute another major class of natural products, see
Table VII. Unsaturated triglyceride oils may have been among the first
monomers polymerized by man. Major commercial products are shown under
items 8 and 9. Originally, triglyceride oils were mixed with driers, and
painted on walls. The driers are actually free radical catalysts; in the
presence of oxygen, the double bonds in the triglycerides become linked
together to form network structures. For oils such as linseed or tung, the
process results in a tight, long lasting, leathery product. In commercial
products, fillers raise the modulus and abrasion resistance, and colorants
add esthetic value.

Castor oil contains an hydroxyl group half-way down each chain. On
reaction with isocyanates such as TDI, important commercial elastomers can
be made. On formation of the IPN's, item 7, tough plastics can be also
prepared.

THE HAUNTED HOUSE

So far, this review has emphasized well-known commercial materials, or
research likely to become commerical in due course. However, there are many
more renewable resource materials from which to choose. There are two
reasons such natural materials should be studied, although these reasons are
not always obvious: (1) The modern polymer scientists can learn to simulate
Mother Nature, and solve problems using her materials as models, and (2)
Many materials exist all around us that might make superior products, but
are rarely considered and are poorly research. Good illustrations of this
are the common spider web and slime.

Spider Webs

Spider webs exhibit great tensile strength, high elasticity, and have
obvious excellent weatherability. Yet, few people have seriously considered
spider webs as realistic engineering fibers (19-26).

Long ago, man learned how to make a delicate cloth from the cocoon of a
caterpillar that grows on mulberry bushes, Bombyx mori. Of course, that
product is silk. But many other insects produce silk, most notably the
spiders. Like the silkworm, the web material of spiders is proteinaceous
and chemically belongs to the fibrins (19). The molecular weight of silk is
commonly $2\text{-}3\times10^{5}$ g/mol. During spinning, a water soluble liquid protein
solution is transformed into an insoluble solid silk thread by orientation
associated with the forces of extrusion. Among other changes,
 intramolecular hydrogen bonds are rearranged into intermolecular bonds, and
the final material becomes water insoluble.

Protein Constituents

Chemically, these macromolecules are high in the short-side-chain amino
acids glycine and alanine, see Table VIII (19). This fact stands out
particularly in view of modern theories of strengths of polymers and fibers
particularly. It is the backbone portion of the chain that supports applied
load; side chains only dilute the system. In addition, long side chains may
actually weaken the polymer further by reducing the effectiveness of
intermolecular hydrogen bonds which transmit the load because then the
chains are physically separated.

The spinnerets of spiders are also of interest. Most spiders possess
three pairs of spinnerets on its abdomen, see Figure 4 (19). The spinning
glands terminate in little spigots on the surface of each spinneret, see

Table VII. Polymeric Products Containing Fats and Oils

No.	Description	Use(s)–Real or Potential	Reference(s)
1	Acetals of aldehyde esters	Plasticizers	1,2
2	Fatty aldehydes (from ozonolytic cleavage of unsaturated fats; or by addition of CO and hydrogen to unsaturated fatty acids by hydroformylation)– convert aldehyde group to acetals, alcohols, acid, amines	General versatile feedstock	3-5
3	Cyclic fatty acids (from alkali treatment of linseed oil and fats or soybean oil soapstock)	Lubricants, in alkyd resins	6-10
4	Polycarboxylic acids	Lubricant, plasticizer,	5,11
5	Epoxidized soybean oil polymerized in situ	Plasticizer/stabilizer for vinyl plastics	12
6	Products from aldehydic acids and formaldehyde	Coatings, plasticizer,s lubricants	13,14
7	Interpenetrating polymer networks based on castor oil	Tough plastics and reinforced elastomers	15-17
8	Free fatty acids plus multifunctional acids and alcohols, polymerized	Alkyd resins	18
9	Crosslinked linseed and tung oils	House paint	19

References

1. E. H. Pryde, D. Moore, J. Cowan, W. Palm, and L. Witnauer, _Polym. Eng. Sci._, 6:60 (1966).
2. R. Awl, E. Frankel, E. H. Pryde, and J. Cowan, _J. Am. Oil Chem. Soc._, 49: 222 (1972).
3. P. Throckmorton and E. H. Pryde, _J. Am. Oil Chem. Soc._, 49:643 (1972).
4. E. H. Pryde and J. Cowan, "Topics in Lipid Chemistry", (F. Gunston, Ed.), Wiley, NY, Chapt. 1, 1971.
5. E. Frankel and E. H. Pryde, _J. Am. Oil Chem. Soc._, 54:873A (1977).
6. J. Friedrich. E. Bell and L. Gast, _J. Am. Oil Chem. Soc._, 42:643 (1965).
7. J. Friedrich and R. Beal, _J. Am. Oil Chem. Soc._, 39(12):528 (1962).
8. W. R. Miller, H. Teeter, A. Schwab, and J. Cowan, _J. Am. Oil Chem. Soc._, 39:173 (1962).
9. E. Bell and L. Gast, _J. Coat. Technology_, 50(636):81 (1978).
10. R. Beal, L. Lauderback, and J. Ford, _J. Am. Oil Chem. Soc._, 52:400 (1975).
11. W. Kohlbase, E. Frankel, and E. Pryde, _J. Am. Oil Chem. Soc._, 54:506 (1977).
12. C. Nevin, B. Moser, _J. Appl. Polymer Sci._, 7:1853 (1963).
13. D. J. Moore and E. Pryde, _J. Am. Oil Chem. Soc._ 45:517 (1968).
14. W. R. Miller and E. Pryde, _J. Am. Oil Chem. Soc._, 55:469 (1978).

15. L. H. Sperling, J. A. Manson, Shahid Qureshi, and A. M. Fernandez, _I&EC Products Research and Development_, 20:163 (1981).
16. N. Devia, J. A. Manson, L. H. Sperling, and A. Conde, _Macromolecules_, 12(3):360 (1979).
17. L. H. Sperling, J. A. Manson, and M. A. Linne, _J. Polym. Mat._ 1:54 (1984).
18. P. Nylen and E. Sunderland, in "Modern Surface Coatings", Wiley-Interscience, New York, 1965.
19. D. H. Parker, "Principles of Surface Coating Technology", Wiley, New York, 1965.

Table VIII. Amino acid composition (%) of the Silk
of the Araneus Diadematus Spider (19)

Amino acid	Total Web	Frame (ampullate gl.)	Cocoon (tubular gl.)	Attachment disc (piriform gl.)
Alanine	27	33	25	29
Glycine	20	24	12	25
Serine	5	6	19	5
Glutamic	9	18	14	15
Proline	13	2	4	5

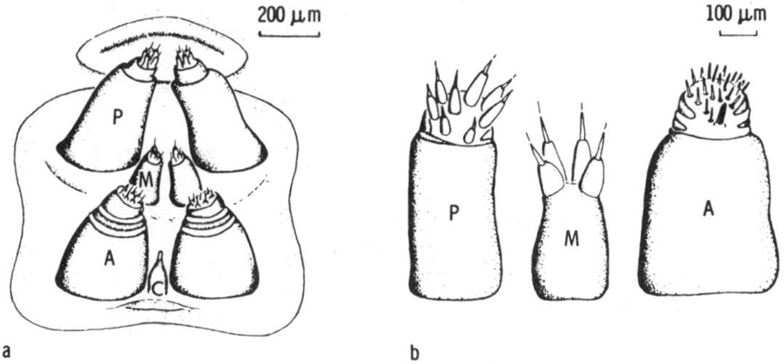

Figure 4. Spinning apparatus of Segestria senoculate spiders. (a)
Relative positions of the three pairs of spinnerets and
colulus. (b) Detail of structure – The different spigots
belong to different kinds of silk glands (19).

17

Figure 4b for detail. Most interestingly, spiders can spin different kinds
of threads.

   A typical spider orb-web is illustrated in Figure 5 (24). Various
kinds of silk are located in different portions of the web. The frame and
radii of the web derive from the large ampullate gland. The viscid spirial
material, the part of the web used to catch insects, comes from the flagel-
liform glands. It is coated with a layer of glue-like material in the form
of microscopic beads.

   Spider web silks are normally partly crystalline and highly elastic.
The crystalline portion of these macromolecules are arranged in antiparalle-
l pleated sheets, a form of the folded chain lamellae familiar to the
physical polymer scientist. The fibers are three to five microns in
diameter, with tensile strengths up to 1.8×10 dynes/cm^2, or 2.3×10 lb/in^2
(20-22). For comparison, Kevlar has a tensile strength of 3.8×10 lb/in^2
(23). Thus, spider webs have tenacities higher than all other natural
fibers and virtually all man-made fibers (20-22).

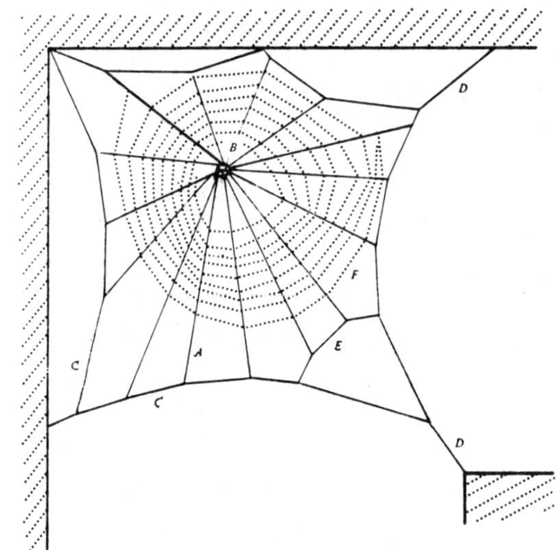

Figure 5. A typical spider orb-web. A, Radii;
          B, hub; C, frame; D, frame attached
          to surrounding structures; E, cord;
          F, viscid spiral. Insects are caught
          by the viscid spirial portion, which
          is covered with sticky droplets (24).

Mechanical Behavior

   Stress-strain curves for various silks are shown in Figure 6 (25).
Like their synthetic counterparts, they vary from rather glassy, Bombyx
mori, to rubbery for the viscid portion of spider webs. A detailed stress-
strain curve for frame silk is shown in Figure 7 (14). Like many polymers,
it appears stiffer at higher rates of extension. The rubber-elasticity
characteristics resemble the segmented polyurethanes in behavior (for
example Spandex^R or Lycra^R).

   Gosline, et al. (26) applied the thermoelastic equation of state (27)
to these materials, and calculated the energetic component of the stress,
f_e/f, see Table IX. For all of these materials, the major component of the
retroactive stress arises through entropic considerations.

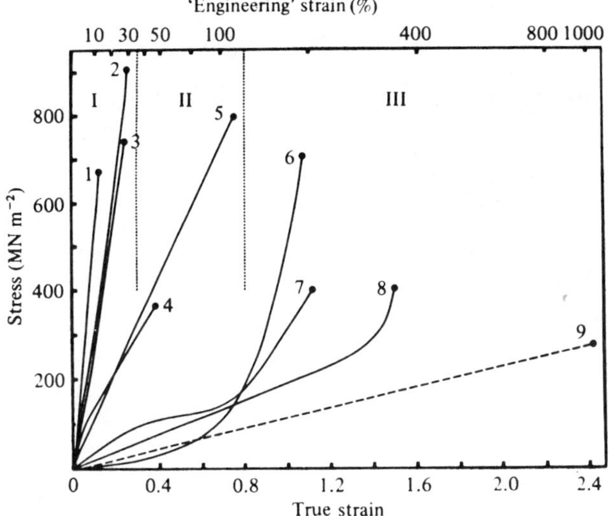

Figure 6. Mechanical behavior of assorted silks. 1, Anaphe
moloneyi; 2, Araneus seratus dropline; 3, Bombyx mori
cocoon; 4, A. Diadenatus cocoon, 5, Galleria mellonella
cocoon; 6, A. sericatus viscid; 7, Apis mellifora larval;
8, Crysopa carnea egg stalk; 9, Meta reticulata viscid.
Note that the viscid portions of the spider web exhibit
rubber elastic behavior (25).

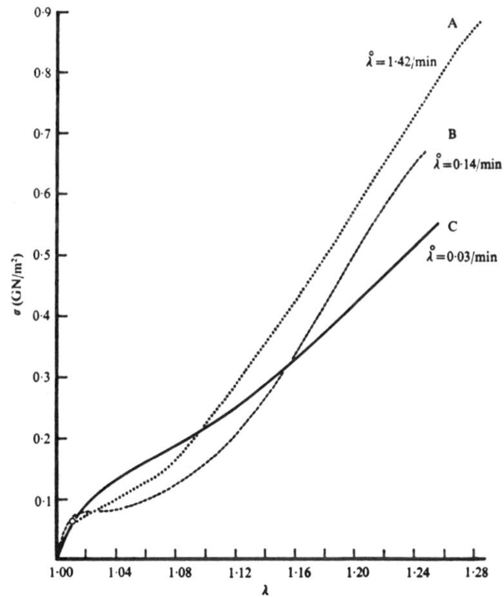

Figure 7. Rubber-elastic behavior of the frame silk of A.
sericatus spider as a function of rate of extension (24).

Table IX. Internal Energy Components of Elastic Force

in Natural and Synthetic Elastomers (26)

Material	fe/f
Elastin	0.26
Natural rubber	0.18
cis-1,4-polybutadiene	0.13
poly(dimethyl siloxane)	0.19
Dragline spider silk	0.14

Denny (24) also studied the stress relaxation behavior of frame spider silk, Figure 7. The rapid initial decay, followed by substantially no stress relaxation is characteristic of crosslinked elastomers. In this case, the crosslinks are provided by the crystalline regions. (It must be emphasized that the reading of Denny's paper (24) will make former Tobolsky student's hearts swell.) Denny concluded that both viscid and frame silks showed a breaking stress of approximately 1 GN/m^2. The breaking extension ratio of viscid silk was at 300%, while that of frame silk was at 125%. Viscid silk exhibits a lower modulus. He further commented that the physical properties of the viscid and frame silks allow them to function effectively as shock absorbers and structural elements, respectively.

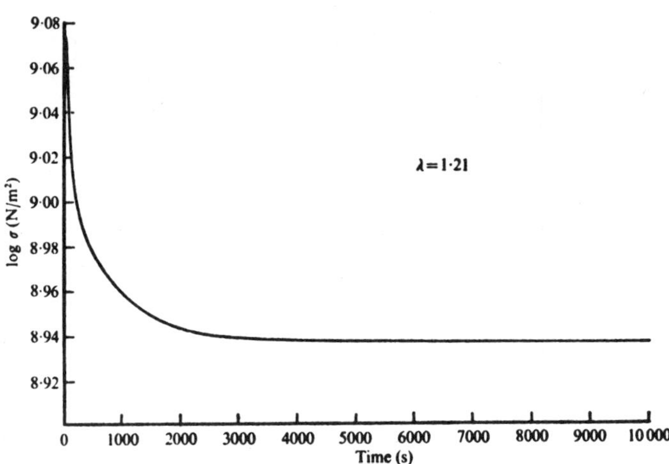

Figure 8. Stress-relaxation behavior of the frame silk of the A. sericatus spider. After a short period of relaxation, the stress is held substantially constant. This behavior is the same as for synthetic elastomers (24).

<u>Applications and Manufacture</u>

Calvert (28) states that in the eighteenth century one M. Bon of
Montpellier made himself stockings and gloves from spider silk, but laments
that it didn't catch on commercially. One may consider how such an operation might be made to work economically in the 1980's.

One might imagine a spider house, with appropriate construction so as
to facilitate web spinning, see Figure 3. The spider house would be
equipped with a paddle-shaped combine collection wheel, operated automatically at the appropriate time intervals to collect the webs. Species of
spider exist that hide in the corner of the web, out of the path of the
collection wheel.

Of course, a fly factory would be required to feed the spiders. The
food can be left over garbage, manure, or other wastes. These are very
inexpensive. The mixed wastes, in trays, would be put first in a "breeder
house" where fly eggs would be deposited. When the adult flies are about to
emerge, automatic equipment would move them to the spider house. When all
the larvae are gone, the waste would be disposed of as before. Such an
arrangement would require minimal space, actually help the environmental problems, and produce web at a low cost. Major actual costs would be
for the automatic equipment. Since further processing would require
equipment now in use in existing industries, ready and presumably
inexpensive transformation to spider-silk end-use products might be
possible.

It must be emphasized that spider webs possess excellent environmental and weathering characteristics besides their strength and elasticity
detailed above. Of course, the sticky droplets on the viscid portion of the
web may have to be removed by washing in organic solvents. On recovery, an
outstanding adhesive may result, as attested to by a carefully controlled
fly population.

<u>The Viscoelasticity of Mucus</u>

"If it is love that makes the world go around, then it is surely mucus
and slime which facilitate its translational motion," R. H. Pain (29). The
mucous glycoproteins have molecular weights of at least 10 Daltons, and a
carbohydrate content of at least 50%, the remainder being protein. The
carbohydrate forms rather short chains, 2-18 residues long, which are
attached in large numbers to a central core of polypeptide.

According to a molecular model described by Pain (29), the mucous
glycoprotein molecule has a spheroidal shape, see Figure 9. The viscosity
of water solutions decreases rapidly with increasing salt concentration. In
the absence of salt, water solutions of 20-30 gm/l will gel. The gelation
mechanism is thought to involve a degree of interdigitation between vicinal
molecules.

Sources of mucus include earthworm surfaces, and slugs provide a pedal
mucus upon which they progress. Pain (29) comments that fish and eels are
covered with the material, the latter rather liberally.

The special lubricant properties of mucus originate in the spherical
nature of the molecular architecture. Literally, slugs and snails move on
spherical biological ball bearings! Potential applications include
specialty lubricants, reversible adhesives, and eye-ball fluid replacements.

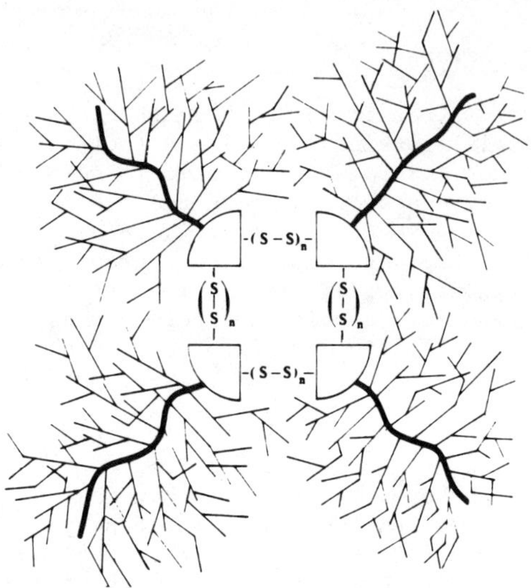

Figure 9.   Spheroid of nature of the gastric mucous glycoprotein molecule. Each of the four disulfide
bridged units has a molecular weight of 5 x
10^5 gm/mol (29).

FIBER BLENDS

A visit to any clothing store will convince even the most hardened
purist of the importance of fiber blends. Many of these blends involve a
natural fiber such as wool or cotton, and a synthetic fiber, see Table X.
Each fiber in a blend contributes its particular characteristics to the
properties of the whole. To obtain a synergism, the good features of both
must dominate, while the poor qualities of each remain somehow hidden.

Wool imparts warmth to its blends because its crimped fiber structure
traps microscopic air pockets. The insulating characteristics follow from
low heat transport coefficients. Because of the hollow nature of cotton
fibers, they "breathe" well. In this case, breathing refers to a high rate
of water vapor transport. Fibers such as polyester and nylon contribute
high strength and abrasion resistance. It must be emphasized that clothing
transmits heat through a combination of conductive and radiative mechanism
(30). Convective heat transfer is negligible, even in light clothing.

A particularly simple yet interesting fiber blend of glass and cotton,
was described by Graham and Ruppenicker (31), see Table X. This blend is
intended for outdoor fabrics such as tentage and tarpaulin. Cotton fabrics
have high wet strength, natural water resistance, and good breathability. A
major disadvantage of cotton resides in its low tensile strength compared
with synthetic fibers.

Glass fibers possess the required high strength, and also offer natural
resistance to environmental degradation from sunlight, mold, and mildew. A
major disadvantage arises from the handling of glass fiber, especially in
the presence of broken ends.

Graham and Ruppenicker propose a solution to the problem by spinning a
yarn with a multifilament glass core and a cotton surface. The fabric has
the appearance and hand of cotton, but increased strength and degradation
resistance, plus added fire retardancy. The synergism in strength can be
explained by mechanical models for two continuous phases in the direction of
pull, for example by the Takayangi upper bound model (32). Thus, a light
weight, high strength fabric results.

22

Table X. Fiber Blends Using Natural Products

Fiber Blend	Advantages/Application	Reference
cotton/blend	flight-crew's uniforms	(a)
wool/polyester	reduced pilling, increased abrasion and wrinkle resistance	(b),(c)
cotton/polyester	longer wear	(d)
cotton/glass fiber	tents and tarpaulin	(e)

References

(a) G. L. Lewis and C. E. Pardo, <u>Textile Res</u>. <u>J</u>. 50:130 (1980).
(b) R. W. Singleton, <u>Textile Res</u>. <u>J</u>. 50:457 (1980).
(c) M. Shiloh, in <u>J</u>. <u>Appl</u>. <u>Sci</u>. <u>Appl</u>. <u>Polym</u>. <u>Sympos</u>. 31:105 (1977).
(d) J. O. Bargenton, H. H. Perkins, Jr., and R. A. Mullikin, <u>Textile Res</u>. <u>J</u>. 48:44 (1978).
(d) C. O. Graham and G. F. Ruppenicker, <u>Textile Res</u>. <u>J</u>. 53:120 (1983).

NON-CELLULOSE WOOD PRODUCTS

Lignin

Wood is composed of cellulose and lignin in roughly equal proportions. While cellulose has become one of the most important polymers known to mankind (see above), lignin remains a poor cousin. In nature, it exists as a three-dimensional network, often called nature's glue. On pulping to free the cellulose, the lignin is degraded significantly. Today, the major application of lignin remains as a fuel for the pulping plants themselves.

Yet other applications do exist or are suggested in the literature. Research on lignin continues, as indeed it must, for the tonnage of available material is huge.

Glasser and coworkers (33-35) made hydroxyalkyl lignin derivatives by reaction kraft and other lignins with propylene oxide and ethylene oxide, and reacting the hydroxy-bearing composition with diisocyanates to make the polyurethane derivative. Depending on the composition, Young's modulus ranged from 1-2 GPa, in the range of soft plastics, and glass transition temperatures ranged from 70 to 190°C. Thus, series of tough thermoset resins were made with lignin serving as the prepolymer.

Tannins

Hemingway and coworkers (36,37) prepared a series of condensed tannin products. Condensed tannin-resorcinol adducts are proposed as adhesives in laminated wood. They concluded (36) that over 60% of the resorcinol required in a room temperature cured wood laminating adhesive can be replaced by extracts from southern pine bark through use of a condensed tannin-resorcinol adduct.

The importance of adhesives in bonded wood products must not be underestimated. Gillespie (38) states that the manufacture of bonded wood products consumes over 37% of the thermoset resins produced annually in

the U.S., besides several percent of the thermoplastic resins. Important
adhesives and types of panel are shown in Tables XI and XII. These uses add
up to 1.90 billion pounds of adhesive. If a sizable fraction of this
adhesive can be made from tannin rather than phenol, resorcinol, etc., the
value of the bark above use as a fuel would be significant.

Table XI. Wood Panel Products Requiring Adhesives (38)

Panel Type	Ft.^{2} x 10^{-6}	Equiv. thick
Softwood plywood	17,000	3/8"
Particleboard	3,000	3/4"
Hardboard	7,000	1/8"
Hardwood plywood	1,400	-
Medium density fiberboard	500	3/4"

Table XII. Use of Thermosetting Resins
in Bonded Wood Products (38)

Thermoset	Pounds, x 10^{-9}, for 1980
Urea-formaldehyde	1.20
Melanmine-formaldehyde	0.07
Phenol-formaldehyde	0.05
Polyester	0.12

Rubber

The largest source of natural rubber, cis-polyisoprene, Hevea
brasiliensis, was originally obtained from Brazil, but is now grown largely
in the Malay Peninsula and the East Indies. However, more than 1000 plants
produce rubber. The more important of these include the lowly dandelion and
goldenrod. Guayule shrubs have long been used as a source of rubber in
Mexico.

Natural rubber has a glass transition temperature near -70°C, while the
vulcanized product is amorphous when relaxed, it crystallizes on extension.
This latter phenomenon is important in the development of high strength in
tires.

24

CONCLUDING REMARKS

It is important to note that research on all of these materials somehow
continues. The funding and publicity given it, however, seems to depend on
world politics and recurring oil crises. When oil prices rise, everybody
becomes interested in renewable resources. Since quality research depends
greatly on sustained efforts, a broader base of interest would do the world
well, for in the long term the oil shortage will get worse.

By way of concluding, a bit on the historical development of polymer
science is in order. Around 1900-1930, many scientists refused to study the
"gunk" in the bottom of their pots, preferring to characterize the smaller,
simpler molecules. Gradually, methods of characterizing the "gunk"
developed, and polymer science arose.

Nowadays simple linear and crosslinked polymers are understood, and
great headway is being made in understanding blends, grafts, blocks, IPN's
and random copolymers. By and large, many of the natural product polymers
appear to be still more complex than the more regularly repeating
snythetics. Thus, a challenge of the first magnitude faces polymer
scientists -- to devise methods of studying nature's own polymeric
structures. Now, the ideas of modern polymer science can and must be
applied in this great new adventure.

REFERENCES

1.   C. E. Carraher and L. H. Sperling, in "Polymer Applications of Renew-
     able Resource Materials", C.E. Carraher, Jr. and L. H. Sperling, Eds.,
     Plenum Press, New York (1983).
2.   E. H. Pryde, L. H. Princen, and K. D. Mukherjee, Eds., "New Sources of
     Fats and Oils", Americal Oil Chemists Society, Champaign, IL (1981).
3.   E. A. MacGregor and C. T. Greenwood, "Polymers in Nature", Wiley,
     New York (1971).
4.   K. V. Sarkanen and C. H. Ludwig, Eds., "Lignins Occurence, Formation,
     Structure, and Reactions", Wiley-Interscience, New York (1971).
5.   D. N. S. Hon, Ed., "Graft Copolymerization of Lignocellulose Fibers",
     American Chemical Society, Washington, D.C. (1982).
6.   J. V. F. Vincent and J. D. Curry, Eds., "The Mechanical Properties of
     Biological Materials", SEB Symposium No. 34, Cambridge Unviersity
     Press, New York (1980).
7.   R. M. Brown, Ed., "Cellulose and Other Natural Polymer Systems",
     Plenum, New York (1982).
8.   T. P. Nevell and S. H. Zeronian, Eds., "Cellulose Chemistry and Its
     Applications", Ellis Horwood, Chicester, England (1985).
9.   E. H. Pryde and F. H. Otey, in "Polymer Yearbook", 1st Ed., H. G. Elias
     and R. A. Pethrick, Eds., Harwood, New York (1984).
10. V. Crescenzi, I. C. M. Dea, and S. S. Stivala, "New Developments in
     Industrial Polysaccharides", Gordon and Breach, New York (1985).
11. T. Shaw and M. A. White, in "Handbook of Fiber Science and Technology",
     Vol. IIB, M. Lewin and S. B. Sello, Eds., Marcell Dekker, New York
     (1984).
12. G. C. Tesoro, in "Handbook of Fiber Science and Technology", Vol. IIA,
     Lewin and S. B. Sello, Eds., Marcel Dekker, New York (1983).
13. F. Lyndon Davies and B. H. Law, Eds., "Advances in the Microbiology
     and Biochemistry of Cheese and Fermented Milk", Elsevier, New York
     (1984).
14. J. Rollings, Carbohydrate Polym. 5:37 (1985).
15. T. E. Creighton, "Proteins Structures and Molecular Properties",
     Freeman, New York (1983).

16. R. L. Whistler, J. N. Bemiller, and E. F. Paschall, Eds., "Starch: Chemistry and Technology", Academic Press, New York (1984).

17. G. Jeronimidis, in "The Mechanical Properties of Biological Materials", J. V. F. Vincent and J. D. Curry, Eds., SEB Symposium No. 34, Cambridge University Press, New York (1980).

18. G. Tsoumis, "Wood as a Raw Material", Pergamon, New York, p. 69 (1968).

19. R. F. Foelix, "Biology of Spiders", Harvard University Press, Cambridge, MA (1982).

20. R. W. Work, Textile Res. J. 46:485 (1976).

21. R. W. Work, Textile Res. J. 47:650 (1977).

22. R. W. Work and N. Morosoff, Textile Res. J. 52:349 (1982).

23. R. E. Wilfong and J. Zimmerman, J. Appl. Polym. Sci, Appl. Polym. Symp. 31:1 (1977).

24. M. Denny, J. Exp. Biol. 65:483 (1976).

25. M. W. Denny, in "The Mechanical Properties of Biological Materials", Cambridge University Press, Cambridge, MA (1980).

26. J. M. Gosline, M. W. Denny, and M. E. DeMont, Nature 307:551 (1984).

27. A. V. Tobolsky and M. C. Shen, J. Appl. Phys 37:1952 (1966).

28. P. Calvert, Nature 309:516 (1984).

29. R. H. Pain, in "the Mechanical Properties of Biological Materials", J. F. V. Vincent and J. D. Curry, Eds., Cambridge University Press, New York (1980).

30. B. Farnsworth, Textile Res. J. 53:717 (1983).

31. C. G. Graham and G. F. Ruppenicker, Textile Res. J. 53:120 (1983).

32. M. Takayanagi, H. Harima, and Y. Iwata, Mem. Fac. Eng., Kyushu Univ. 3:1 (1963).

33. L. C. F. Wu and W. G. Glasser, J. Appl. Polym. Sci. 29:1111 (1984).

34. W. G. Glasser, C. A. Barnett, T. G. Rials, and V. P. Saraf, J. Appl. Polym. Sci 29:1815 (1984).

35. V. P. Saraf and W. G. Glasser, J. Appl. Polym. Sci. 29:1831 (1984).

36. R. E. Kreibich and R. W. Hemingway, Forest Products J. 35(3):23 (1985).

37. R. W. Hemingway, G. W. McGraw, J. J. Karchesy, L. Y. Foo, and L. J. Porter, J. Appl. Polym. Sci., Appl. Polym. Symp. 37:967 (1983).

38. R. H. Gillespie, J. Adhesion, 15:51 (1982).

39. A. Pizzi, Ed., "Wood Adhesives, Chemistry and Technology", Marcel Dekker, N.Y. (1983).

SECTION II - SACCHARIDES AND POLYSACCHARIDES

NATURAL VEGETABLE FIBERS: A STATUS REPORT

M.B. Amin, A.G. Maadhah and A.M. Usmani

Research Institute
University of Petroleum and Minerals
Dhahran 31261, Saudi Arabia

INTRODUCTION

Natural fibers are a major renewable resource material through the
world, specifically in the tropics. For the past 60 years they have
faced competition with the man-made fibers. Research on natural fibers
was expanded to minimize market losses and this has resulted in
modification of natural fibers to give improved properties e.g.,
wrinkle resistance, flame resistance, enhanced dyeability, and
increased resistance to heat and microorganisms. The natural fibers,
particularly those that are cellulose based, have some advantages.
Cellulose is abundant, renewable, inexpensive, and derives its carbon
from the air instead of petroleum or natural gas. Additionally
cellulose is amenable to chemical and mechanical modification.
Finally, cellulose is the most abundant organic material on our planet
and is produced to the extent of about 10^{15}lb annually by plants.

The full commercial potential of natural fibers has not been
achieved however due to lack of research and development of high
technology applications. In this work we shall describe commercially
important vegetable fibers, their composition and properties, fiber
processing, applications, and newer research areas.

COMMERCIALLY IMPORTANT VEGETABLE FIBERS

There are about 2,000 species of useful fiber plants grown in
various parts of the world and they are used for many applications.
Many fibers are used because they are the best available locally for
specific needs. Some countries, specifically developing, utilize locally
grown fibers because they are acceptable substitutes for expensive
imported fibers.

Natural fiber, leaf or bast is composed of fibrils glued together
with natural resinous materials of the plant tissue. Essentially,
fibers are composed of cellulose with associated gummy binder and woody
tissue. The cell is primarily cellulose and the related carbohydrate
xylan. The woody tissue, specifically present in the leaf fibers,
contains lignin which is phenolic in nature.

Important vegetable fibers of commercial interest and their major producers are shown in Table 1.

Table 1. Important Vegetable Fibers

Fiber	Botanical name	Major producers
Bast (soft)		
jute	Corchorus capsularis; C. olitorius	Bangladesh, India, Thailand
flax	Linum usitatissimum	Belgium, Luxembourg, The Netherlands, USSR, France, Ireland
sunn	Crotarlaia juncea	India
hemp	Cannabis sativa	Yugoslavia
ramie	Boehmeria nivea	Philippines, Brazil, China, Japan, Taiwan
kenaf	Hibicus cannabinus	India, Iran, USSR, Latin America, Pakistan
Leaf (hard)		
sisal	Agave sisalana	Brazil, Eastern Africa, Haiti, Indonesia, Mexico
henequen	Agave fourcroydes; Agave letonae	Mexico, Cuba, El-Salvador, Australia
abaca	Musa textilis	Philippines, Singapore, Indonesia
istle (tampico)	Agave; Amaryllis	Mexico
Palm-Type		
coir	Cocos nucifera	Sri Lanka, Mexico, Jamaica, India
crin vegetal	Chamaerops humilis	Morocco
piassava	Leopoldinia piassava	Western Africa, Nigeria, Brazil
Seed		
cotton	Gossypium hirsutum, barbadense, arboreum, and herbaceum	USSR, China, USA, India, Pakistan, Brazil, Turkey, Egypt
kapok	Ceiba petandra	Thailand, Indonesia, India

Cotton has been the principal textile fiber for almost 200 years. Current annual world production of cotton is about 32 billion lb. Rayon and acetate made from cellulose, the principal component of the plant fibers, total to about 7.4 billion lb whereas about 22 billion lb of non-cellulosic man-made fibers are annually manufactured. Every year about 2.3 billion lb of jute, 900 million lb of sisal, 300 million lb of henequen, 130 million lb of abaca, and 400 million lb of coir are produced in the world.

Some important vegetable fibers are briefly discussed below.

Jute. The color of jute fiber ranges from creamy white to reddish brown turning into dingy brown after aging. Fibers are soft with silky luster and are grouped into strands 5-10 ft long. The ultimate cells composing the fibers are 1-5 mm long and 14-20μ in diameter. Since chemically jute has a higher percentage of lignin than any other commercial soft fiber and a lower cellulose content it lacks in strength and durability.

Flax. Flax fibers vary in color from creamy white to dark brown, 12 to 36 inches in length and 2-20 mils in width. Flax fibers are strong, low stretching with high water absorption.

Ramie. The degummed ramie fiber is a multiple-celled long fiber and the ultimate cells are considerably longer and thicker (1-20 in. long and 20-70 μ in diameter) than other bast fibers. Ramie fiber is superior to all other bast fibers in strength and versatility because of extermely high cellulose content.

Kenaf. Kenaf is quite comparable to jute. The fiber is silky, soft, light cream to tan in color, 5-9 ft in length and slightly stronger than jute.

Sisal. Commercially, African sisal, Indonesian sisal and Haitian sisal are available. They range in color from almost white to creamy color. Commercial fibers range from 24 to 64 in. in length and from 1/8 to 1/2 mm in diameter. The cells are polygonal in outline with an average major diameter of 16 μ and the minor diameter of 11 μ.

Abaca. Commercial abaca fibers are 12 ft or more in length with a diameter of 0.2-1 mm. The ultimate cells are 3-12 mm long and 16-32 μ in diameter.

Coir. Coir is a hard and tough fiber. It is multi-cellular with a central pore. Each cell is polygonal or round in shape. Coir consists of 32-43% cellulose, 40-45% lignin and 3-4% pectin.

Cotton. Cotton is the most important vegetable fiber used in spinning. Its origin, cultivation, morphology, and chemistry have been described in innumerable publications. Cotton is a member of the mallow family, a plant of the genus Gossypium, and is widely grown in warmer climates all over the world. Cotton is the most important natural fiber cash crop (Figure 1).

COMPOSITION AND PROPERTIES

The chemical composition of various vegetable fibers is given in Table 2. In general, chemical composition varies greatly between plants and within specific fibers depending on genetic characteristics, part of the plant, growth, harvesting and fiber preparation conditions.

Cotton is essentially 95% cellulose. Several typical fibers are shown in Figure 2.

Leaf fibers are multi-celled and not easily split into component cell whereas bast fibers are easily broken down, permitting spinning. The microfibrillate structure of cotton includes pores, channels, and cavities that play an important role in its chemical modification. The arrangement of fibrils follows a spiral pattern and at times reverses itself. Scanning electron micrographs of sisal, banana, talipot palm, and coir (longitudinal view) are shown in Figures 3 to 6.

Mechanical properties of certain vegetable fibers are shown in Table 3.

Table 2. Chemical Composition of Select Vegetable Fibers, wt%

Fiber	Cellulose	Moisture	Ash	Lignin & pectins	Extrac- tives
Bast fibers					
hemp	77.07	8.76	0.82	9.31	4.04
jute	63.24	9.93	0.68	24.41	1.42
kenaf	65.7	9.8	1.0	21.6	1.9
ramie	91	0.65			
sunn	80.4	9.6	0.6	6.4	3.0
Leaf fibers					
abaca	63.72	11.83	1.02	21.83	1.6
henequen	77.6	4.6	2.2	13.1	3.6
sisal	77.2	6.2	1.0	14.5	1.1
istle	73.48	5.6	1.65	17.37	1.9
Palm-type					
coir	32-43			43-49	
Seed					
cotton	88-96	4-8	0.7-1.6	0.7-1.2	0.3-0.5

FIBER PROCESSING

Vegetable fibers, except those used as filler, or in papermaking, are processed into twisted yarns prior to manufacturing operations. There are two processing stages, namely, fiber preparation and fiber spinning. In fiber preparation individual fibers are converted into ribbons of parallel overlapping fibers, cleaned of pithy materials, softened and lubricated. In spinning, the ribbon of aligned fibers is drawn down to size and twisted (Figure 7).

Figure 1.   Cotton is an Important Cash Crop in many African
Countries.

Figure 2.   Several Typical Vegetable Fibers.

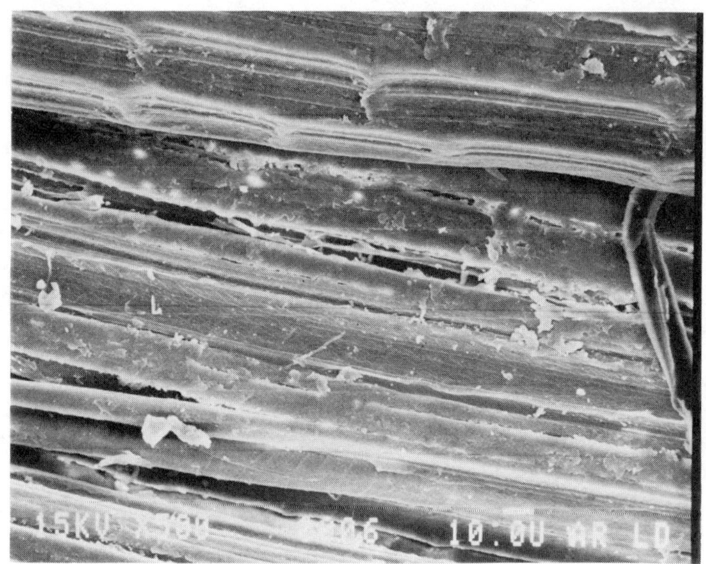

Figure 3.   Scanning Electron Micrograph (SEM) of Sisal (Longitudinal) at 500X.

Figure 4.   SEM of Banana Fiber (Longitudinal) at 500X.

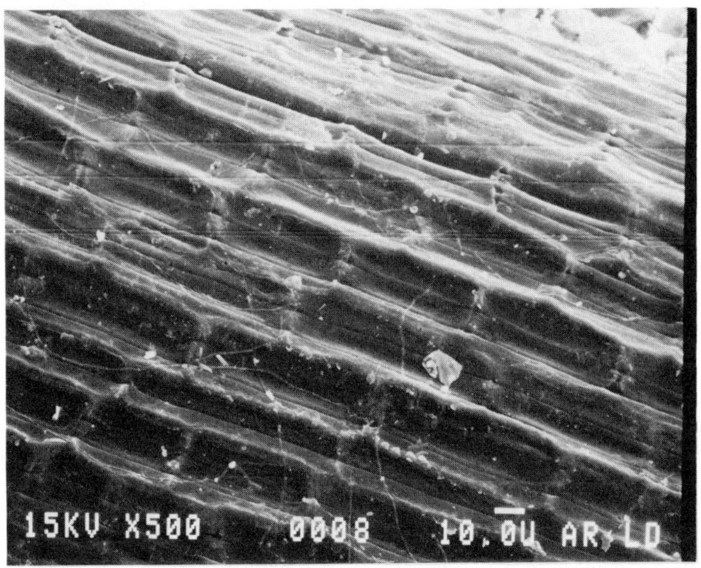

Figure 5. SEM of Talipot Palm (Longitudinal) at 500X.

Figure 6. SEM of Coir (Longitudinal) at 100X.

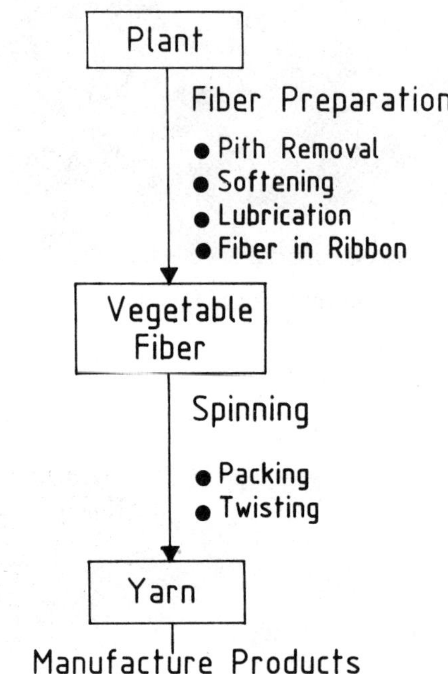

Figure 7. Process Outline for Fiber Processing.

Table 3. Mechanical Properties of Certain Vegetable Fibers

Fiber	Tensile strength, psi x 10^{-3}	Relative weight for comparable strength	Elongation at break, %
jute	57	167	1.5
flax	96	100	1.5
hemp	76	125	2.0
ramie	97	100	4.0
sisal	74	125	3.0
henequen	56	167	5.0
abaca	93	100	3.0
istle	46	200	5.0

CURRENT APPLICATIONS

Current applications of certain vegetable fibers are indicated in Table 4. There is a pressing need for diversifying the utilization of natural fibers because they are a renewable resource and because of competition from synthetic fibers. Therefore processes must be developed including non-woven technologies.

Table 4. Principal Applications of Select Vegetable Fibers

Fiber	Applications
jute	Burlap, sacking, backing for linoleum and rugs, webbing, twine, packing
flax	Shoe-stiching thread, textiles
sunn	High-quality tissue paper, cordage, twine
hemp	Fabrics, strong twines, packing
ramie	Textiles
kenaf	Coarse textiles
sisal	Hard fiber cordage, upholstery padding, paper
henequen	Sacking for coffee bags, cordage
abaca	Ropes, cordage, twine, manila paper, fine strong tissue paper
istle	Brushes, upholstery, coarse twines
coir	Mats, rugs and carpets, ropes, upholstery filling
kapok	Insulating material in refrigeration and sound insulation
cotton	All grade of textiles, cordage

SUGGESTED RESEARCH AREAS

Some potential applications of natural fibers are now discussed. Certain natural vegetable fibers can be used as filler and reinforcer in plastics, clay, cement, rubber and adhesive compositions, specifically for roofing and building material applications. Molded parts could find application in gears and transportation industries. Certain vegetable fibers can be upgraded by grafting with suitable modifying monomers by chemical or radiation bombardment. Because of the high electrical resistance and easy fabrication into complicated shapes the fibers can be used as insulators. Fibers can be used to manufacture specialty papers, hard board, and plastics. Certain fibers, e.g., sisal, can be dyed to prepare artificial hair for wigs. Carbonization of natural fibers may produce products that could find newer applications.

To develop newer applications research in the following areas are suggested.

- Dyeing characteristics of fibers should be investigated along with their standardization.
- Structure-property relationships should be developed for natural fibers under various environmental conditions. Degradation should be studied so that service temperatures for various fibers become known to composite scientists.
- Processes and products for production of roofing and building materials using wastes, e.g., coir dust and sisal dust, should be developed in combination with suitable binders.
- Sisal fiber cottonization by blending with cotton for use in wall decoration should be developed. Similarly sisal and glass fiber may be used in composites for making water tanks and grain silos.
- Sorption characterization of fibers should be studied. This will help in determining compatibility of natural fibers with polymers, clay, cement, and rubber.
- Flammability should be studied and efficient, inexpensive flame proofing compositions should be developed.
- Carbonization and graphitization of natural fibers should be investigated. The ultimate objective should be development of newer refractory composites for high temperature applications.

FUTURE PROSPECTS

The future prospects of vegetable fibers are encouraging in terms of quantities but a slight decrease in percent of the total textile market is foreseen. Most of the developing nations of the world are committed to improve their quality of life and therefore increase in usage of fibers are indicated. The effect of high petroleum and energy cost in interfiber competition is not fully known. Lower petroleum and energy cost should favor the man-made fibers somewhat however. The crude oil prices are soft now and therefore natural fiber scientists should develop newer high technology applications.

The per capita consumption and textile markets are expanding the world over and thus there is room for growth of both natural and man-made fibers.

CONCLUSIONS

Natural fibers are a major renewable resource materials. In this work we have described commercially important vegetable fibers, their composition and properties, fiber processing, applications and needed newer areas of research. The prospects of these fibers are good due to increased per capita consumption and an expanding textile market.

ACKNOWLEDGEMENT

Thanks to Mohammed Riazuddin for typing this work. The SEM work, done at Arabian American Oil Company, Dhahran, is acknowledged.

REFERENCES

1. H.B. Brown and J.O. Ware, "Cotton," 3rd ed., McGraw-Hill, New York, 1958.
2. R.E. Perdue, Jr., "Fiber," in Encyclopedia American, Vol.11, 1977.
3. R.K. Warner and D.B. Saku, "Literature of the Natural Fiber," Advances in Chemistry Series 10, American Chemical Society, Wasington, D.C., 1954.
4. J.N. McGovern, "Vegetable Fibers," in Kirk-Othmer Encylopedia of Chemical Technology, 3rd ed., Vol.10, Wiley, New York, 1980.
5. K.G. Satyanarayana, A.G. Kulkarni and P.K. Rohatgi, J. Sci. Ind. Res.,40, 222 (1981).
6. A.G. Kulkarni, K.G. Satyanarayan and K. Sukumaran, J. Mater. Sci., 16, 905 (1981).
7. A.G. Kulkarni, K.G. Satyanarayan and P.K. Rohatgi, J. Mater. Sci. Lett.,16, 1719 (1981).
8. K.G. Satyanarayan, C.K.S. Pillai, K. Sukumaran and G.K. Pillai, J. Mater. Sci., 17, 2453 (1982).
9. A.G. Kulkarni, K.A. Cheriyan, K.G. Satyanarayana and P.K. Rohatgi, J. App. Polym. Sci., 28, 625 (1985).
10. R.M.V.G.K. Rao, N. Balasubramanian and M. Chanda, J. App. Polym. Sci., 26, 4069 (1981).
11. V.M. Murty and S.K. De, J. App. Polym. Sci., 27, 4611 (1982).
12. R.N. Mukherjea, S.K. Pal and S.K. Sanyal, J. App. Polym. Sci., 28, 3029 (1983).
13. A. Nagaty, A.B. Mustafa and O. Mansour, J. App. Polym. Sci., 23, 3263 (1979).
14. V.M. Murty, S.K. De, S.S. Bhagawan, R. Sivaramakrishnan and S.K. Athithan, J. App. Polym. Sci., 28, 3483 (1983).
15. R.M. Kishore, M.K. Shridhar and R.M.V.G.K. Rao, J. Mater, Sci. Lett., 2, 99 (1983).
16. T.M. Aminabhavi, N.S. Biradar and R.M. Holennavar, J. Macromol. Sci. Chem., A20(4), 515 (1983).
17. K. Gopakumar, T.P. Murali and P.K. Rohatgi, J. Mater. Sci., 17, 1041 (1982).
18. V.M. Murty and S.K. Dey, J. Rubber Chem. Technol., 55, 287 (1982).
19. K.G. Satyanarayana, A.G. Kulkarni and P.K. Rohatgi, Proc. Ind. Acad. Sci., India, 4(4), 419 (1981).
20. A.G. Kulkarni and K.G. Satyanarayana, Fib. Sci. Technol., 19,(1), 59 (1983).
21. K.G. Satyanarayana, C.K. Manglakumar, A.G. Kulkarni and P. Koshy, Bull. Elec-Micro. Soc. India, 1-4, 179 (1983).

POLYSACCHARIDES FROM LICHENS: 13C-NMR

STUDIES ON (1-6) - BETA-D-GLUCAN (PUSTULAN)

Arthur J. Stipanovic, P. J. Giammatteo and S. B. Robie

Texaco Research Center
P. O. Box 509
Beacon, New York   12508

INTRODUCTION

The ubiquitous character of polysaccharides in nature provides the historical basis for the study and utilization of these renewable materials.[1-3] Considerable research has been focused on the chemical structure, molecular conformation, crystalline morphology, and physical properties of commercially important polysaccharides such as cellulose, starch, chitin, and xanthan gum while lesser attention has been directed to those carbohydrate-based polymers which serve as naturally occurring "specialty chemicals". In the present study we have investigated the unique molecular properties of pustulan, a water-soluble, gel-forming polysaccharide produced by the lichen Umbilicaria papullosa (pustulata). Lichens are rather unusual organisms which represent a symbiotic relationship between a photosynthetic algae and a fungus which develops for the mutual benefit of both microorganisms in response to environmental stress such as dehydration[4].

Pustulan was first isolated by Drake[5] from Umbilicaria pustulata and was later identified by Lindberg[6] as a linear, (1-6)-Beta- linked glucan (Figure 1). In the native state, pustulan has been shown to contain acetate substituents on approximately 10% of its glucose residues[7]. Upon removal of these acetate groups by mild, base-catalyzed hydrolysis, pustulan becomes capable of forming thermally reversible gels in aqueous solvents[8-11]. Pustulan gelation has previously been monitored by ultraviolet circular dichroism spectroscopy[8,9] and solution 13C-NMR[10]. It has been postulated that the solution/gel transition involves a "random-coil" to "pseudo-helical" conformation transformation leading to crosslinks and gel formation.[13] To gain further insight into this mechanism, we have employed 13C-NMR, both conventional solution techniques and Cross Polarization/Magic Angle Spinning (CP/MAS), to compare the conformational and crystal structure of pustulan in the solution, solid, and gel states. In doing so the utility of solid-state CP/MAS 13C-NMR in defining structural aspects of polysaccharide solids and gels is clearly demonstrated.

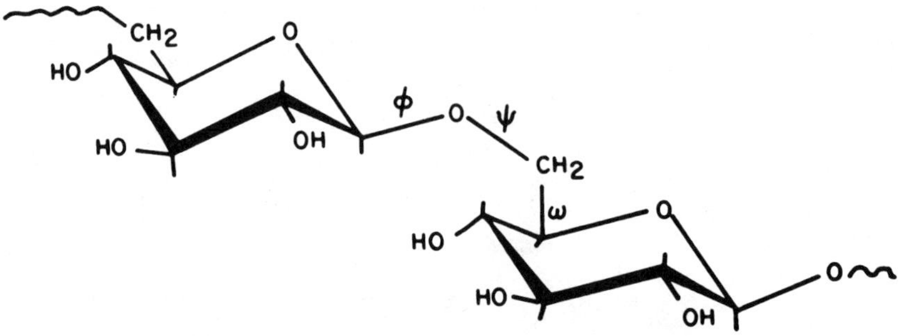

FIGURE 1.   PUSTULAN: (1-6)-BETA-D-GLUCAN

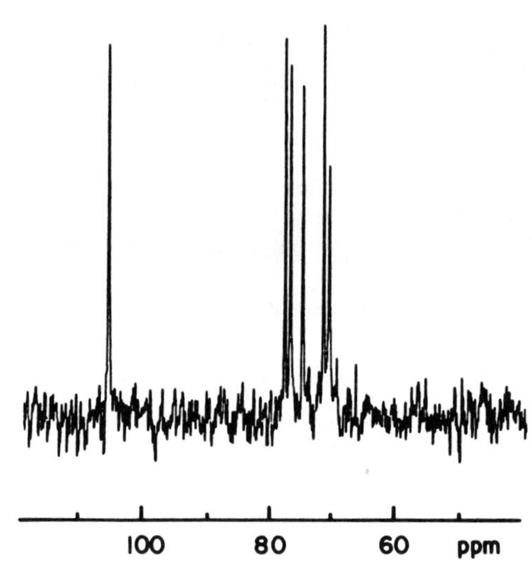

FIGURE 2.   13C-NMR SPECTRUM OF PUSTULAN IN D_2O at 85°C

<u>Experimental</u>

Partially acetylated native pustulan, extracted from the lichen <u>Umbilicaria papullosa (pustulata)</u>, was purchased from Calbiochem-Behring. This material was gently deacetylated in dilute sodium carbonate by a procedure outlined elsewhere.[12] Pustulan solutions/gels were prepared by dispersing 5-10% (wt/v) of deacetylated material in $H_2O(D_2O)$ and heating the mixture for five minutes at 95-100°C to effect dissolution. Gels were found to set at room temperature only after an aging period of 1-16 hours. Gelation was prevented to record solution NMR spectra by maintaining pustulan solutions above 80°C, by using non-gelling, slightly acetylated, native pustulan, or alternatively, by dissolving pustulan in 4M urea or DMSO (Aldrich Chemical Co.). ^{13}C-NMR solution spectra were recorded at 20 MHz on a Varian FT80A spectrometer. Chemical shifts in D_2O solutions were referenced to either TMS placed in an external capillary tube or to internal TSP. CP/MAS ^{13}C-NMR spectra were recorded for solid samples at 15 MHZ on a JEOL FX-60QS spectrometer equipped with a Chemagnetics Solids Accessory. X-ray diffraction analysis was performed on a Scintag PAD V powder diffractometer with a broad focus copper anode X-ray tube (Rich. Seifert and Co.) operated at 45 kV and 40 mA. Detection was by a scintillation detector and graphite monochromator combination. Further details on the NMR and X-ray methods employed are outlined elsewhere.[11]

RESULTS AND DISCUSSION

Solution ^{13}C-NMR of Pustulan

The use of ^{13}C-NMR spectroscopy to identify the monomeric composition and linkage stereochemistry of carbohydrates and polysaccharides in solution is well documented[13-18]. More recent studies have used this technique to monitor changes in molecular conformation which occur with alterations in temperature and solvation[10,19,20]. ^{13}C-NMR data recorded for pustulan solutions are summarized in Table I. Resonances were assigned to specific carbon atoms according to Bassieux[21]. All non-gelling conditions were found to yield similar chemical shifts and relative peak intensities (not shown) indicating that the molecular conformation of pustulan is similar in all cases. Estimates of the spin-spin relaxation time, T_2, are consistent with randomly-oriented macromolecules ($\sim$ 50 ms).[22] As previously concluded,[8-10] it is likely that pustulan adopts an unordered, random-coil conformation in solution prior to the onset of gelation or when gelation is suppressed by alternative solvents and/or elevated temperatures.

CP/MAS ^{13}C-NMR of Solid Pustulan

Cross Polarization/Magic Angle Spinning ^{13}C-NMR has become a valuable technique for structural characterization of organic solids such as polymeric materials. Detailed descriptions of the experiment as well as applications have been described elsewhere.[23-26] Three principle interactions become significant problems in performing a high resolution NMR experiment on a solid. These are: 1) Chemical shift anisotropy, 2) [13]C-^{1}H dipolar interactions and 3) Long [13]C-^{1}H, T_1 relaxation times.

TABLE I

^{13}C-NMR Chemical Shifts of Pustulan Solutions

Pustulan Sample/Solvent	Chemical Shifts (ppm)[a,b]					
	C(1)	C(2)	C(3)	C(4)	C(5)	C(6)
Deacetylated/D_2O - 85°C	105.7	76.0	78.6	72.7	77.8	71.8
Deacetylated/4M Urea	106.0	76.1	78.8	72.6	77.9	71.8
Deacetylated/DMSO[c]	106.0	76.2	78.7	72.7	78.5	71.2
Acetylated/D_2O	106.0	76.1	78.7	72.6	77.9	71.8

[a]Assignment based on Reference 21.
[b]Referenced to internal TSP.
[c]Referenced based on C(1) = 106 ppm.

TABLE II

CP/MAS ^{13}C-NMR of Pustulan

Pustulan Sample (Figure No.)	Chemical Shifts (ppm)[a,b]					
	C(1)	C(a)	C(b)	C(c)	C(d)	C(6)
Slightly Crystalline (3A)	104	NR	--	74	--	70[c]
Moderately Crystalline (3B)	103.7	81.4	76.7	74.6	73.1	70.0
Most Crystalline (3C)	103.7	81.6	77.0	74.6	72.8	70.0
10% Gel (5A)	104	81.5	77	74	72	70[c]
Dried Gel (5B)	103.5	81.4	76.5	74.1	72.3	69.4
DMSO Freeze Dried	106.1	NR	--	75	--	70[c]

[a]Referenced to external TMS.
[b]No assignment for C(2)-C(5); NR = no resonance.
[c]Unresolved shoulder.

Chemical shift anistropy line broadening occurs due to small magnetic field inhomogeneities that arise from specific molecular bond orientations with respect to the applied external magnetic field (See Fig. 5.1, Ref 23). In liquids, rapid molecular tumbling averages all bond orientations to a single average orientation giving an isotropic value for the chemical shift. Since molecular motion is severely restricted in a solid, many possible bond orientations result in a dispersion of the chemical shift. The resultant spectrum is a broad line typically 50-200 ppm wide for randomly-oriented amorphous solids. Rapid spinning (2-5,000 Hz) of the sample at an angle of 57.4° (the magic angle) off the external magnetic field averages the anisotropic chemical shift towards its isotropic value, narrowing the solid sample NMR line to widths from several Hz to 1-2 ppm depending on the sample.[23]

Heteronuclear proton decoupling is routinely applied in liquid ^{13}C-NMR analysis. In the solid state, the ^{1}H-^{13}C dipolar interactions still exist but are now intensified by intermolecular interactions as well as intramolecular ones. In order to remove these ^{1}H-^{13}C dipolar interactions, much higher proton decoupling with magic angle spinning is required to further reduce solid ^{13}C linewidths[24].

Finally, the problem of long ^{13}C-^{1}H, T_1 relaxation times is addressed via cross-polarization. Basically, the principle of the experiment involves bringing the proton (abundant) magnetization spin system into thermal contact with the ^{13}C (dilute) magnetization spin system. This is accomplished by finding a "matching frequency" in the rotating frame that sets the effective Larmor precessional frequencies of ^{1}H and ^{13}C equal to each other. Thermal contact of the magnetization during this spin - locked time achieves an approximate four-fold increase in ^{13}C signal intensity and an approximate 10 to 100 fold decrease in the effective ^{13}C-^{1}H, T_1 relaxation.[25].

A typical CP/MAS ^{13}C-NMR spectrum of solid pustulan powder is presented in Figure 3C; Chemical shifts relative to TMS are given in Table II. A comparison of solution and solid-state ^{13}C-NMR data (Figure 2 and 3C, Tables I and II) differs most significantly near 82 ppm where only solid pustulan exhibits a distinct resonance. Although the specific carbon atom responsible for this resonance cannot be readily established, a 2-5 ppm downfield shift of a glucose ring atom [C(2)-C(5)] induced by intermolecular packing interactions could account for this signal. Pfeffer et al have shown for α and β-glucose model compounds that solution and solid NMR chemical shifts may, typically, vary ±2 ppm for a particular carbon atom.[27] Since the magnetic environment of the C(6) linkage atom is influenced primarily by the inductive effects of chemical bonds and torsion angle considerations rather than intermolecular interactions, we do not believe that the 82 ppm resonance arises from C(6) which normally occurs in the 60-70 ppm range.

To further investigate the origin of the unique resonance characteristic of solid pustulan, CP/MAS ^{13}C-NMR and X-ray diffraction techniques were used to study samples of varying degrees of crystallinity. CP/MAS studies on cellulose have shown that increasing crystallinity results in an improved signal resolution and a downfield shift for C(4) and C(6) resonances.[28,29] For solid pustulan, we have determined that crystallinity increases directly with increasing degree of sample hydration as calculated from C,H,O elemental analysis. Diffractograms shown in Figures 4A and 4D result from pustulan samples

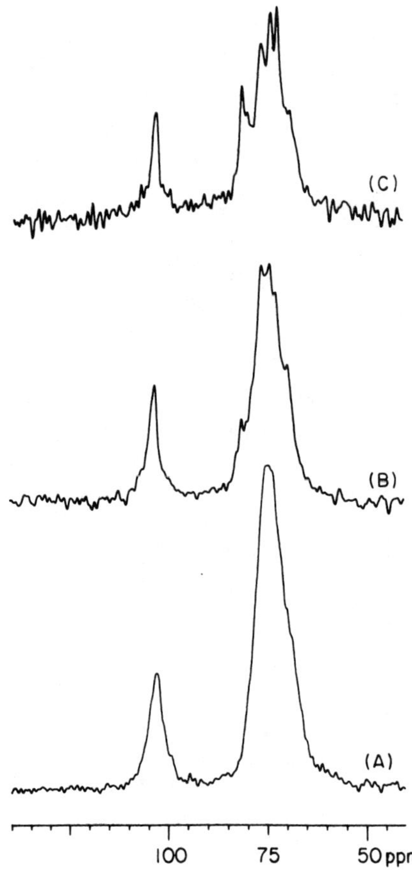

FIGURE 3. CP/MAS ^{13}C-NMR SPECTRA OF SOLID PUSTULAN: (A) SLIGHTLY CRYSTALLINE, (B) MODERATELY CRYSTALLINE, (C) CRYSTALLINE PUSTULAN HEMIHYDRATE

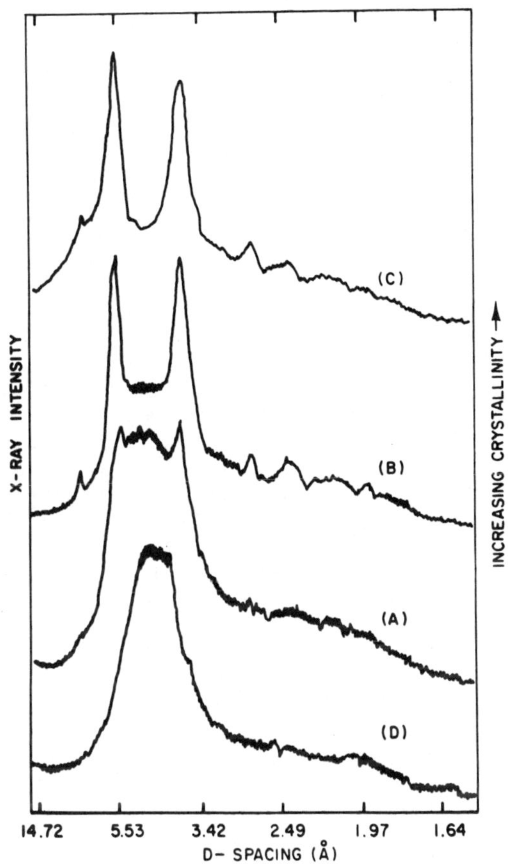

FIGURE 4.   X-RAY DIFFRACTOGRAMS OF: (D) AMORPHOUS PUSTULAN, (B) SLIGHTLY CRYSTALLINE PUSTULAN, (C) MODERATELY CRYSTALLINE PUSTULAN, (A) CRYSTALLINE HEMIHYDRATE PUSTULAN

containing no detectable water of hydration while the sample shown in 4C contains approximately one water molecule per two glucose residues (pustulan hemihydrate). A comparison of Figures 3 and 4 reveals that the intensity and resolution of the 82 ppm CP/MAS NMR resonance increase for pustulan samples of higher crystallinity which clearly display the two strong X-ray reflections located at 5.8 A and 3.8 A. Maximum signal resolution was observed for the pustulan hemihydrate sample shown in Figure 5B which approaches the practical limit (1-2 ppm) expected by CP/MAS for incompletely crystalline polymer materials.[23,26] The data above suggest that the 82 ppm signal results from a very specific crystal field effect possibly involving the interaction of pustulan with a water molecule.

Additional insight into the crystal structure of pustulan is obtained by comparing the CP/MAS spectrum of this material to that of gentiobiose, the (1-6)-Beta linked dimer of pustulan. Rohrer et al have demonstrated that the crystal structure of gentiobiose is unique among glucose-containing carbohydrates in that no intramolecular hydrogen bonds are observed.[30] From the packing arrangement of gentiobiose in the crystal lattice, these authors propose a possible crystal structure for pustulan which also contains no intramolecular hydrogen bonds. The CP/MAS ^{13}C-NMR spectrum of crystalline gentiobiose, shown in Figure 6, does not exhibit the 82 ppm resonance characteristic of crystalline pustulan suggesting a fundamental difference in crystal packing exists between the dimer and polymer. It may be speculated that these differences originate from intramolecular hydrogen bonds in the crystal structure of pustulan which is more typical of glucose-containing polysaccharides. Recent potential energy calculations have indeed shown that the pustulan molecule can exist in a minimum energy conformation characterized by a hydrogen bond between the O(2) and O(4) atoms of contiguous glucose residues.[31] Alternatively, the absence of an 82 ppm resonance in the CP/MAS spectrum of gentiobiose compared to pustulan may be related to the effect of hydration on the crystallinity of pustulan. The dimer was found to contain no water in its crystal structure.[30]

CP/MAS ^{13}C-NMR of Pustulan Gels

It has been demonstrated[10] that pustulan gelation results in a loss of liquid NMR signal intensity due to viscosity-induced line broadening effects resulting from the short spin relaxation times typical of crosslinked molecules in a gel network[20]. As a result, CP/MAS ^{13}C-NMR was applied to pustulan since this technique is capable of obtaining high resolution spectra from solids and gels. Figure 5A contains a typical CP/MAS spectrum for a 10% (w/v; D_2O) pustulan gel; the spectrum of a crystalline, solid pustulan sample resulting from gel dehydration is shown in Figure 5B for comparison. Significantly, the gel spectrum contains the 82 ppm resonance shown to reflect crystallinity in solid pustulan. The poorer spectral resolution observed for pustulan gels compared to crystalline solids may arise from a lower degree of crystallinity and smaller crystalline size for the gel. The weakly crystalline nature of these gels was confirmed, however, by X-ray diffraction.

MECHANISM OF PUSTULAN GELATION

The CP/MAS ^{13}C-NMR and X-ray diffraction results reported above provide evidence that aqueous pustulan gels contain crystalline regions

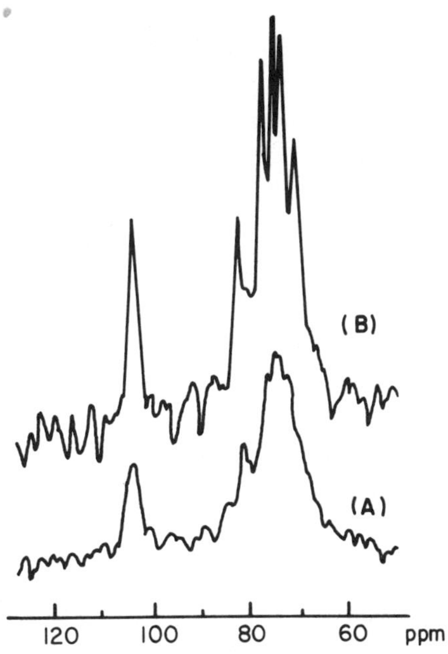

FIGURE 5. CP/MAS ^{13}C-NMR SPECTRA OF: (A) 10% PUSTULAN GEL
IN D$_2$O, (B) CRYSTALLINE PUSTULAN

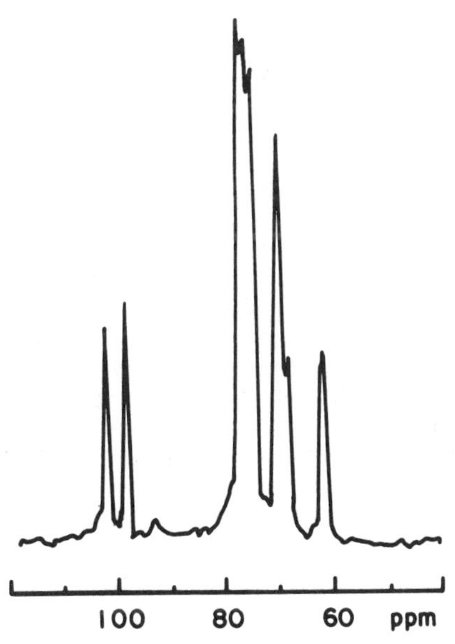

FIGURE 6. CP/MAS ^{13}C-NMR OF GENTIOBIOSE

which presumably act as "junction-zones" to establish a three-dimensionally crosslinked network structure. We have determined that crystallization can be disrupted by elevated temperatures, the presence of urea or DMSO, or by partial acetylation of the pustulan molecule as it exists in nature. These factors apparently prevent the development of intra- and intermolecular hydrogen bonds which lead to crystallization. It has also been observed that crystallization rates in water can be enhanced by freeze/thaw cycling which ultimately results in a complete precipitation of crystalline pustulan (upon thawing) without the formation of a gel network.[6,12] This result suggests that the gel state of pustulan may be "metastable" and results only when there exists an appropriate equilibrium between the concentration of water-soluble, random-coil molecules (or molecular segments) and those molecules (segments) bound in water-insoluble crystallites which provide junction zones. This equilibrium is apparently optimized for pustulan gel formation in H_2O in the temperature range of 5-25°C. At lower temperatures, crystallite sizes become sufficiently large that they are not adequately solubilized by random-coil segments resulting in complete precipitation of the solid matrix prior to gel formation. On the other hand, higher temperatures apparently yield an insufficient number of crystallites to provide adequate crosslinking.

Although the presence of crystallinity in pustulan gels is somewhat analogous to the situation observed for amylose gels, differences between these systems exist, however. Amylose gels (1-2% w/v) will develop spontaneously if an aqueous suspension of this biopolymer is heated above its gelation temperature (70°C) while pustulan gels are established in a time-dependent fashion only upon cooling. Further, pustulan gels can readily be remelted while amylose gels are not totally thermally reversible. It has been postulated that amylose gels are established by the development of multiple helical conformations which join molecules in aqueous solution followed by a slow aggregation of multiple helical segments to form crystallites. In the case of pustulan, we cannot unambiguously determine if the initial random-coil conformation transforms to an "ordered" or "helical" conformation prior to gelation and the development of crystallinity. Thermodynamic considerations, however, favor the intermolecular association of "ordered" molecules in solution since conformation entropy losses are minimized.[22] There also exists an interesting parallel between the gelation of pustulan and the polypeptide derivative, gelatin, which is obtained from the partial hydrolysis of collagen, a protein which exists in the skin of animals.[32-34] In the case of α-gelatin, it is believed that solutions form gels upon cooling by first adopting short-segments of a helical, "collagen-fold" conformation followed by crosslink formation through intermolecular hydrogen-bonds which ultimately results in gelation if concentrations are in the range where molecular overlap can occur. At high concentrations ($\sim$2.0%), gelatin has been observed to develop crystallinity through the interaction of aggregated helices in the gel.[35]

It should also be noted that the thermoreversible gelation of pustulan by crystallization appears mechanistically similar to processes reported for isotactic polystyrene[36] and chlorinated polyethylenes[37]. In these cases, the initial phase of gelation involves a "fringed micelle" type crystallite which provides a nucleation site for further development of larger crystalline regions which provide the crosslinks necessary for gel network formation.

REFERENCES

1. Marchessault, R. H. (1984) Chemtech, 542-552.

2. Davidson, R. L. (1980) Handbook of Water-Soluble Gums and Resins, McGraw-Hill Book Co.

3. Whistler, R. L. (1973) Industrial Gums, Academic Press.

4. Mish, L. B. (1953), Ph.D. Thesis, Harvard University.

5. Drake, B. (1943) Biochem. Z. 313, 388-395.

6. Lindberg, B. and McPherson, J. (1954) Acta Chemica Scandinavica 8, 985-988.

7. Nishikawa, Y., Takeda, T., Shibata, S. and Fukuoka, F. (1969) Chem. Pharm. Bull. 17(9), 1910-1916.

8. Stipanovic, A. J. and Stevens, E. S. (1980) Int. J. Biol. Macromol 2, 209-213.

9. Stipanovic, A. J. and Stevens, E. S. (1981) ACS Symp. Series No. 150, 303-315.

10. Stipanovic, A. J. and Stevens, E. S. (1981) Makromol. Chem. Rapid Commun. 2, 339-341.

11. Stipanovic, A. J., Giammatteo, P. J. and Robie, S. B. (1985) Biopolymers, in Press.

12. Shibata, S. Nishikawa, Y., Takada, T. and Tanaka, M. (1968) Chem. Pharm. Bull. 16(12), 2362-2369.

13. Dais, P. and Perlin, A.S. (1982) Carbohydr. Res., 103-116.

14. Colson, P., Jennings, H. J., and Smith, I.C.P. (1974) J. Amer. Chem. Soc, 8081-8087.

15. Perlin, A. S. and Hamer, G. K. (1979) ACS Symposium Series 103, 123-141.

16. Heyraud, A., Rinaudo, M., Vignon, M. and Vincendon, M. (1979) Biopolymers, 167-185.

17. Inoue, Y. and Chijo, R. (1978) Carbohydr. Res., 367-370.

18. Rinaudo, M., Milas, M., Lambert, F. and Vincendon, M. (1983) Macromolecules 16, 816-819.

19. Matsuo, K. (1984) Macromolecules 17, 449-452.

20. Saito, H. (1981) ACS Symposium Series 150, 125-147.

21. Bassieux, D., Gagnaire, D., and Vignon, M. (1977) Carbohydr. Res. 56, 19-33.

22. Drake, A., Morris, E. R., Rees, D. A. and Welsh, E. J. (1970), Carbohydr. Res. 66, 133-144.

23. Fyfe, C. A. (1983) Solid-State NMR For Chemists, CFC Press, Guelph.

24. Lambert, J. B. and Riddell, F. G. (eds) (1983) The Multinuclear Approach to NMR Spectroscopy, D. Reidel Publishing Co., Dordrecht, 111-131.

25. Randall, J. C. (ed), (1984) NMR and Macromolecules, ACS Symposium Series 247, Washington, 21-41.

26. Fukushima, E. and Roder, S. B. W. (1981) Experimental Pulse NMR Addison-Wesley Pub. Co., 218-220.

27. Pfeffer, P. E., Hicks, K. B., Frey, M. H., Opella, S. J. and Earl, W. L. (1984) J. Carbohydrate Chem. 3(2), 197-217.

28. Maciel, G. E., Kolodziejski, W. L., Bertran, M. S., and Dale, B. E. (1982) Macromolecules 15, 686-687.

29. Fyfe, C. A., Dudley, R. F., and Stephenson, P. J. (1983) J. Macromol.Sci.-Rev. Macromol. Chem. Phys. C23(2), 187-216.

30. Rohrer, D. C., Sarko, A., Bluhm, T. L. and Lee, Y. N. (1980) Acta. Cryst., Section. B. B36, 650-658.

31. Sathyanarayana, B. K. and Stevens, E. S., (1983) J. Biomolecular Structure and Dynamics 1,947-959.

32. Finch, C. A. (ed), Chemistry and Technology of Water Soluble Polymers (1983), Plenum Press, New York.

33. Eagland, D., Pilling, G. and Wheeler, R. G. (1974) Discussions Faraday Soc. 181-200.

34. Henderson, G. V. S., Campbell, D. O., Kuzmicz, V. amd Sperling, L. H. (1985) J. Chem. Ed. 62(3), 269-270.

35. Boedtker, H. and Doty, P. (1954) J. Phys. Chem. 58, 968-984.

36. Atkins, E.D.T., Hill, M. J., Jarvis, D.A., Keller, A., Sarhene, E., and Shapiro, J. S. (1984) Colloid and Polymer Science 262, 22-45.

37. Tan, H. M., Chang, B. H., Baer, E. and Hiltner, A. (1983) Eur. Polym. J. 1021-1025.

SUGAR CONTAINING POLYMERS DERIVED FROM ORGANOSTANNES AND BIS-(CYCLOPEN-

TADIENYL) TITANIUM DICHLORIDE

Yoshinobu Naoshima[a], Charles E. Carraher, Jr.[b,c], Satomi Hirono[a], Tamara S. Bekele[b], and Phillip D. Mykytiuk[b]

Department of Chemistry, Okayama University of Science[a], Ridai-Cho, Okayama 700, Japan and Department of Chemistry, Wright State University[b], Dayton, Ohio 45435, USA, and Department of Chemistry Florida Atlantic University[c], Boca Raton, Florida 33431, USA

ABSTRACT

Crosslinked metal-containing networks are formed through condensation of Cp_2TiCl_2 and organotin dichlorides with sucrose. The presence of the metal-containing moiety is indicated by color (in the case of Ti), elemental and thermal analyses, mass spectroscopy and infrared spectroscopy. The crosslinked nature of the product is indicated by solubility studies. The formation of the M-O-R linkage is consistent with previous studies.

INTRODUCTION

Carbohydrates represent the most abundant (weight and molecular-wise) renewable resource. About 400 billion tons are photosynthetically produced yearly. Sucrose is referred to as sugar. In the USA it is obtained mainly from sugar beets (about 15% sucrose; 3.1 million metric tons-1980). About 100 million metric tons are produced (under cultivation) worldwide. This represents well less than 1% of the total sucrose produced synthetically. Primary uses for cultivated sucrose are as a sweetner and as the major feedstock in the fermentation process that produces ethanol.

The initial major source of sugar was the sugar cane dating in written records to the fourth century B.C. in northern India. The name sugar is derived from the Sanskrit word "sarkara" and the Sanskrit word for solidified sugar is "khanda" from which the word candy originates.

The second major source of sugar today is the sugar beet, Beta vulgaris. It is interesting to note the complementary nature of the sugar beet and the sugar cane. The sugar beet is grown in temperate to cold climates whereas sugar cane is grown in tropical regions. Further, the sugar is stored in the root for the beet and the stalk portion for the cane.

The isolation of the sugar from the beet occurs through a series of more technology-intensive steps, compared to sugar isolation from the cane. This is due to at least two factors. First, the more recent emergence of the beet as a source of sugar. Second, the fact that much

of the cane production occurs in rural, less technologically developed
areas counter to beet production. Of importance here is that the isolation and refining of sugar is well known and widely spread throughout
both developed and less developed countries. Major sugar manufacturers
are listed in Table 1.

Sugar is a high energy (about 1,800 kcal/lb) food source that both
improves the palatibility of other foods and it can act as a preservative.
The amount of sugar produced for the market place is controlled in a
general manner by the International Sugar Agreement where quotas are
assigned with periodic adjustments to reflect the demand (price) of
sugar. Thus the capacity for sugar production is greater than the actual
production.

Reasons for employing sucrose as a feedstock include the ready
availability and low cost. Further, unlike polysaccharides such as
cellulose and dextran, sucrose represents a chemically well defined,
pure material where structure-property relationships are more easily
reproducible regardless of the source.

Sucrose, alpha-D-glucopyranosyl beta-D-fructofuranoside, is a disaccharide containing one glucose and one fructose unit. It is the most
widely occurring disaccharide, found in all photosynthetic plants.

The reaction of simple sugars and nucleosides with various metal-
containing reactants has focused on the use of monofunctional metal-
containing reactants with little work done with difunctional reactants
(for instance 1-7). In all cases emphasis focused on the synthesis
of the monomeric species. For instance, Moffatt and co-workers (5)
synthesized in good yield, a number of cyclic organotin derivatives
of nucleosides (2) from reaction with dibutylin oxide. The cyclic products are soluble in polar solvents such as methanol, ethanol and DMF.

Several points should be made. First, almost all of these studies
involved nonacid chloride reactants. Even so the reaction characteristic
of metal-containing reactants as dialkyltin oxides should be similar
to that of dialkyltin dichlorides. Second, the ring hydroxyl groups
are more reactive then are the non-ring hydroxyl group (5'-OH) with
respect to nucleophilic substitution. Third, almost all of the studies
involved simple sugars that have only the 2' and 3'-hydroxyl groups
free, leading to the synthesis, in good yield, of the cyclic, monomeric
product. As will be described, the present study involves the use of
sucrose that contains five ring and three non-ring hydroxyls increasing
the opportunity for interconnection of the sugar rings through condensation with metal-containing bifunctional reactants.

EXPERIMENTAL

Sucrose (Fisher Scientific, Fairlawn, N.J. and Wako Pure Chemical
Industries Ltd., Japan), bis(cyclopentadienyl)titanium dichloride (Aldrich,
Milwaukee, Wis.), diphenyltin dichloride (Aldrich) and di-n-butyltin
dichloride (Aldrich) were employed as received. Reactions were conducted
employing a one quart Kimex emulsifying jar placed upon a Waring Blendor
(Model 7011G) with a "no-load" stirring rate of 20,000 rpm. The product
was obtained as a precipitate. Water and organic liquid washes assisted
in the purification of the product.

A second procedure was employed where the solid precipitate was
allowed to remain with the aqueous phase. Reaction procedures are similar
to that described above except the solid and aqueous phase are allowed

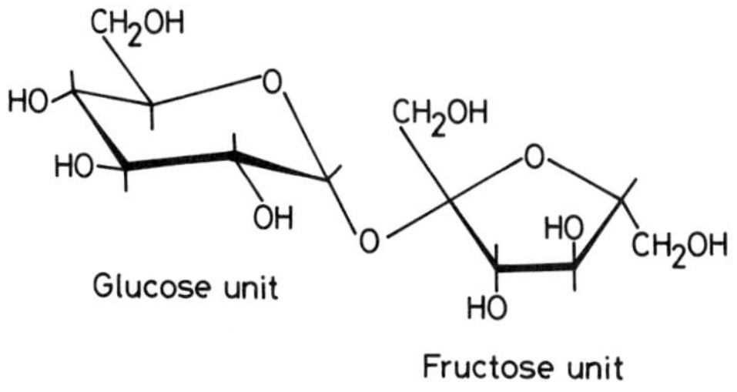

Structure of the disaccharide (+)-sucrose

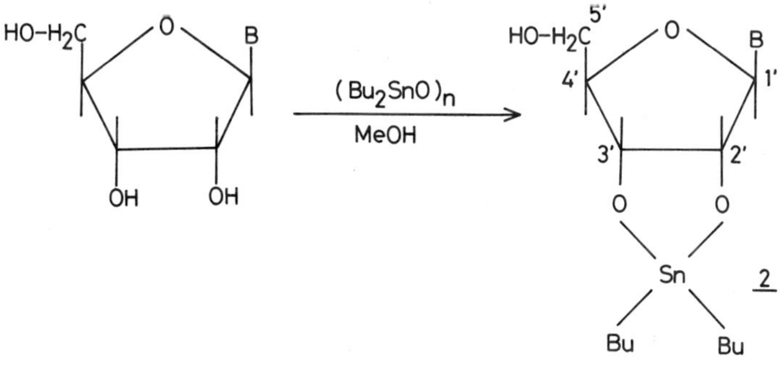

TABLE 1. Leading Commercial Sugar Producing Countries

Country	Total Production[a]	Sugar Beet Production[b]	Sugar Cane Production[b]
Brazil	8.2		90
Russia	7.0	66	
India	7.0		140
Cuba	6.0		54
USA	5.2	26	26
France	4.3	23	
Australia	3.3		22
Mexico	2.9		32
West Germany	2.9	17	
China	2.8	6.7	34
Phillipines	2.6[b,c]		24
Pakistan	2.3[b,c]		21
Poland	2.1[b,c]	15	
Colombia	2.1[b,c]		19

a. Source: USDA, 1980; millions of metric tons of sucrose production.

b. Source: Production Yearbook, 1975, FAO; millions of metric tons of raw beet or cane produced.

c. Assuming sugar beets give 14% by weight sugar and cane gives 11% by weight sugar.

to evaporate to dryness. The solid is washed with carbon tetrachloride
to remove unreacted metal-containing organohalides. A sample run is
described following. The aqueous layer (100 ml) contained sucrose (1.075
mmoles) and sodium hydroxide (1.075 mmoles) while the organic phase
consisted of carbon tetrachloride (200 ml) and Cp_2TiCl_2 (0.33 mmoles).
The system was blended for 180 seconds. The two phases were separated
and the aqueous phase allowed to dry and washed with carbon tetrachloride
to remove unreacted Cp_2TiCl_2. The product was a brown tacky, solid
material.

Common heavy microscope slides were used in quantitative tests.
The glass surfaces were scrubbed with soapy water to remove the oily
film and allowed to dry. The product was smeared onto the glass surfaces
employing a simple glass spatula.

Infrared spectra were obtained employing Hitachi 260-10 and 270-
30 spectrometers and a Digilab FTS-IMX FT-IR. The EI (70eV) and CI
(isobutane) mass spectra were obtained using a JEOLJMS-D300 GC mass
spectrometer with a JAI JHP-2 Curie Point pyrolyzer (injector temperature-
250°C; ion source temperature-200°C; pyrolysis temperature-315°C).
Additional mass spectra were obtained employing a Kratos-MS-50 mass
spectrometer operating in the El mode with probe temperatures varying
from 300 to 350°C. Analysis for titanium and tin were carried out employing the usual wet analysis procedures using $HClO_4$. DT and TG analyses
were carried out employing a SINKU-RIKO ULVAC TGD-500M or a DuPont 990
TGA and 900 DSC.

RESULTS AND DISCUSSION

This paper concentrates on the brief identification of the structural
units of the condensation product between sucrose and the metal-containing
dichlorides and on the preliminary description of properties of products
resulting from simple evaporation of the aqueous layer. The product
contains a mixture of crosslinking units as those depicted in 3-9, unreacted
units and probably includes internally cyclized units such as 2.

Historical

The reaction of hydroxyl-containing reactants forming the Sn-O-
R and Ti-O-R moieties is well known including diols and hydroxyl-containing
synthetic (polyvinyl alcohol) and natural (dextran and cellulose) polymers
(for instance 8-13). It is reasonable to consider the proposed reactions
as extensions of these reactions.

Elemental Analysis

Elemental analyses were conducted on samples derived from a variety
of reaction conditions. Titanium content was typically in the range
of 9 to 19%. Products depicted in 3-6 have calculated titanium percentages
of 9 to 20%. Thus for the product derived employing 3 mmoles of Cp_2TiCl_2,
1 mmole of sucrose and 8 mmoles of Et_3N, a titanium content of 14.5%
was formed.

Tin content was typically within the range of 20 to 40% depending
on the specific stannane. For products derived from dibutylin dichloride,
the percentage tin was generally within the range of 21 to 27% over
a wide range of reaction conditions. The calculated values for tin
of form 9 (one tin per sucrose unit) is 20%, for 7 (two tin atoms per
sucrose unit) is 29% and for 8 (four tin atoms per sucrose) is 34%.
The percentage titanium and tin indicate products with two to three
metal-containing moieties per sucrose unit.

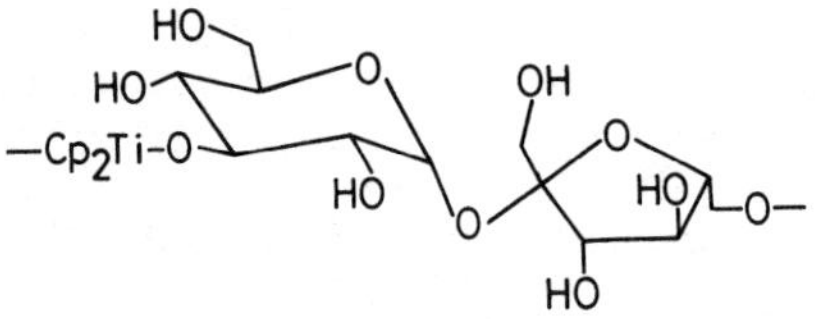

—Cp₂Ti—O
3 – %Ti=9 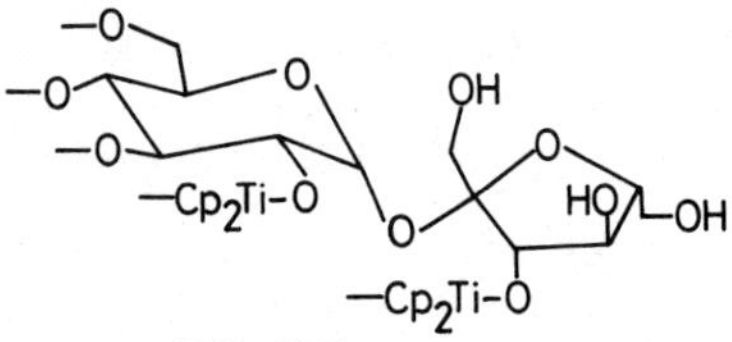
4 – %Ti=13.5 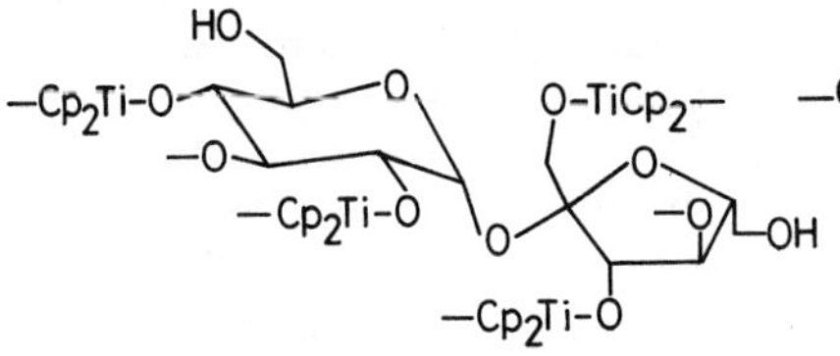
—Cp₂Ti—O
5 – %Ti=18 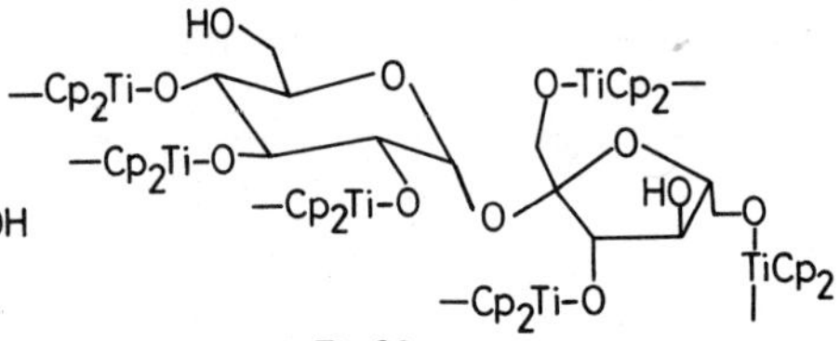
6 – %Ti=20
—R₂Sn—O
7
R₂SnCl₂ +
8
9

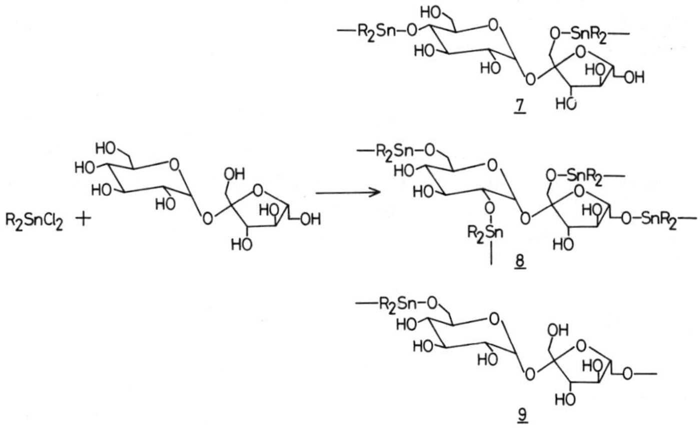

<u>Color</u>

Products derived from the organostannanes are white consistent
with the known lack of color sites for both sucrose and the employed
organostannanes. The titanium-containing products are light to dark
brown in color consistent with the presence of the Ti-Cp brown-color
site. The lack of red coloration is consistent with the lack of significant amounts of Ti-Cl moieties (a red color site) being present.

<u>Infrared Spectroscopy</u>

Analysis of the infrared spectra of the products is also consistent
with products of forms <u>3-6</u>. Bands characteristic of the Cp_2Ti moiety
are present at 1400 and 1040 cm^{-1}. Bands characteristic of the sucrose
moiety are present (within the 1600 to 1000 cm^{-1} region) at 1370, 1340,
1245, 1170, 1160, 1070 and 1000 cm^{-1}. A band assigned as arising from
the Ti-O stretching in titanium ethers may be present at 1130 cm^{-1}.
Products derived from triethylamine contain bands about 2695-2825 cm^{-1}
and 2500-2825 cm^{-1} attributed to the presence of the triethylamine
moiety. This is verified by MS studies reported following.

The stannane products show the presence of bands about 660 to 690
cm^{-1} consistent with the formation of the Sn-O-R ether linkage (assymmetric
stretch) and a doublet about 550 to 600 cm^{-1} attributed to the symmetric
stretch of tin ethers. For the dibutylin dichloride product bands characteristic of the methylene deformation are present at 1470 and 1150 cm^{-1};
bands characteristic of methyl groups are present at 1420 (assymmetric
stretching) and 1380 (symmetric stretching) cm^{-1}; bands characteristic
of the C-H stretch in n-butyl groups are present at 2910, 2880 and 2860
cm^{-1}; and bands at 668, 598, and 585 cm^{-1} are assigned to the presence
of the Sn-O-R moiety.

<u>Mass Spectroscopy</u>

Mass spectral data are consistent with a product containing units
derived from the metal-containing moiety and sucrose (specifically ion
masses at m/z=17(OH), 18(H_2O), 27, 29, 39, 40, 42) with some ion masses
assignable to both moieties (Table 2).

For the products obtained from the regularly derived Cp_2TiCl_2 there
are masses at m/z 36, 38 which could be assigned to HCl (Cl=35) and
HCl (Cl=37). The natural abundance ratio of Cl 35/37 is about 3:1,
while the m/z ratio of 36:38 is 1:12. Further both masses can be derived
from the sugar. Thus, the mass spectral data does not indicate the
presence nor absence of TiCl endgroups. For the products derived from
evaporation of the aqueous layer, the ratio of 36:38 is 2.8 and the
ratio of 35:37 is 3.2 both near the known isotopic abundance for Cl(35)
to Cl(37) consistent with the presence of at least some TiCl endgroups
or entrapped chloride ion. While there are masses at m/e 49, 51 (possible
CH_2Cl) and 84, 86, 88 (possible CH_2Cl_2), the intensities are not present
at the ratios predicted by the isotopic abundances of Cl (35=75%; 37=25%).
Thus, while entrapped solvent may be present, it is not supported by
MS data.

For the products isolated through evaporation of the aqueous phase,
there are no ion fragments near m/e 117 (CCl_3 (C35)) or 152 (CCl_4).
There are ion fragments at 82, 84 and 86 m/e in the ratios of 2.3:1.5:1
(predicted ratios (9:3:1) not the predicted ratios. Further there are
m/e ion fragments at 85, 83, and 81. This group of ion fragments (m/e
81-86) is generally assigned as being derived from the sugar moiety.
Again, while entrapped organic solvent may be present, it is not supported
by MS data.

TABLE 2. Mass spectral fragmentation pattern results for the condensation products of Cp_2TiCl_2, diphenyltin dichloride and sucrose.

<u>Triethylamine</u> (Cp_2TiCl_2)

m/z	86	58	101	56	100	87	70	72	57	54	55
standard	1000	262	157	85	66	56	38	33	22	19	10
Present-Et_3N	1000	354	184	62	74	61	33	53	20	7	6

<u>Cyclopentadiene</u>

m/z	66	65	39	40	38	63	67	62	31	61
standard	1000	473	316	273	82	82	63	58	49	45
Et_3N Product	1000	520	314	282	81	90	53	54	38	43
NaOH Product	1000	648	715	-	163	99	204	64	16	55

<u>Benzene</u>

m/z	78	51	52	50	39
standard	1000	205	196	179	141
Present-NaOH	1000	313	325	156	638[a]

a. Derived from both the stannane and sucrose moieties.

The mass spectral data also shows the presence of triethylamine
(Table 2) for products derived employing triethylamine as the added
base. In previous studies employing dextran instead of sucrose it was
found that the triethylamine was present as an endgroup as depicted
by <u>10</u> (12).

$$R-O-Cp_2TiNEt_3^{+}Cl^{-}$$

<u>10</u>

Thermal Analysis

The DT thermograms in air of the products derived from employing
NaOH or Et_3N as the added base are similar. A small endotherm occurs
between 50° to 100°C followed by a similar exotherm to 175°C. A much
more highly (net) degradation occurs from 200 to 550°C with peaks occurring
at 320° and 490°C. The initial exotherm coincides with onset of weight
loss, corresponding to an 8% weight loss. The endothermic region to
175°C corresponds to an additional 8% weight loss. The highly exothermic
region between 200 to 550°C corresponds to an addtional 24% weight loss.
From 550°C through 1000°C 3% weight loss occurs and an additional 2%
weight loss occurs as the sample is held at 1000°C for 30 minutes.
The total weight loss is 45% corresponding to a percentage titanium
(assuming the residue is only TiO_2) of 27%. The sites of highest exothermic behavior (at 320 and 490°C) coincides with regions of accelerated
weight loss. The occurrence of stability plateaus, retention of significant weight past 500°C (carbohydrates, including sucrose, degrade yielding
only small amounts of char past 500°C) and highly exothermic degradative
pathways are all characteristic of organometallic polymeric materials
(11).

Solubility

Solubility tests were carried out employing about 1 mg of sample
to 3 ml of liquid. A wide variety of liquids were examined including
chloroform, dichloromethane, acetone, dioxane, pyridine, ether, ethyl
acetate, benzene, hexane, DMSO, and water. The product is insoluble
in the afore liquids consistent with product being crosslinked. It
swells and becomes tacky in HMPA.

General Properties of Modified Isolation Product

One general aim of this work is the synthesis of potentially useful
coatings and/or adhesives. The products isolated in the usual manner
are typically powdery solids. It was found that a sticky, tacky material
was forthcoming if the product is isolated through evaporation of the
aqueous phase. As cited previously in this chapter, the physical characteristics of this product are generally the same as those derived from
employing the usual isolation procedure. A number of qualitative experiments were performed using the product derived from the modified isolation
procedure. The product is brown. It remained a tacky, viscous solid
capable of giving cold drawn fiber-like strands even after three years
of exposure to the atmosphere. The material chars when heated to 130°C
but forms a good bond when adminstered prior to heating and heating
to 130°C. When placed between glass slides, it retains its tacky nature,
also after about three years. When heated to 50°C for three minutes
it forms a flexible, tacky coating on glass. Further heating produces
a tough, abrasive resistant (to the fingernail) glass coating.

About one-half of the product is water soluble. The water soluble
portion gives a brown solution. The water soluble portion forms a brown
precipitate on addition to benzonitrile. This fraction is believed
to be uncrosslinked product. Again, it forms a tough tacky glass film.
It also forms a good, stiff bond when placed between glass plates and
heated to about 100°C for five minutes. The water insoluble portion
forms a tough, rough abrasive resistant (to the fingernail) coating
on glass when heated to 50 to 60°C.

Clearly much remains to be done in both describing the product
and its properties but preliminary results indicate that this product
may offer good long term tacticity, flexible glass bonding and when
heated, decent glass coatings.

REFERENCES

1.   J. Zuckerman (Ed.), 'Organotin Compounds': 'New Chemistry and Appli-
     cations', ACS, Washington D.C., 1976.
2.   A. Crowe and P. Smith, J. Organomet. Chem., 110, C57 (1976).
3.   J. Alais and A. Maranduba, Tetrahedron Letters, 24, 2383 (1983).
4.   Y. Tsuda, M.E. Haque and K. Yoshimoto, Chem. Pharm. Bull., 31(5),
     1612(1983).
5.   D. Wagner, J. Verheyden and J.G. Moffatt, J. Org. Chem., 39, 24(1974).
6.   A. Husain and R.C. Poller, J. Organometal Chem., 118, C11 (1976).
7.   R. Munavu and H.H. Szmant, J. Org. Chem., 41, 1832(1976).
8.   C. Carraher and G. Burrish, J. Macromol. Sci.-Chem., A10, 1457(1976).
9.   C. Carraher and S. Bajah, Br. Polym. J., 7, 155(1975).
10. C. Carraher and S. Bajah, Polymer, 15, 9(1974).
11. C. Carraher and J.D. Piersma, J. Macromol. Sci.-Chem., A7, 913(1973).
12. Y. Naoshima, C. Carraher, and G. Hess, Polymeric Materials, 49,
     215 (1983).
13. C. Carraher, J. Macromol. Sci.-Chem., A17, 1293 (1982).

SYNTHESIS OF TITANIUM, ZIRCONIUM AND HAFNIUM MODIFIED POLYSACCHARIDES

Yoshinobu Naoshima[a], Charles E. Carraher, Jr.[b], and Koichi
Matsumoto[a]

Okayama University of Science[a], Department of Chemistry, Ridai-
cho, Okayama 700 Japan and Wright State University[b], Department
of Chemistry, Dayton, Ohio 45435 and Florida Atlantic University[b],
Department of Chemistry, Boca Raton, Florida 33341

ABSTRACT

Dextran has been modified employing the classical interfacial poly-
condensation technique through reaction with Group IV B bicyclopenta-
dienylmetallocene dichlorides. Product yields, both percentage and
weight, follow the general trend Hf>Zr>Ti consistent with the Hard-Soft
Acid Base theory. Dextran is employed as a representative polysaccharide.

INTRODUCTION

About 400 billion tons of saccharides are produced annually through
natural photosynthesis. The vast majority decompose within less than
a single growing season. The saccharides range in structure from simple
sugars to polymers which offer their own tertiary dimensional structures.
As a group, saccharides represent a large source of under-used feedstock
that should be investigated as an alternative feedstock to largely non-
renewable sources as petroleum, natural gas or coal.

Dextran was chosen as the representative polysaccharide for at
least the following reasons. First it is water soluble allowing three
dimensional modification employing aqueous solution and classical interfacial condensation routes. Second, it is readily available in industrial
quanities. Third, it is available in a range of molecular weights allowing
product modification to be studied as a function of dextran chain size.
Fourth, it is generally considered to be an under-utilized natural feedstock.

We recently reported the initial synthesis and structural characterization of titanium-containing dextran (1-3). Here we report the initial
synthesis of the analogous zirconium and hafnium-containing products
and an expanded study involving the synthesis of the titanium-containing
product.

Dextran is the collective name of extracellular bacterial poly-
alpha-D-glucopyranoses linked largely by 1,6 (1-6) bonds. Physical
properties vary according to the amount and manner of branching, nature
of endgroup, molecular weight and molecular weight distribution,
processing, etc.

Dextran is produced from sucrose by a number of bacteria the major
ones being the nonpathogenic bacteria Leuconostoc mesenterodes and Leuconostoc dextranicum. As expected, the structure (and consequently
the properties) of the dextran is determined by the particular strain
that produces it.

Dextran is the first microbial polysaccharide produced and utilized
on an industrial scale. The potential importance of dextran as a structurally (and property) controlled feedstock is clearly seen in light
of the recent emphasis of molecular biologists and molecular engineers
in the generation of microbes for feedstock production. Dextran is
employed as pharmaceuticals (additives and coatings of medications),
within cosmetics, as food extenders, as water-loss inhibitors in oil-
well drilling muds and as the basis for a number of synthetic resins.

Svenska Sockerfabriks Aktiebolaget for Aktiebolaget Pharmacia in
Sweden began large scale production of dextran about 1942 (4). By 1947,
Dextran Ltd. (East Anglia Chemical Co, England) began production of
dextran and in 1949 Commercial Solvents Corp. (USA) began production
(5,6). In 1952, the R.K. Laros Co. (6) began the enzymic production
of dextran in the presence of living cells. In an effort to standardize
the dextran produced, by 1952, the majority of companies employed the
L. mesenteroides NRRL B-512-(F) in the production of dextran. The production of dextran involves merely mixing the appropriate quantities of
sucrose and enzyme under prescribed conditions. In actuality, the resultant "pot" contains a variety of products that must be purified. Figure
1 lists a brief flow chart for the industrial production and purification
of dextran.

Following describes briefly general properties of dextran. (Again,
while the singular term dextran is used, it must be remembered that
the term dextran applies to a wide range of 1,6-polysaccharides.) Dextran is a white, tasteless solid with the specific properties dependent
on both the bacteria strain employed and the specific conditions for
synthesis and purification.

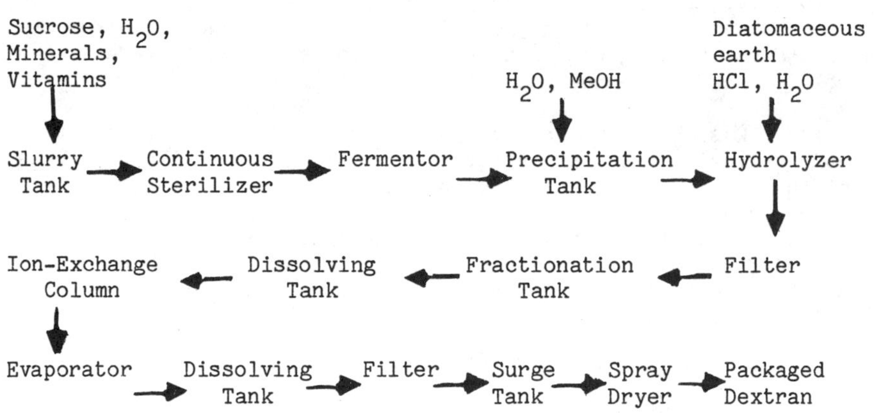

Figure 1. Flow sheet for the production of dextran.

Generally, solubility decreases with increase in chain size and
extent of branching. Typically the solubility of dextran can be divided
into four groups--those that are readily soluble at room temperature
in water, DMF, DMSO and dilute base; those that have difficulty dissolving
in water; those that are soluble in aqueous solution only in the presence
of base; and, those that are soluble only under pressure, at high temper-

atures ($>100^{\circ}$C) and in the presence of base. Dextran B-512 readily
dissolves in water and 6M, 2M glycine and 50% glucose aqueous solutions.

Dextran B-512 in water approaches a form of compact spherical-like
helical coils (7-9). Streaming, birefringence measurements indicate
that the dextran has some flexibility. The B-512 dextran shows a refractive index increment, dn/dc, of 0.154 ml/g at 436 mu in water and an
apparent radius of gyration in water on the order of 2×10^{3}Å. Branching
occurs through about 5% of the units through the 1,3 linkages with about
80% of these branches being only one unit in length. There exists a
few, less than 1%, long branches in B-512 dextran (9).

Chemically dextrans are similar to one another. The activation
energy for acid hydrolysis is about 30-35 kcal/mole (8). The C-2 hydroxyls
appear to be the most reactive in most Lewis base and acid-type reactions.
A wide variety of esters and ethers have been described as well as car-
bonates and xanthates (10,11). In alkaline solution, dextran forms
a varying complex with a number of metal ions (12).

The biological properties of dextran again vary with strain. The
B-512 is digestible in mammalian tissues as the liver, spleen, and kidneys,
but not in the blood (13). It appears not to stimulate formation of
antibodies in man upon intravenous infusion (4). Counter, subcutaneous
injection in humans led to skin sensitivity and to the formation of
precipitating antibodies (14). Antibody formation and precipitation
with antiscrums decreases as chain length increases (14,15).

Linear, water-soluble dextrans have many uses. One dextran is
employed in viscous water-flooding for secondary recovery of petroleum
with a potential market of about 2×10^{5} tons per year. This dextran
is superior to carboxymethyl cellulose when employed in high calcium
drilling muds (16,17). These dextran-muds show superior stability and
performance at high pH and in saturated brines (18-20). Other dextrans
show good resistance to deterioration in soil and the ability to stabilize
aggregates (21) in soils. They can also be used in binding collagen
fibers into surgical sutures (22).

Dextran is employed in food processing based on their ability to:
a. stabilize syrup confections against crystallization, b. stabilize
texture in ice creams and sherbets, c. act as adhesives and humectants,
and d. form viscous liquid sugars (10,23). Water insoluble dextran
flavored with fruit liquids constitute the Philippine dessert hata (24).
Table 1 lists additional uses of dextran.

EXPERIMENTAL

Biscyclopentadienlytitanium dichloride (BCTiCl, Aldrich, Milwaukee),
biscyclopentadienylzirconium dichloride (BCZrCl, Aldrich, Milwaukee),
biscyclopentadienylhafnium dichloride (BCHfCl, Alfa, Danvers) and dextran
(USB, Cleveland OH and Wako Pure Chemical Industries; molecular weight
= 2 to 3×10^{5}) were used as received. Reactions were conducted using
a one quart Kimex emulsifying jar placed on a Waring Blendor (Model
7011 G) with a "no load" stirring rate of 20,000 rpm. The product is
obtained by suction filtration. Repeated washings with the organic
liquid and water assisted in the purification of the product.

Infrared spectra were obtained employing Hitachi 260-10 and 270-
30 spectrometers and a Digilab FTD-IMX FT-IR. The EI (70eV) and CI
(isobutane) mass spectra were obtained employing a JEOL JMS-D300 GC
mass spectrometer with a JAI JHP-2 Cure Point pyrolyzer. Analysis for

metals was carried out employing the usual wet analysis procedures using $HClO_4$. DT and TG analysis was carried out employing a SINKU-RIKO ULVAC TGD-5000M.

RESULTS AND DISCUSSION

Structural characterization is based on solubility, thermal analysis elemental analysis, infrared spectroscopy and mass spectroscopy. Results will be detailed in a subsequent paper. Characterization was in agreement with the product containing units as depicted in 1 to 4 including unreacted units.

Following is a brief description of several pertinent points related to structural analysis. First, all products exhibit infrared spectral bands at 1405, 1030, and 855 cm^{-1} assigned as characteristic of the Cp-M moiety and bands between 3000 and 3600 cm^{-1} characteristics of unreacted hydroxyl groups. Second, mass spectral analysis is consistent with the presence of both moieties derived from metallocene and dextran. Ion fragmentation patterns for cyclopentadiene are given in Table 2 and are consistent with the presence of the cyclopentadiene moiety. Third, products derived utilizing triethylamine exhibit infrared and mass spectral ion fragments characteristic of the presence of endgroups identified previously for the titanium-containing products as being of the form 5.

TABLE 1. Uses of dextran as a function of chain length.

Molecular Weight (10^{-4})	Applications	References
200-3	Molecular sieves for separating watersol substances	11,25
200,50,18,4	Macromolecular separations in two-phase systems	26
200-3	Fractionation of nucleic acids, proteins, polysaccharides by ion-exchange	11,25
74,23,18	Medium for suspension polymerization of unsaturated monomers	27
50	Dialysis	28
50,25	Concentration and separation of blood	29
30	Photosensitive resists for pre-coating deep etch process plates	30,31
20,7,2.5	Suspend contrast agent for roentgenographic examination of body cavities, blood vessels and organs	32
10-7	Blood anticoagulant, antilipemic agent	33
7	Treatment of hephrotic, hypoproteinemia and cerebral edema	4
7	Blood plasma substitute	4
3-1	Stabilizer in freeze-drying sensitive materials such as bacteria and vaccines	35

Table 2. Ion fragmentation patterns for cyclopentadiene.

Ion fragment (m/e)	66	65	39	40	38	63	67	62	31	61
Standard	100	47	32	27	8	8	7	6	5	5
Ti, NaOH	100	52	73	22	13	8	8	6	3	3
Ti, Et_3N	100	53	43	–	10	8	6	6	3	4
Zr, NaOH	100	53	79	10	17	8	11	–	–	–
Zr, Et_3N	100	56	53	–	11	8	8	5	1	4
Hf, NaOH	100	49	59	22	13	7	9	5	–	4
Hf, Et_3N	100	52	46	29	10	8	9	5	7	4

For the present system the percentage yield trend is not always clear but in general follows the order Hf>Zr>Ti. For the synthesis of polyethers (6), the yield trend is not clear, with a mild trend also of the order Hf>Zr>Ti (for instance 36-39).

Table 3 contains results as a function of metal and molar ratio.
Product yield is given in both actual weight of precipitate obtained
and percentage yield based on the repeat unit 1 assuming a one-to-one
reaction ratio. Percentage yield can be varied by varying the basis
for its calculation, but the general trends are independent of this
assumption. The trends are better detailed in figures 2-5. The dashed
tie-line represents the presence of an equal molar amount of Lewis base
and acid reactants.

Yield, both actual weight and percentage, are greatly diminished
at extremes of metallocene dihalide concentrations with decent (greater
than 50%) yields occurring over a molar ratio of reactant groups of
about 2:9 through 1:1 MCl:ROH. The maximums of percentage yield do
not occur at 1:1 ratios of reactants indicating that factors in addition
to mass abundance factors are important. The reaction system appears
to be less dependent on the amount of dextran present with decent (greater
than 35%) yields occurring over the entire range tested (ratios of MCl:ROH
from 4:1 to 1:2).

The complex dependency of reaction results with reactant molar
ratio is consistent with a number of other interfacial condensations
and indicates the complexity of the interfacial reaction systems.

Table 3.   Results as a Function of Cp_2MCl_2 and Reactant Molar Ratio.

Cp_2MCl_2 (mmole)	Dextran (mmole)	Cp_2TiCl_2			Cp_2ZrCl_2			Cp_2HfCl_2		
		Yield (g)	Yield (%)	Ti (%)	Yield (g)	Yield (%)	Zr (%)	Yield (g)	Yield (%)	Hf (%)
0.50	3.00	0.0261	15	7	0	–	–	0	–	–
1.00	3.00	0.142	42	9	0.1843	48	20	0.0499	11	32
2.00	3.00	0.3282	49	12	0.5263	69	24	0.2943	31	31
3.00	3.00	0.338	33	21	0.7142	62	30	0.9592	68	37
4.00	3.00	0.0136	1	18	0.8544	75	33	1.4275	102	48
5.00	3.00	0	–	–	0.0733	6		1.6411	116	49
6.00	3.00	0	–	–	0	–	–	0.0183	1	49
3.00	0.50	0.2435	140	21	0.5428	284		0.7707	320	
3.00	1.00	0.3023	90	15	0.5165	135		0.8986	192	49
3.00	2.00	0.231	34	18	0.4845	64	30	1.0118	72	39
3.00	3.00	0.338	33	21	0.7142	62	30	0.9592	68	37
3.00	4.00	0.3381	33	17	0.3121	27	21	0.5507	39	37

Reaction conditions – Dextran and triethylamine (9.0 mmole) dissolved in 50 ml water and added to stirred (20,000 rpm, no load) solutions of Cp_2MCl_2 in 50 ml $CHCl_3$ with 30 secs. stirring time at 25°C.

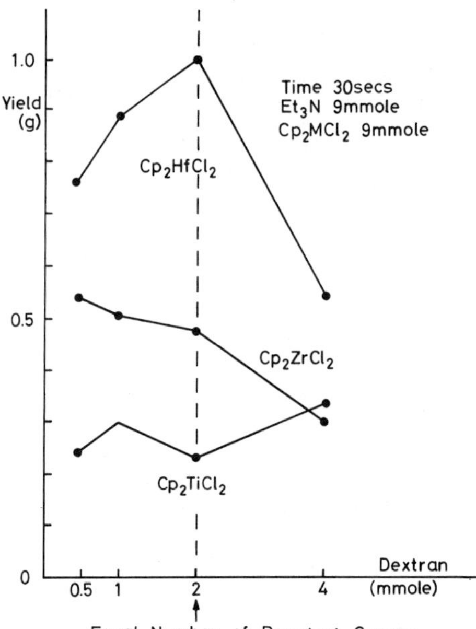

Figure 2. Weight yield of product as a function of molar amount of dextran (holding Et$_3$N (9.00 mmoles) and Cp$_2$MCl$_2$ (3.00 mmoles) constant.)

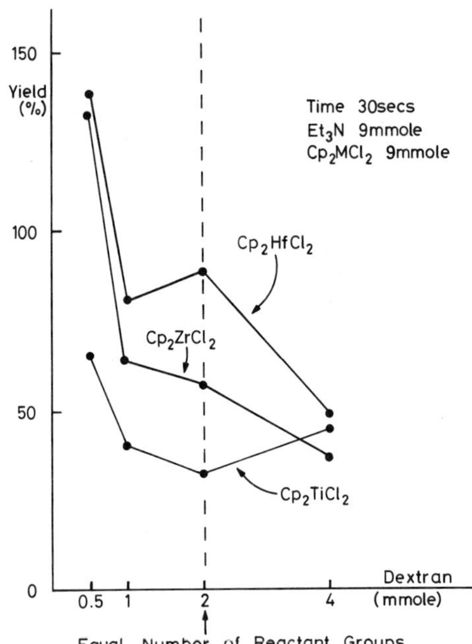

Figure 3. Percentage yield of product as a function of molar amount of dextran.

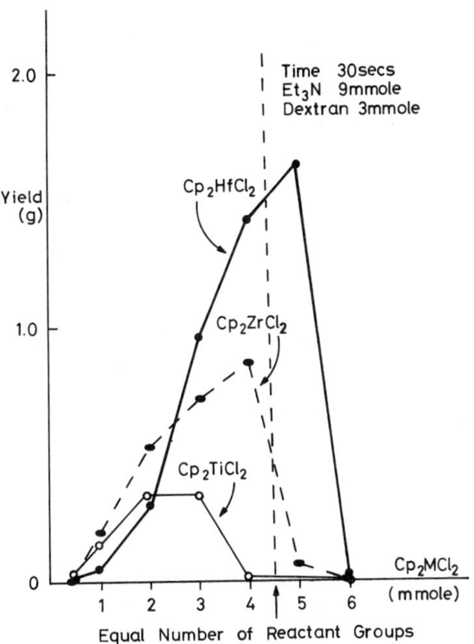

Figure 4. Weight yield of product as a function of molar amount of Cp_2MCl_2 (holding dextran (3.00 mmoles) and Et_3N (9.00 mmoles) constant.)

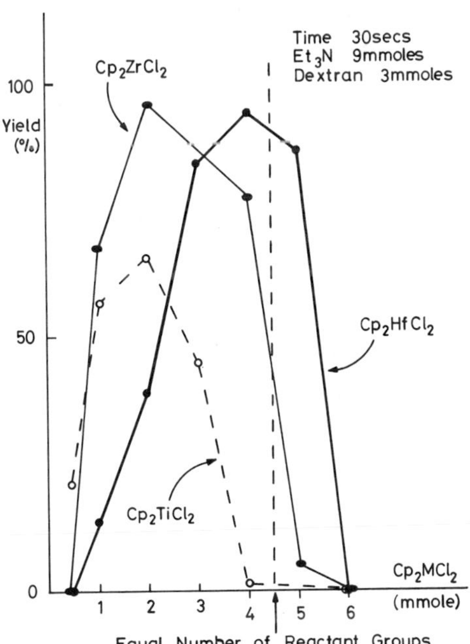

Figure 5. Percentage yield of product as a function of molar amount of Cp_2MCl_2.

$$R_2MCl_2 \; + \; HO-R-OH \longrightarrow \{M-O-R-O\}$$

with R substituents on M (structure **6**).

Due to the lanthanide contraction, both Zr and Hf have similar
sizes and since they are within the same chemical family, they are found
to exhibit similar physical and chemical properties (40). Though the
sizes are similar, the electron density distributions are different.
The outer orbitals of Hf have one additional radial "node" with the
center of electron density further from the metal nucleus compared with
Zr. This represents a more disperse (or softer acidic) outer shell
for Hf compared to Zr and Ti. Thus the Lewis base may more easily and
at longer distances "distort" the orbitals of Hf being seen as an increase
in the rate of reaction for Hf, giving as an overall reactivity trend
of Hf Zr Ti, which is consistent with that observed. Other factors
may also be (partially or wholly) responsible for the observed trend.
These include relative solubility of M-modified dextran, tendency to
form cyclic and crosslinked products, and diffusion rates.

Reaction between Group IVB Cp_2MCl_2 and diols occurs within the
organic phase, but quite near the interface, being seen as a high dependency of the reaction system on the nature of both phases. Results given
in Table 1 are consistent with this, with yield and extent of modification highly dependent on both the nature and amount of the Lewis acid
and on the amount of dextran.

Extent of modification (i.e. inclusion of Cp_2M moiety) is generally
high and, as expected, increases as the relative ratio of Cp_2M to dextran increases, though no product is formed at the higher ratios of
Cp_2M to dextran.

The products from Cp_2Ti are light brown to yellow brown (due to
the presence of the yellow-brown Cp-Ti color site), while the products
from Cp_2ZrCl_2 and Cp_2HfCl_2 are white to light gray.

REFERENCES

1.   Y. Naoshima, C. Carraher, G. Hess and M. Kurokawa, "Metal-Containing
     Polymeric Systems," (J. Sheats, C. Carraher and C. Pitman, Eds.,)
     Plenum Press, New York (1985).
2.   Y. Naoshima and C. Carraher, Polymeric Materials, 50 403 (1984).
3.   Y. Naoshima, C. Carraher and G. Hess, Polymeric Materials, 49,
     215(1983).
4.   A. Gronwall, "Dextran and Its Use in Colloidal Infusion Solutions,"
     Academic Press, N.Y., 1957.
5.   Mfg. Chemist, 23(2), 49(1952).
6.   Chem. Eng. 59,215(Sept. 1952) and 240(Dec. 1952).
7.   H. Van Oene and L. H. Cragg, J. Polymer Sci., 57, 175(1962).
8.   E. Antonini, L. Bellelli, M. Bruzzesi, A. Caputo, E. Chiancone
     and A. Rossi-Fanelli, Biopolymers, 2, 35(1964).
9.   F. Bovey, J. Polymer Sci., 35, 167 and 183 (1959).
10. P.J. Baker, "Dextrans," Academic Press, N.Y., 1959.

11. P. Flodin, Dextran Gels and Their Applications in Gel Filtration, Halmstad, Uppsala, 1962.
12. C.E. Rowe, Ph.D. Thesis, Univ. Birmingham, 1956.
13. E. Fischer and E. Stein, "Dextranases," Academic Press, N.Y., 1960.
14. E. Kabat, Bull. Soc. Chim. Biol., $\underline{42}$, 1549(1960).
15. E. Hehre, J. Sugg and J. Neill, Ann. N.Y. Acad. Sci., $\underline{55}$, 467(1952).
16. P. Monaghan and J. Gidley, Oil Gas J., $\underline{57}$, 100(1959).
17. G. Dumbauld and P. Monaghan, U.S.Pat. 3,065,170(1962).
18. E. Mueller, Z. Angew. Geol., $\underline{9}$, 213(1963).
19. W. Owen, Sugar $\underline{47}$(7), 50(1952).
20. W. Owen, U.S. Pat. 2,602,082(1952).
21. L. Novak, E. Witt and M. Hiler, Agr. Food Chem., $\underline{3}$, 1028 (1955).
22. L. Novak, U.S. Pat. 2,748,774(1956).
23. D. Wadsworth and M. Hughes, U.S. Pat, 2,409,816(1946).
24. F. Adriano, J. Oliverso and E. Villanueva, Philippine J. Ed., $\underline{16}$, 373(1933).
25. "Sephadex," Pharmacia Fine Chemicals, N.Y., 1964.
26. P. Albertsson, "Partition Methods for Fractionation of Cell Particles and Macromolecules," Interscience, 1962.
27. J. Kearney, U.S. Pat. 2,857,367(1958).
28. F. Aly, Biochem. Z., $\underline{325}$, 505(1954).
29. V. Essman, Scand. J. Clin. Lab. Invest., $\underline{13}$, 134(1961).
30. G. Schwarz, U.S. Pat. 2,942,974(1960).
31. F. Pautard, Nature, $\underline{171}$, 302(1953).
32. A. Allen, U.S. Pat. 2,677,645(1954).
33. A. Gronwall, B. Ingelman and H. Mosimann, Swedish Pat., 418,014(1947).
34. C. Ricketts, Brit. Pat. 695,787(1953).
35. P. Muggleton and J. Ungar, U.S. Pat. 2, 908,614(1959).
36. C. Carraher, L. Jambaya and S. Bajah, Polymer Preprints, $\underline{18(2)}$.
37. C. Carraher and L. Jambaya, Angew. Makromolekulare Chemi, $\underline{52}$, 111(1976) and $\underline{39}$, 69(1974).
38. C. Carraher and L. Jambaya, J. Macromolecular Sci.-Chem., $\underline{A-8}$, 1249(1974).
39. C. Carraher and S. Bajah, Polymer (Br.), $\underline{15}$, 9(1974) and $\underline{14}$, 42(1973).
40. F. Cotton and G. Wilkinson, "Advanced Inorganic Chemistry," 2nd Ed., Interscience, N.Y., 1980, pp. 822-831.

ACKNOWLEDGEMENT

Portions of this work were supported by ACS-PRF Grant 13084-B3.

NEWER APPLICATIONS OF BAGASSE

M. B. Amin, A. G. Maadhah, and A. M. Usmani

Research Institute
University of Petroleum & Minerals
Dhahran 31261, Saudi Arabia

INTRODUCTION

Together with molasses, bagasse is a by-product of the sugar cane industry. Bagasse is composed of the sheath and pith material from sugar cane stalk and consists of two distinct cellular constituents. The first is a thick-walled, relatively long, fibrous fraction derived from the rind and fibrovascular bundles dispersed throughout the interior of the stalk. The second is a pith fraction derived from the thin-walled cells of the ground tissues.

Bagasse is basically cellulose but contains hemicellulose, lignin, and extractives as well, with a high fiber content and a minor quantity of amorphous pith. A small percentage of inorganic silica is also present. Finally, bagasse also contains some residual sugars, pentosans, hexosans, and other reactive low molecular weight products.

Large quantities of bagasse are available in tropical countries, e.g., India, Pakistan, Indonesia, Philippines, Zambia, and Jamaica. It is also available in south-eastern states of the USA. In this chapter, we will describe the chemistry and newer applications of this important renewable resource.

CHEMISTRY OF BAGASSE

The chemistry of bagasse reveals that it contains $\sim$40% cellulose, $\sim$30% hemicellulose, and $\sim$15% lignin. Bagasse fiber resembles cotton fiber in that it has spiral structure (length = 1-4mm, width = 0.01-0.04mm). Cellulose is a linear polyglucose with a DP of about 7,000-10,000 and is highly hydrogen bonded. This makes bagasse a highly crystalline, stable polymer and extremely resistant to solvent and chemical attack. Least resistant to crystallinity and attack are the hemicelluloses. They are also polymers, but may contain several different sugar units and also their degree of polymerization is low ($\sim$200). Lignins are complex, cross-linked polymers of phenylpropanoid units joined by benzylic and phenolic ether linkages and C-C linkages. The DP of lignin is only several hundred. Both lignin and hemicellulose can be used to make resin intermediates. Hemicellulose hydrolysis can be done either with 0.1-1% H_2SO_4 at 95-120°C for short reaction times, or with 2-4% acid at $90^{\circ}C$ for about 4 h to yield primarily pentoses.[2]

The internal bonds of cellulose are protected by the formation of crystalline structure, assuming cellulose as polyglucose.[3] Further protection against hydrolysis of cellulose is made by the formation of a seal of lignin around it. Thermal treatment of bagasse above $200^{\circ}C$ for a long period will result in loss of fiber structure. At $150^{\circ}C$ minor degradation or cross-linking can be observed. The reactions of cellulose with various monomers have been earlier described.[4]

PROPERTIES AND AVAILABILITY

The color of bagasse is generally gray-yellow to pale green. It is bulky and quite nonuniform in particle size. Bagasse is found to be of comparable composition regardless of origin.

Conventionally, bagasse is used as fuel to power the sugar mill. The burning of 70% of the bagasse produced furnishes enough steam for power heat to run the mill.[5] The amount of bagasse usually available is equal to the sugar yield. Large-scale production of sugar cane produces excess bagasse which must be either utilized or, if disposed of, will create environmental problems. The utilization of bagasse in pulp, paper, and board has not been too successful. Economical analysis has been made for furfural extraction, single cell protein production, and paper pulp production from bagasse.[6] Recent research indicates that bagasse can be upgraded by bonding with resins to produce composites suitable for building materials.[7]

APPLICATIONS

Bagasse is widely used as a fuel in sugar cane factories. The manufacture of mulch and litter appears to offer a very promising outlet for large amounts of bagasse. Bagasse can also be employed to manufacture roofing and building materials. It can be hydrolyzed to produce fuel and chemicals. Lignin, the phenolic polymer matrix in bagasse upon chemical modification may find use in adhesives, foams, films, coatings, and plastics.[8]

Roofing and Building Materials

Usmani and coworkers examined a large number of agricultural and other low cost residues as fillers in composites with various resin binders. Typical fillers examined in raw form were bagasse, jute sticks, rice straw, rice hulls, coconut husks, palm fronds, water hyacinths, balsa wood, wood shavings, sawdust, and excelsior. The evaluation was done by determining the effects of the raw fillers, at high volume percent loading, on the mechanical properties of the composite, initially and after 1,000 hours accelerated aging in a weatherometer.

The bagasse filled composites retained a higher percentage of their initial strength, after weatherometer accelerated aging tests than any other filler materials tested. Sawdust was second best in overall merit. Furthermore, bagasse has other inherent features which make it especially attractive including: renewable natural resource; high fiber content; readily "powdered" and fibrillated; high silica content and good fire retardant properties; compatibility with binder resins in

high volume percentages; availability in quantity; "gathered" to a
central place; and low cost.

Six composite building material systems that utilized major percentages of bagasse filler and lesser amounts of phenolic or other
resin binders were developed. The six materials developed are bagasse-
phenolic (BP), bagasse-rubber (BR), bagasse-thermoplastic (BT), oriented bagasse-phenolic (OP), random bagasse-phenolic (RP), and bagasse
in-situ generated (BI) composites.[1,4,7] The process technology for
these six systems are shown in Figure 1.

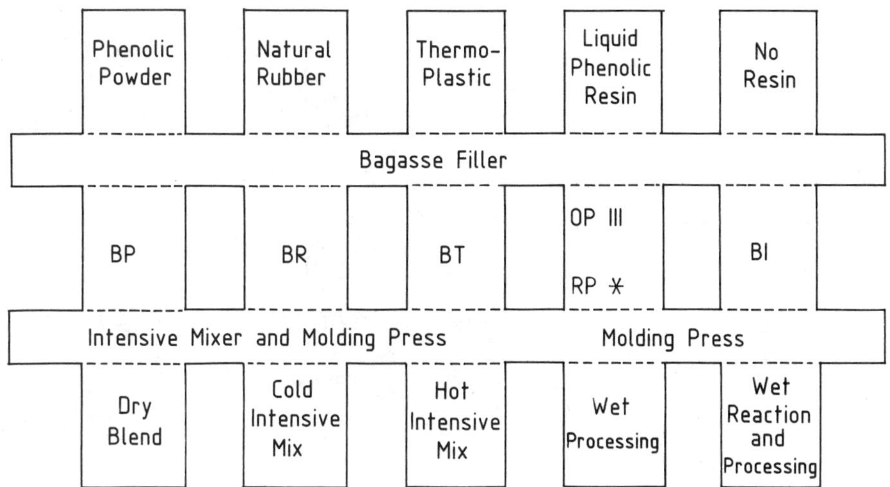

Figure 1. Process Technology for Six Bagasse-based Composites.

Product properties (Table 1), highlight of composition (Table 2) and
comparison of processing and raw materials costs (Table 3) are summarized below.

Table 1. Product Properties of Bagasse-based Composites

Composite	Properties
BP	Strong, durable, heat- and moisture resistant, colorful (pigmented red) panel; rigid, insulating, lightweight, and easily transportable; sawable and nailable for easy installation; excellent outdoor durability.
BR	Rigid, tough dark panel of desired element shapes; nailable and sawable for easy installation; good outdoor durability.
BT	Rigid, tough, dark panel of desired element shapes; amenable to easy installation; good outdoor durability.
OP	Durable, buff colored, flat or corrugated panel with direc-tional mechanical properties.

(continued)

Table 1 (continued)

Composite   Properties

RP   Durable, buff colored, flat or corrugated panel.

BI   Buff colored flat or corrugated panel; water resistance marginal.

Table 2. Highlights of the Composition of Bagasse-based Composites

Composite	Composition, wt %
BP	Phenolic Binder/Whole Bagasse/Processing Aid (30/61/9)
BR	Natural Rubber/Sulfur/Bagasse/Processing Aid (20/10/55/15)
BT	SAN or ABS/Bagasse/Fe_2O_3, Clay (27/63/10)
OP	Phenolic/Processed Bagasse Fibers/Product Improvers (4/80/16)
RP	Phenolic/Bagasse Fibers/Product Improvers (5/85/10)
BI	Bagasse/Phenol or Melamine (95/5)

Table 3. Comparison of Processing and Raw Material Costs

Composite	Raw Material	Processing
BP	High resin cost	Easiest and low cost
BR	High rubber cost	Easy and low cost
BT	High resin cost	Easy and low cost
OP	Low cost	Difficult and some environmental problem
RP	Low cost	Easy and low cost
BI	Lowest cost	Easy and low cost

All bagasse-based composites are suitable as building material. We believe only BP and BR are suitable for roofing application. Examples of other potential applications include wall panels, ceilings, flooring, counter tops, fences, siding, acoustical panels, sinks, furniture, door, and shutters.

<u>Hydrolysis of Bagasse and Value-Added Products Therefrom</u>

The concept of hydrolyzing cellulose to glucose is more than 100 years old. Hydrolysis can be done either by acid or by enzyme. Major shortcomings of acid hydrolysis are acid recovery when concentrated acids are used and the formation of dehydration products that limit sugar yield to 50 percent when dilute acids are used. Recently, a twin-screw extruder device has been used in high temperature (450°F) acid hydrolysis (sulfuric acid) process for conversion of wood pulp to glucose.[9] The combination of high temperature and 20 seconds hydrolysis time resulted in 60 percent yield of glucose while avoiding acid-catalyzed degradation. Dilute acid two-step hydrolysis looks quite attractive and this process can be adapted to production of value-added products.[10]

In enzyme hydrolysis, side products are not formed and the enzyme-catalyzed reactions can be conducted at nearly ambient conditions. The enzymes that promote saccharification of cellulose are β-(1$\rightarrow$4) glucan-glucanohydrolases (EC 3.2.1.4) and β-glucosidases (EC 3.2.1.21). These enzymes acting together degrade cellulose to cellodextrins, which are water soluble glucose polymer of about six repeating units, and glucose.[11,12] Of the many microorganisms examined, the best appears to be <u>Trichoderma reesei</u> mutant, a mold found in the soil.[13,14]

Enzyme catalyzed saccharification of cellulose may be economically viable on a large scale. But certain problems such as the inability of cellases to rapidly and totally hydrolyze native cellulose[15] and the inhibition effect that glucose and cellobiose (dimer) have on enzyme action are problem areas.[16]

The wide commercial use of renewable agricultural plants as typified by bagasse and microorganisms as raw material and energy sources is a possibility for the developing countries of the world with tropical climates. Genetically manipulated microorganisms that could ferment alcohol directly from cellulose is an attractive alternative but the technological realization is still far off. Should such a development gain momentum it would be necessary to transfer technology to developing nations.

Purdue researchers[17] have recently advanced the cellulose conversion technology. Their work is based on conversion of the hemicellulose and α-cellulose fractions of native cellulose to pentose (C_5) and hexose (C_6) sugars, fermentation of C_5 and C_6 sugars to fuel ethanol and some limited chemical intermediates, use of lignin fractions to power the ethanol concentration process, and utilization of biomass materials, e.g., agricultural residues or municipal waste, as raw materials for the process. We shall now briefly describe the Purdue process.

The hemicellulose extraction step is straightforward and involves extraction of native cellulose with dilute acid. Pentose is quantitatively extracted and can be fermented to ethanol or butanediol. The extracted residue now consists mostly of lignin and α-cellulose. The next step is cellulose pretreatment and recovery. Lignin seals microfibrils of highly crystalline linear cellulose. This situation can be reversed by using small amounts of solvents that swells, disrupts, and ruptures lignin seals to make the cellulose more accessible to hydrolysis.[18] Cellulose pretreatment solvents that have been used are Cadoxen[18] (4-5 percent CdO in solution with 25-30 percent ethylene diamine), chelating metal cellulose swelling agent[19] (aqueous solution of 17 percent sodium tartrate, 6.6 percent ferric chloride, 7.8 percent sodium hydroxide, and 6.2 percent sodium sulfite), and concentrated sulfuric acid.[20] After pretreatment, a nonsolvent, e.g., methanol and/or water is added. α-Cellulose is precipitated and the washing can be concentrated to recover the pretreating solvent.

The pretreated residue can be hydrolyzed by <u>Trichoderma reesei</u> or by acid. The in-situ precipitated α-cellulose having greater surface area allow the <u>Trichoderma reesei</u> access to the cellulose. The in-situ precipitated α-cellulose is sterilized and fed into an enzyme flask where it is mixed with nutrients and air and innoculated with the fungus. High yields of sugar were obtained with bagasse. Following hydrolysis, lignin fraction was separated by filtration. Part of the enzyme can be recovered by ultracentrifugation and recycled.

The C_5 sugar[21] obtained from the hemicellulose can be fermented to 2,3-butanediol by mutant strains of <u>Klebseilla</u>[22] and <u>Aeromonas genera</u>[23]. The C_6 sugar can be mixed with yeast in a controlled environment and fermented to ethanol.

A dilute acid hydrolysis (two-step process) is schematically shown in Figure 2. Possible value-added products that can be potentially obtained from bagasse are shown in Figure 3.

<u>Lignin</u>

Because of recent advances in analytical technology, the structure-performance relationship of lignin is better understood now. Thus lignin utilization in development of engineering plastics is possible. Lignin-based prepolymers are already in commercial use in phenolics. Newer markets for lignin point the way to adhesives, thermosets, foams, coatings, and films.

CONCLUSIONS

Bagasse is an underutilized renewable resource material. Potentially it is attractive to produce from bagasse value-added products. These include composites for building materials as well as chemical and resin intermediates.

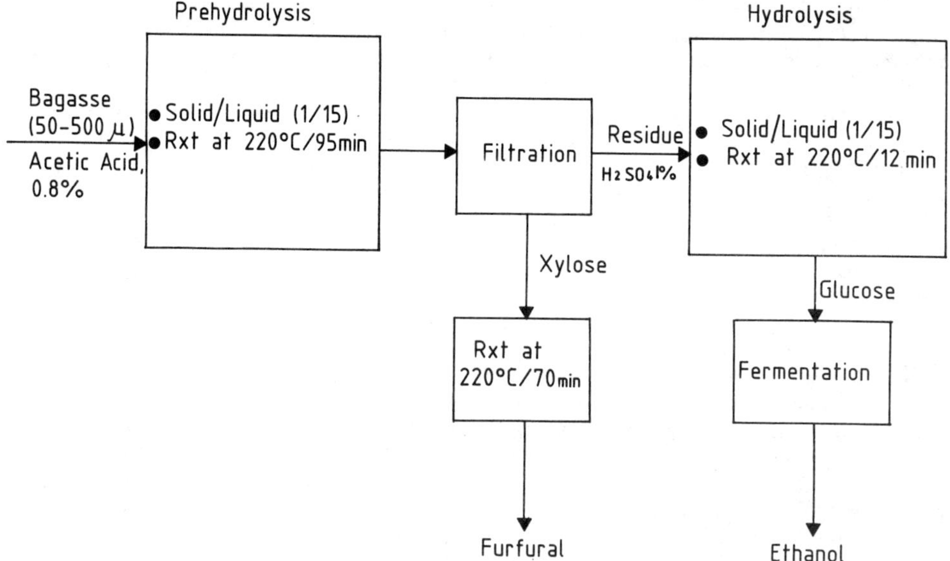

Figure 2. Two-Step, Dilute Acid Hydrolysis of Bagasse.

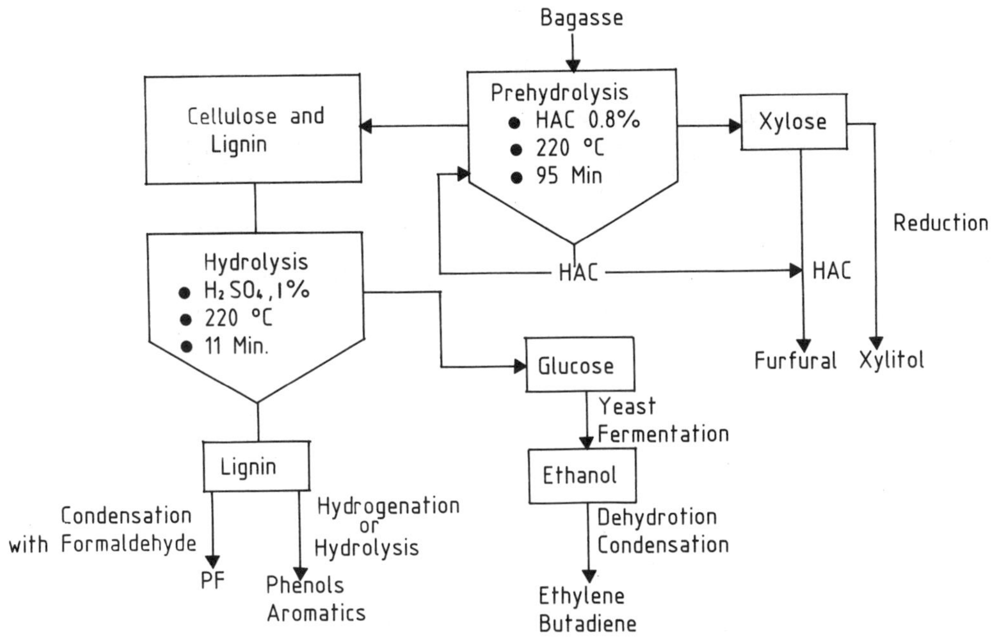

Figure 3. Value-Added Products from Bagasse.

REFERENCES

1. I.O. Salyer and A.M. Usmani, Ind. Eng. Chem. Prod. Res. Dev., 21, 17 (1982).
2. H.F. Wenzel, "The Chemical Technology of Wood," Academic Press, New York, 1970.
3. M. Chang, J. Polym. Sci., Part C, 36, 343 (1971).
4. A.M. Usmani and I.O. Salyer, "Bagasse-Rubber Composite Technology," in Use of Renewable Resources for Polymer Applications (C.E. Carraher and L.H. Sperling, eds.), Plenum, New York, 1982.
5. R.N. Shreve, "Chemical Process Industries," McGraw-Hill, New York, 1967.
6. G.D. Agrawal, Indian J. Environ. Health, 16(2), 159 (1974).
7. S.H. Hamid, A.G. Maadhah and A.M. Usmani, Polym. Plast. Technol. Eng., 21(2), 173 (1983).
8. News Item, Chem. Eng. News (Sept. 24, 1984).
9. News Item, Chem. & Eng. News (Oct. 8, 1979).
10. A. Singh, K. Das, and D.K. Sharma, Ind. Eng. Chem. Prod. Res. Dev., 23, 257 (1984).
11. G.H. Emert, E.K. Gum, Jr., T.A. Lang, T.H. Lieu, and R.D. Brown, Adv. Chem. Ser. 79 (1979).
12. T.K. Ghose, in "Advances in Biochemical Engineering," Springer Verlag, Berlin, 6, 39 (1977).
13. For example, see; "Enzymatic Conversion of Cellulosic Materials: Technology and Applications," Proceedings of Symposium, Wiley, New York, 1976.
14. For example, see; "Advances in Enzymic Hydrolysis of Cellulose," Pergamon Press, New York, 1963.
15. T.K. Ghose and A.N. Pathak, Process Biochem., 20 (May 1975).
16. F. Bisset and D. Sternberg, Appl. Environ. Micro. 35(4), 750 (1978).

17. M.R. Ladisch, M.C. Flickinger, and G.T. Tsao, Energy, $\underline{4}$, 263 (1979).
18. G. Jayme and K. Neuschaffer, Makro. Chem., $\underline{28}$, 71 (1957).
19. B.E. Dale, M.R. Ladisch, T. Hamilton, and G.T. Tsao, Paper presented at the 176th ACS Meeting (1976).
20. G.T. Tsao, M.R. Ladisch, C. Ladisch, D. Hsu, B.E. Dale, and T. Chou, Annual Reports in Fermentation Processes, Academic Press, New York, $\underline{2}$(1), 1 (1978).
21. M.C. Flickinger and G.T. Tsao, Annual Reports in Fermentation Processes, $\underline{2}$(2), 23 (1978).
22. W.A. Wood and R.P. Motlock, J. Bac., $\underline{88}$(4), 838 (1964).
23. R.Y. Starier and G.A. Adams, Canadian J. Res., $\underline{38}$, 168 (1944).

SECTION III - GRAFT POLYSACCHARIDES

USE OF GRAFTED WOOD FIBERS IN THERMOPLASTIC COMPOSITES
V. POLYSTYRENE

Bohuslav V. Kokta, Famakan Dembélé, and Claude Daneault

Centre de recherche en pâtes et papiers
Université du Québec à Trois-Rivières
C.P. 500
Trois-Rivières (Québec) Canada   G9A 5H7

INTRODUCTION

Fibrous wood flour[1], the product of attrition or hammer mill grinding of relatively resin-free fibrous soft woods (i.e. pine, fir, spruce), and to a lesser extent hardwood in the form of cellulose fibers, has been used as fillers and reinforcement in polymer composites. In a recent research by Kokta et al.[2], it has been shown that the use of grafted wood fibers has improved mechanical properties of polystyrene composites as compared with the use of non-grafted substrates. In the research, wood fibers from aspen (hardwood), spruce or balsam fir (softwoods) have been used to fill and reinforce the polystyrene matrice. The effect of fiber modification by the xanthate method of grafting on resulting composite properties has also been studied.

EXPERIMENTAL

Materials

In this experiment, pulp was used in the form of wood flour prepared on a laboratory «ARTHUR» grinder (fir), of mechanical pulp (spruce) or thermo-mechanical pulp (aspen). Styrene (Eastman Chemical) was purified by distillation. The central cut was collected and then stored in dark bottles in a refrigerator. All other chemicals were used as supplied by the manufacturer.

Synthesis

The xanthate method of grafting using hydrogen peroxide-ferrous ion initiation system has been applied to graft polystyrene on wood fibers. The procedure in pulp conditioning and the xanthate method of graft copo-

Table 1. Fibre Aspect Ratio L/d

	$\overline{L}$ (mm)	$\overline{d}$ (mm)	Average L/d
Spruce (Mesh 100) (mechanical pulp)	0.37	0.12	3.08
Spruce (Mesh 100) (grafted)	0.37	0.146	2.53
Spruce (Mesh 40)	1.02	0.30	3.4
Spruce (Mesh 40) (grafted)	1.02	0.251	4.08
Fir (Mesh 40) (grafted)	0.8	0.20	4.0
Fir (Mesh 40) (grafted)	0.8	0.251	3.64
Fir (Mesh 100)	0.41	0.13	3.15
Fir (Mesh 100) (grafted)	0.41	0.14	2.93
Aspen (Mesh 100) (TMP)	0.14	0.09	1.56
Aspen (Mesh 100) (grafted)	0.14	0.10	1.40

lymerization have been described in previous papers[3-7].

Polymer loading of grafted styrene on wood fibers was defined as follows:

$$\text{Polymer loading (\%)} = (A-B)/(B) \times 100$$

and total polymer deposition (%) on wood fibers (grafted styrene + homopolymer) was defined as:

$$(D-B)/(B) \times 100$$

Where A corresponds to the weight of product after copolymerization and extraction with acetone, B the weight of the fiber, and D the total weight of the product after copolymerization.

The preparation of composite samples was identical to what was described recently[2]. In brief, polystyrene powder was mixed with wood fibers and then molded in a Carver laboratory press for 30 minutes at a temperature above T_g of PS (150°C) and a 4 MPa pressure. Annealing of samples was achieved in an oven at 105°C for 16 hours.

Mechanical measurements were made on an Instron tester (Model 1131) at 23°C and a 50% level of relative humidity. The rate of elongation was 10 mm per minute for a shoulder shape sample of 3.175 mm in width and 6.4 cm in length (1.7 cm between the grips). The thickness of all samples was measured with a micrometer. All experimental data represent an average of at least four measurements (and in most cases, six).

RESULTS AND DISCUSSION

It is a well known fact that the ratio between fibre length and cross section defined as Fiber Aspect Ratio, plays an important role in reinforcing the thermoplastic matrice. In Table 1, the Fiber Aspect Ratio is shown for all species used in this study. It is clear that wood fibers prepared by mechanical means all have very low L/d values (below 5) compared to fibre of chemical or chemithermomechanical pulps (above 15). In addition, grafting of polystyrene on wood fiber does not modify in any consistent way this Aspect Ratio since the grafting reaction takes place on the fiber surface, mostly.

Figure 1 represents typical relations between property (in this case modulus) and the weight fraction of fibers. The weight fraction of fibers was expressed in percentage points, and varied from 10 to 40 percent. In general, mechanical properties improved consistently up to 40% but in some cases the maximum value was reached at the 30% filling level. Consequently, for all other results shown here, mechanical properties were compared at 30% of the weight fraction of the fiber present in the composite. This behavior is similar to what was described in one of our previous studies[2].

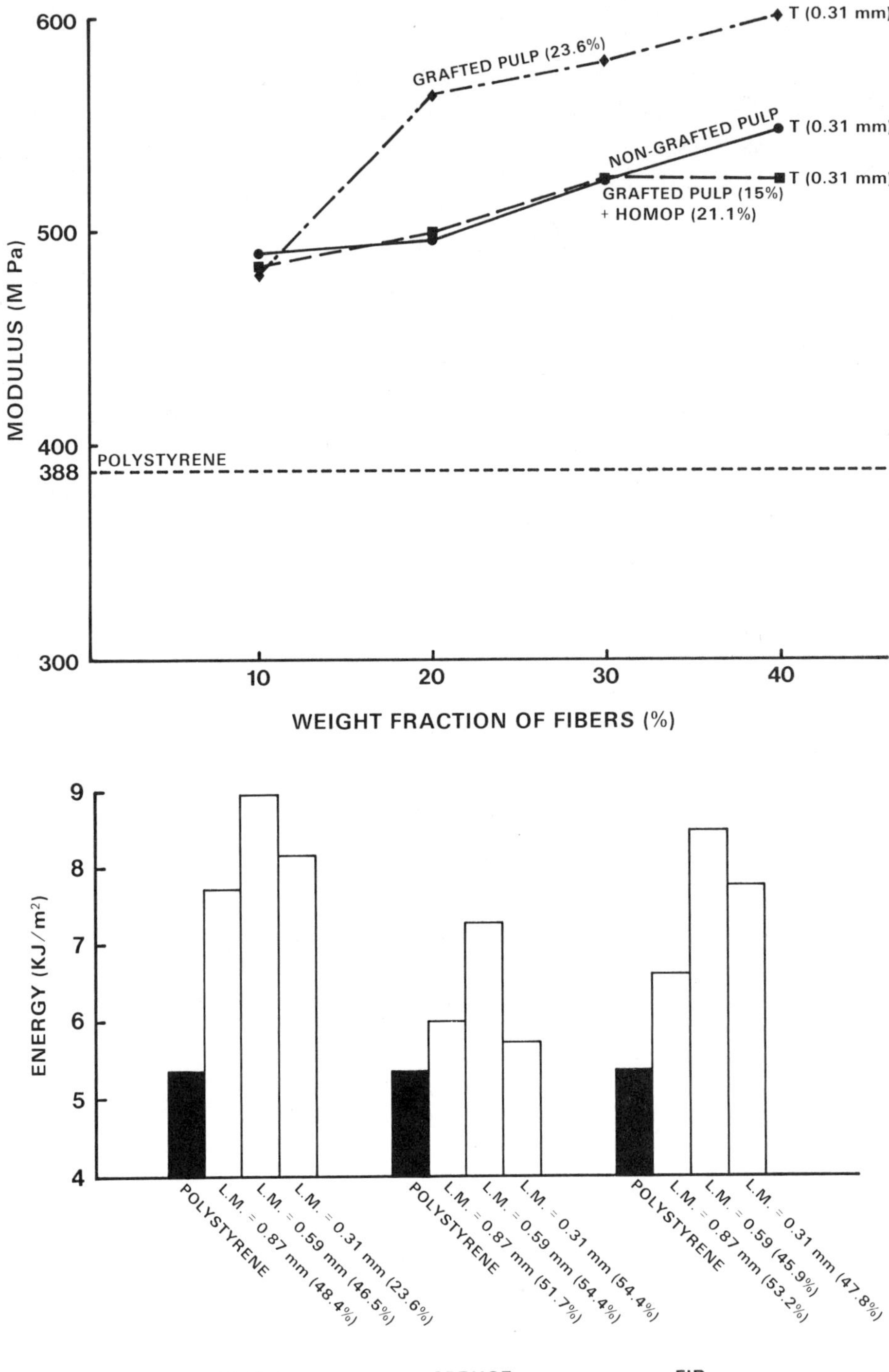

Fig. 1 Effect of weight fraction of fibers on modulus

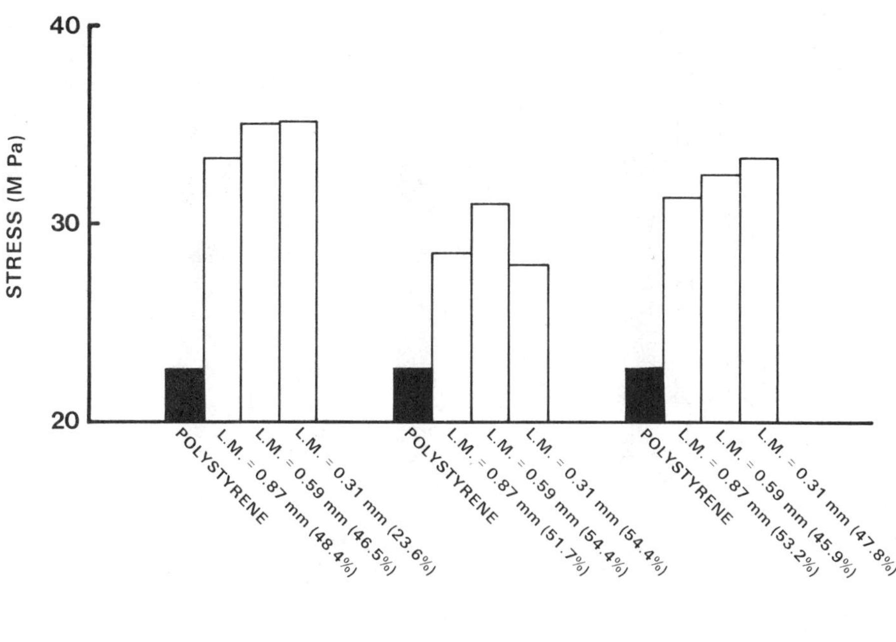

Fig. 3 Effect of fiber length on stress. Fiber content: 30%

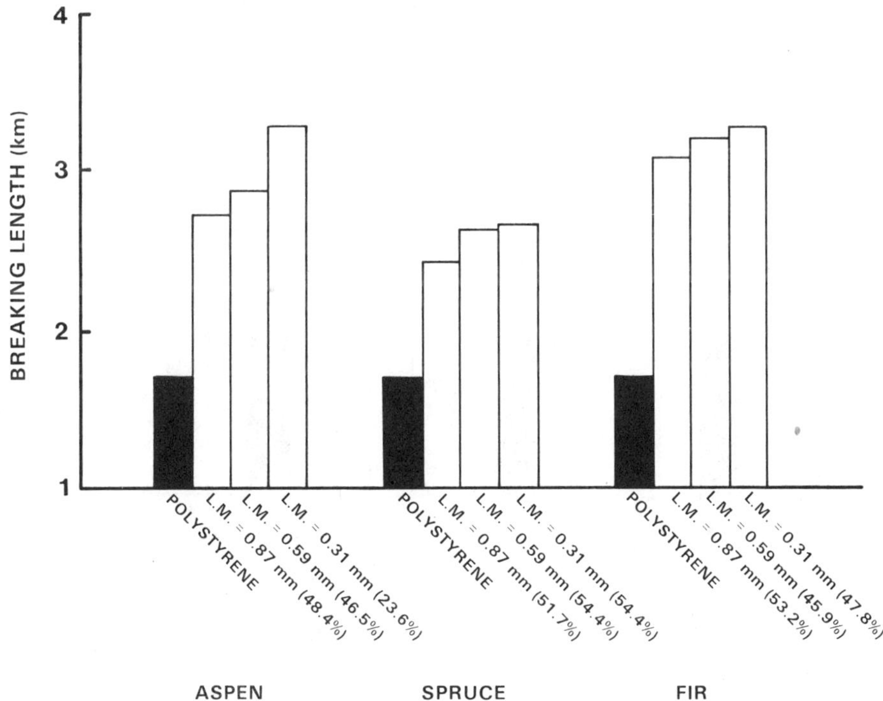

Fig. 4 Effect of fiber length on breaking length. Fiber content: 30%

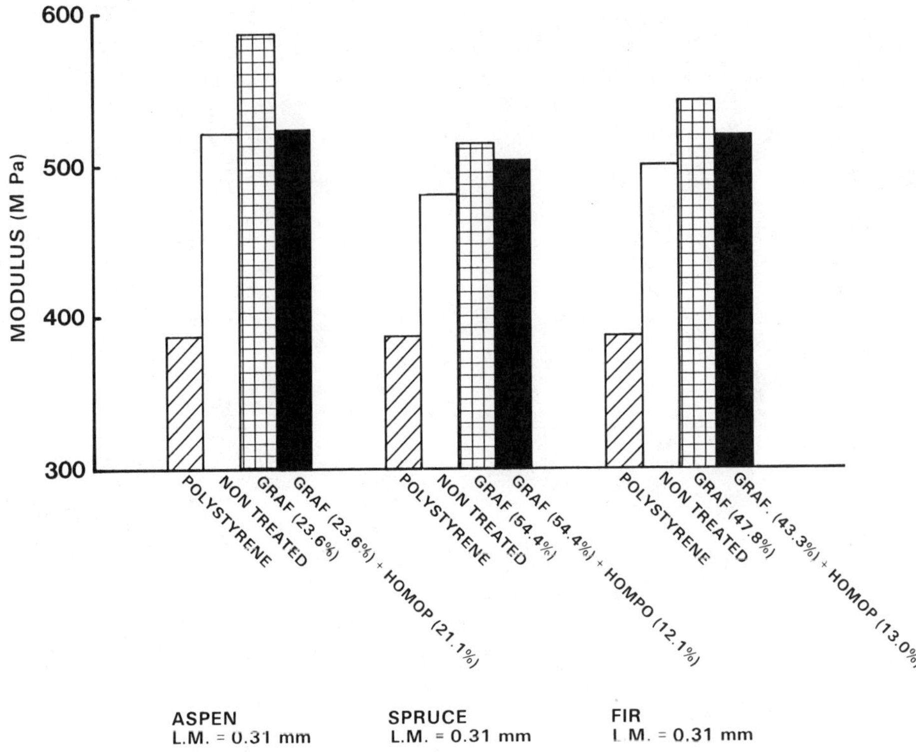

Fig. 5 Effect of fiber length on modulus. Fiber content: 30%

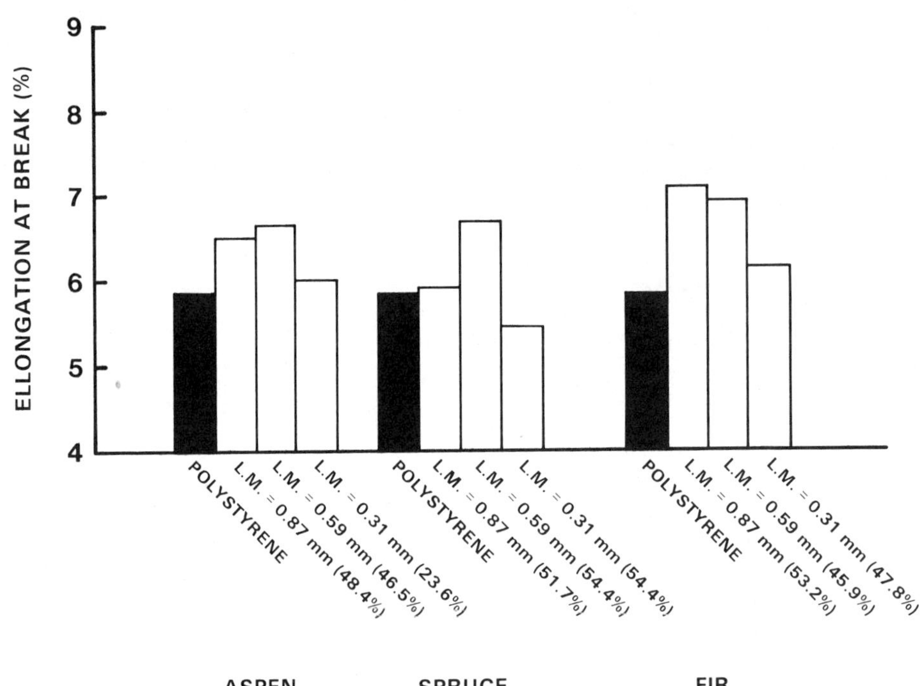

Fig. 6 Effect of fiber length on elongation. Fiber content: 30%

In Figures 2-6, the effect of fiber length on the following mechanical properties is shown energy at break (Fig. 2); stress (Fig. 3); breaking length (Fig. 4); modulus (Fig. 5) and elongation at break (Fig. 6). Composite properties of all three fiber species (aspen, spruce and fir) are being simultaneously compared to those of the original polystyrene. The following conclusions can be drawn from Figures 2-6: a) in almost all cases, there has been a significant improvement in all mechanical properties of composites due to the presence of grafted fibers (polymer loading levels indicated in brackets) for all three wood species; b) with the exception of the energy at break and elongation at break where the average fiber length of 0.59 mm gave superior values, the best results were obtained with the shortest fiber length: 0.31 mm; this strengthens our previous study which indicated the same trend[2]; c) grafted aspen wood fibers gave composites with the best improvement in mechanical properties, with the exception of elongation where grafted fir had the best performance. Grafted aspen fiber composites showed various increases: 50% in modulus, 67% in energy, 92% in breaking length and 55% in breaking stress. In addition, even elongation at break increased (by 14%). The fact that graft-

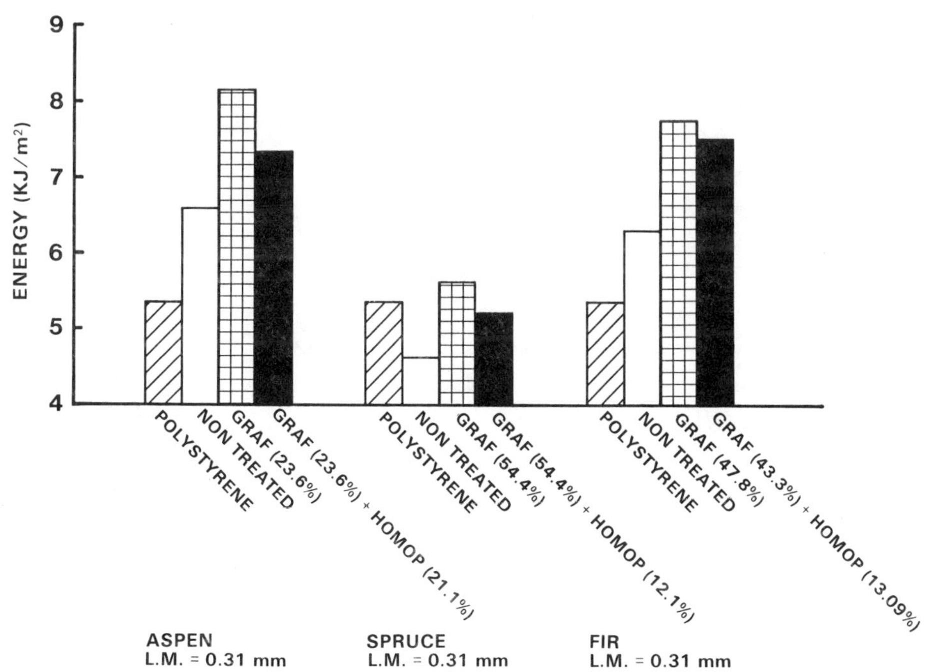

Fig. 7 Effect of homopolymer on energy. Fiber content: 30%; fiber length: 0.31 mm

ed wood fibers can increase (or preserve) elongation of original polymer
represents its significant advantage over either glass fiber or other
inorganic filler composites which become brittle and usually lead to a
considerable decrease in elongation at break [8,9]. The increase in elongation also clearly indicates good interface contact between grafted wood
fibers and polymeric matrice.

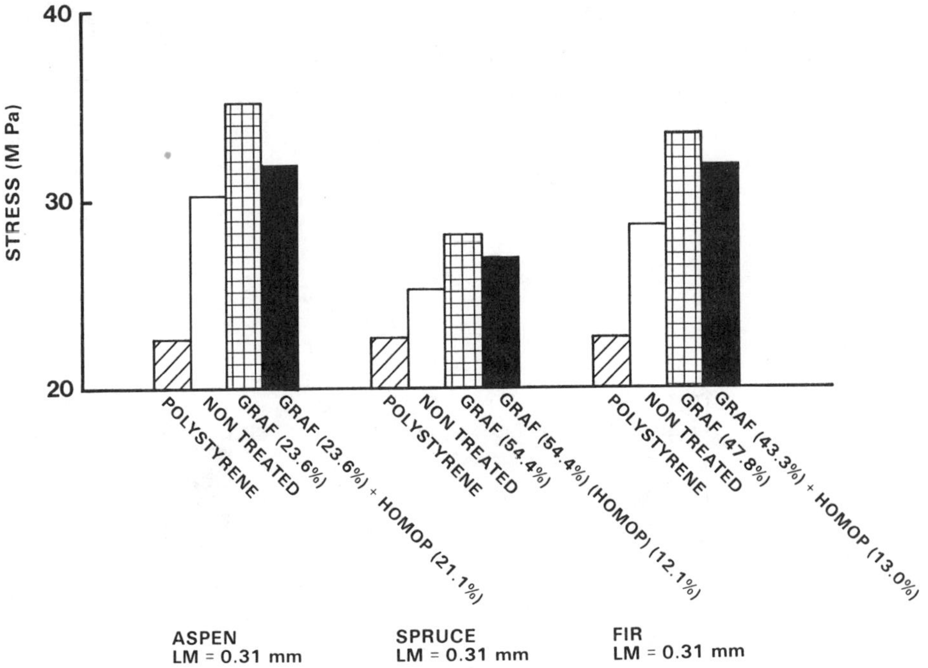

Fig. 8   Effect of homopolymer on stress.   Fiber content: 30%; fiber
length: 0.31 mm

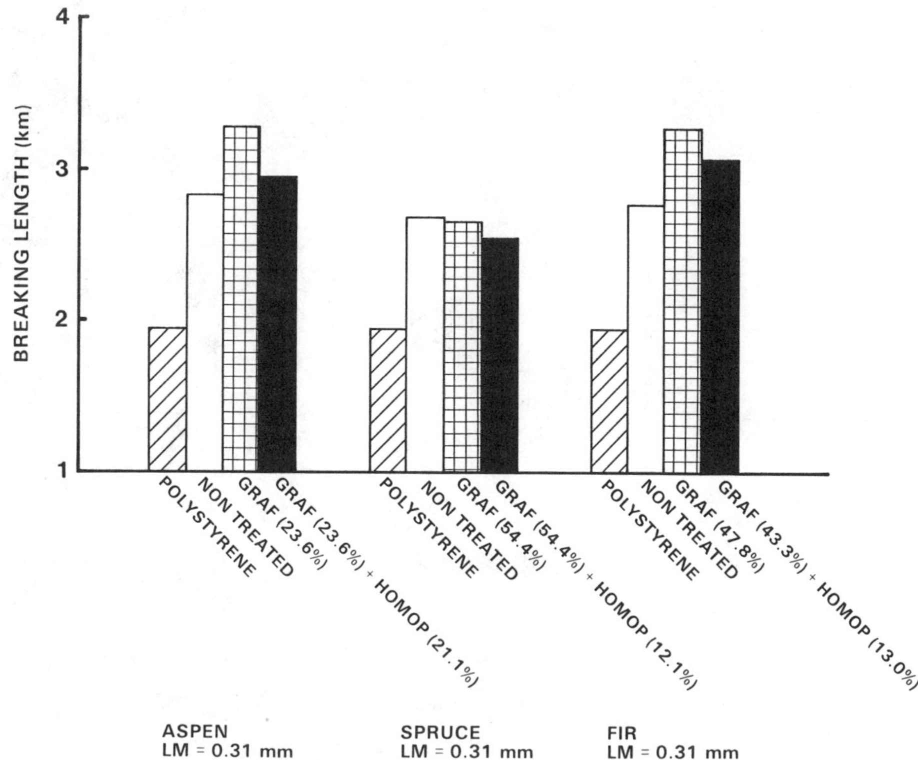

Fig. 9 Effect of homopolymer on breaking length. Fiber content: 30%; fiber length: 0.31 mm

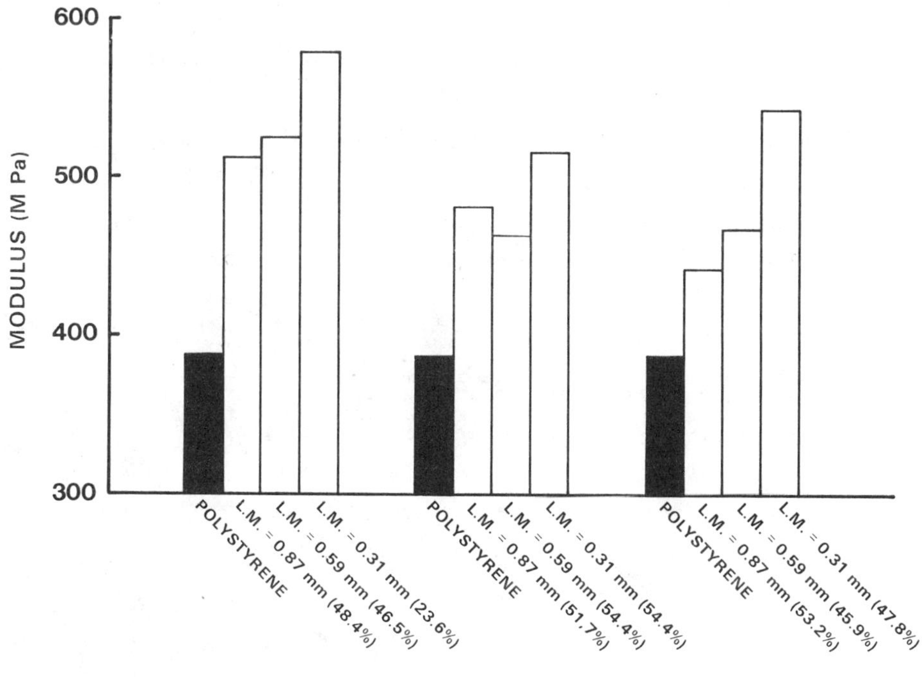

Fig. 10 Effect of homopolymer on modulus. Fiber content: 30%; fiber length: 0.31 mm

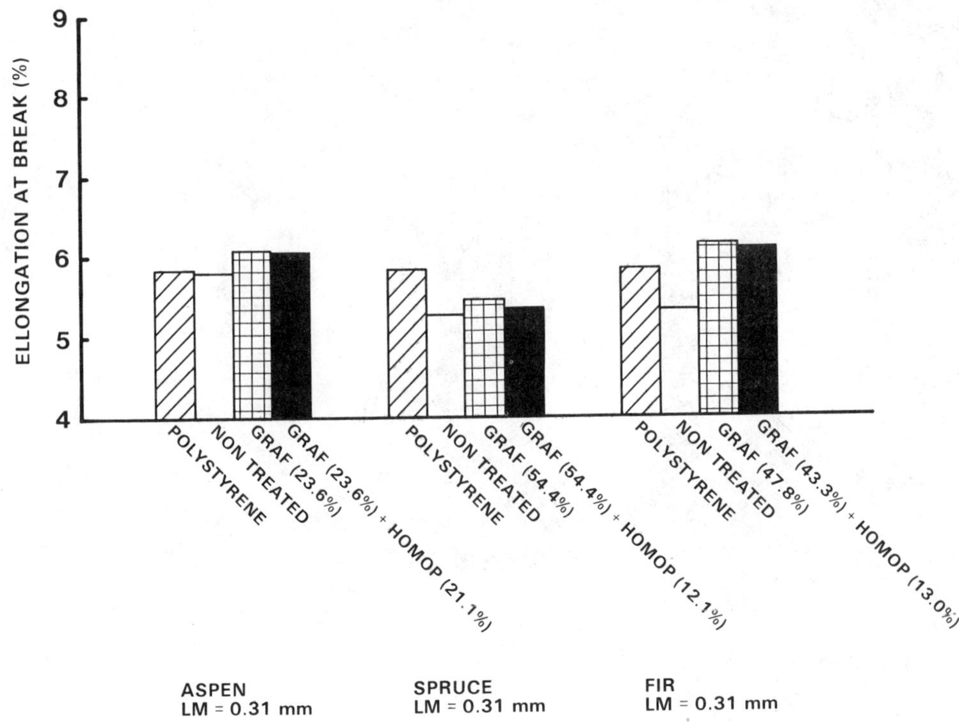

Fig. 11   Effect of homopolymer on elongation.   Fiber content: 30%;
fiber length: 0.31 mm

Figures 7 to 11 compare the mechanical properties of composites made
of three different wood species at a 30% fiber addition level for an average fiber length of 0.31 mm.   The effect of different fiber states (i.e.
non-grafted fiber; grafted fiber; grafted fiber + homopolymer) on composite
mechanical properties is also compared with respect to the original polystyrene.

Based on the results indicated in Figures 7 to 11, one may conclude
that: a) grafted fibers showed the best mechanical properties; b) the
presence of homopolymer displayed some negative effects on the final properties; c) non-grafted wood composites showed an improvement of properties as compared to polystyrene; and d) aspen fibers performed better than
spruce or balsam fir fibers.

It is clear that grafting substantially improves mechanical properties of composites because of a good interphase contact (fiber-polymer-
matrice) and no phase separation under stress.   This explains the superior

behavior of grafted fibers as opposed to those impregnated with polymer
which displays only a physical fiber-polymer bond[9].

The above-mentioned results indicate that sawdust may well be used
to decrease the price as well as reinforce the polystyrene matrice. It
has been shown recently[8] that grafted hardwood fiber-filled poly (methyl)
methacrylate) composites outperformed both mica or glass fiber PMMA. This
is also true for grafted aspen-filled polyethylene even under extreme
conditions[9].

CONCLUSION

Grafted fibers of aspen, spruce or birch can improve mechanical properties of polystyrene and produce stronger composites. The improvement
of properties of composites may be compared as followed:

	Elastic Modulus	Energy To Break	Stress To Break	Breaking Length	Elongation To Break
	%	%	%	%	%
Aspen	50	67	55	92	14
Spruce	33	58	32	56	14
Fir	40	36	47	92	21

In addition, the best mechanical properties can be reached with an
aspen fiber length of 0.3 to 0.6 mm and fiber loading of 30% to 40%.

Finally, the mechanical properties of composites made from grafted
wood fibers were superior to those of non-grafted fibers, and slighty
better than those of made with non-extracted homopolymer. Preliminary
results indicate that grafted wood-fiber composite properties could be
much further increased by the optimization of mixing and molding conditions
as well as by the use of fibers with a higher L/d ratio.

ACKNOWLEDGEMENT

The authors wish to thank the Natural Sciences and Engineering Research Council of Canada and the FCAC Fund (Québec) for their financial
support.

REFERENCES

1. M. H. Fisher, Modern Plastics 14 (2), 50 (1936).

2. B.V. Kokta, R. Chen, C. Daneault and J.L. Valade, Polymer Composites, 4, (4), 229 (Oct. 1983).

3. V. Hornof, B.V. Kokta and J.L. Valade, Journal of Applied Polymer Science, 20, 1543 (1976).

4. V. Hornof, B.V. Kokta and J.L. Valade, Journal of Applied Polymer Science, 21, 477 (1977).

5. P. Fuertes and B.V. Kokta, Pulp and Paper of Canada, Transaction, TR53 (September 1981).

6. B.V. Kokta, R. Chen and C. Daneault, ACS Symposium Series 187, 269, (1982).

7. B.V. Kokta and J.L. Valade, Tappi, 55, 3, 366 (1972).

8. B.V. Kokta, P.D. Kamdem, A.D. Beshay and C. Daneault, World Poplar Commission Symposium, Proceeding of Papers, 19 pp. Ottawa (October 1984).

9. A.D. Beshay, B.V. Kokta and C. Daneault, NRCC-IGM Symposium, «Composites '84», Proceeding of Papers, 8200-8-1.4, pp. 10.0-10.64. Boucherville (November 1984).

THE XANTHATE METHOD OF GRAFTING:

XI.   GRAFTING OF METHYLMETHACRYLATE ON HARDWOOD PULP

Claude Daneault and Bohuslav V. Kokta

Centre de recherche en pâtes et papiers
Université du Québec à Trois-Rivières
C.P. 500
Trois-Rivières (Québec) Canada   G9A 5H7

INTRODUCTION

From an industrial point of view [1-3] the xanthate method of grafting
seems to be one of the most practical methods in use. This method is
not limited to pure cellulose, solely. It can also be applied for the
grafting of chemical as well as mechanical softwood pulps [4-7]. The above-
mentioned method has recently been applied to the grafting of acrylonitrile to aspen [8], or acrylonitrile and/or styrene to birch [9].

In the present study, conditions which influence grafting parameters
using the xanthate method of copolymerization have been studied. Methyl
methacrylate (MMA) was applied as a monomer, and chemithermomechanical
pulp (CTMP) of aspen as a substrate. In addition, the effect of the presence of either styrene (STY) or ethyleneglycoldimethacrylate (EGDMA) in
the mixture with MMA on grafting efficiency, conversion and polymer loading have been tested.

EXPERIMENTAL

Materials

The chemithermomechanical pulp from aspen (Populus Tremuloids Michx)
was prepared in a Sund Defibrator: temperature: 126°C; retention time:
5 min; pressure: 0.12 MPa; refining energy: 5.26 MJ/kg; Na_2SO_3: 5%;
NaOH: 5%; pH: 12.9; yield: 92%; lignin: 17.9; average fiber length:
0.75 mm; drainage index (CSF): 119; breaking length: 4.46 km; tear index:
7.2 mN·m²/g. Monomer: Methyl methacrylate (Eastman Chemical Grade) was
purified by distillation and stored in dark bottles. All other chemicals

were of an analytical grade and used as supplied by manufacturers.
Polymer: Poly(methyl methacrylate) (Polyscience Inc.), $\overline{M}w$ 60600; $\overline{M}n$ 33200.

Procedure

<u>Xanthation</u>: Experimental procedures pertaining to grafting (xanthate method)
and extraction and fiber pre-treatment used in the present study were similar to those previously published [9]. Specific conditions were as follow:
pulp conditioning: Pulp 18 g + 0.02 g (oven-dried weight); partial mercerization at 25°C; 45 min in 300 ml 0.75N NaOH (solubility in NaOH: 3.5%);
xanthation, 2 hrs exposed to CS_2 vapors, 25°C; washing, 600 ml of acidified (pH: 4.5) distilled water; ion exchange, 2 min in 2000 ml of 0.006%
$(NH_4)_2\ Fe(SO_4)_2$; washing, 500 ml of distilled water (ph: 6.5).

<u>Copolymerization</u>: Monomer, methyl methacrylate, 4.5 g; impregnation, 15 min,
25°C; surfactant (Tween-80), 1.8 g; distilled water, 450 ml, H_2O_2 (conc.
30 g/1), 5 ml; H_2SO_4 to pH: 5; reaction time: 19 hrs; atmosphere: nitrogen; temperature, 25°C; quenching of polymerization: washing of product
with distilled water and 1% $K_2S_2O_5$, 5 min; washing with 1000 ml of distilled water.

<u>Extraction</u>: The apparent level of grafting, polymer loading and conver-
sion was determined by extraction with Soxhlet for 4 hrs with the use of
acetone. The oven-dried weight was corrected for pulp loss during the
mercerization stage. Grafting parameters were defined as follows: Polymer
loading, % (A-B/B) x 100; grafting efficiency, % (A-B/D-B) x 100; Degree
of conversion, % (D-B/C) x 100, where A is the weight of the product after
copolymerization and extraction, B the weight of the pulp, C the weight of
the monomers charged, and D the weight of the product after copolymerization.

RESULTS AND DISCUSSION

In a recent study [7], we have shown that the most important parameters
controlling the grafting reaction (in the case of softwood) are the following: concentration of sodium hydroxide during mercerization; concentration of hydrogen peroxide; time; monomer concentration; initial pH and
concentration of ferrous ammonium sulfate.

The variables mentioned above have been examined for hardwood spe-
cimen-aspen and monomer MMA. Furthermore, the monomer mixtures (MMA +
STY) or (MMA + EGDMA) have been copolymerized under optimum grafting conditions in order to discover their impact on final results.

Results, dealing with optimisation of the xanthate method are presented in Figures 1 to 7.

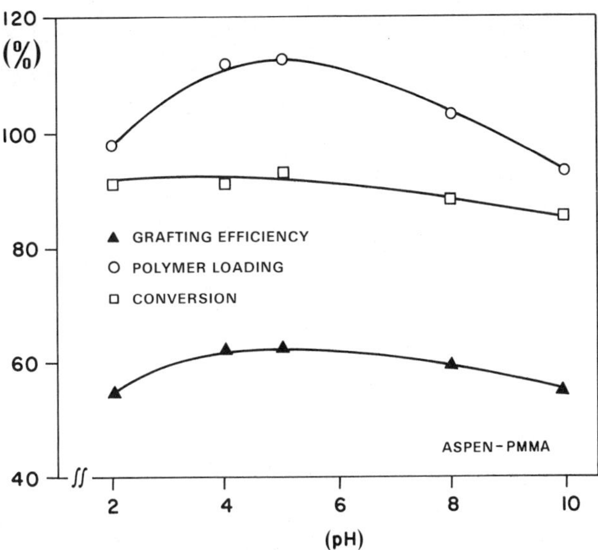

Fig. 1 Influence of initial pH on grafting parameters. (NaOH: 0.75N; time 19 hrs; 25°C; F.A.S.: 0.006%; pulp: 4.52 g; monomer: 9 g; peroxide: 1.5 g)

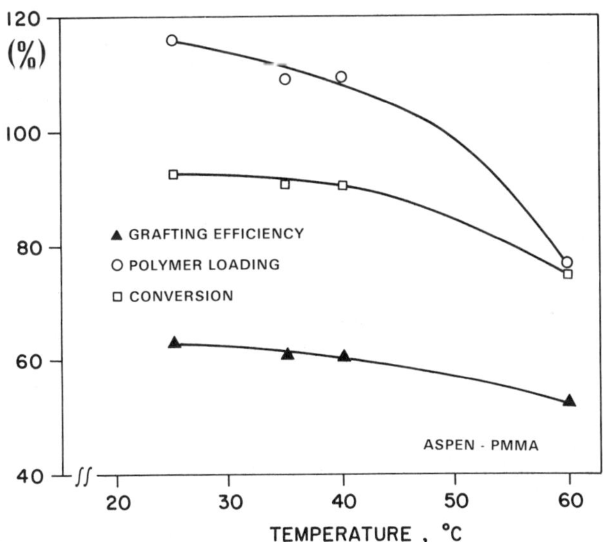

Fig. 2 Influence of temperature on grafting parameters. (NaOH: 0.75N; time 19 hrs; F.A.S.: 0.006%; pulp: 4.5 g; monomer: 9 g; peroxide: 1.5 g; pH = 5)

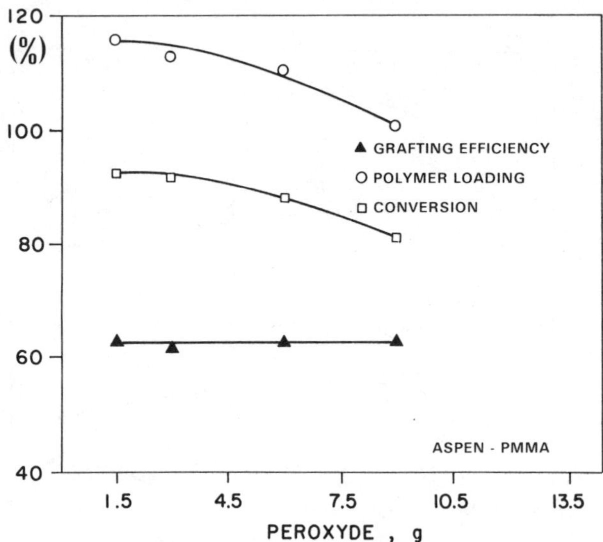

Fig. 3   Influence of hydrogen peroxide on grafting parameters (NaOH:
0.75N; time: 19 hrs; 25°C; monomer: 9 g; pulp: 4.52 g; pH
= 5; F.A.S. = 0.006%)

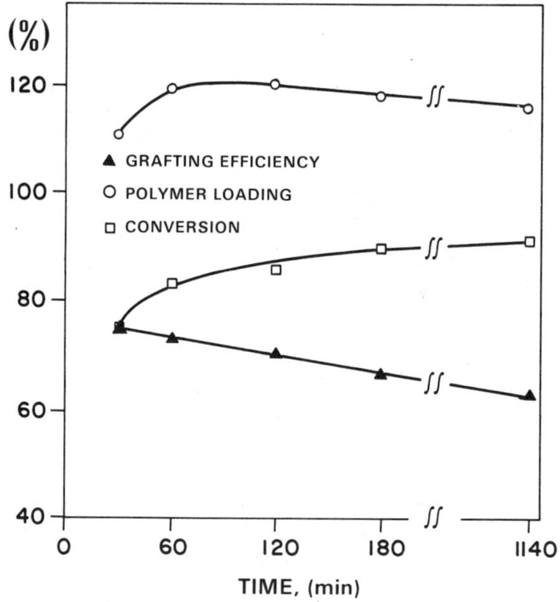

Fig. 4   Influence of reaction time on grafting parameters (NaOH:
0.75N; F.A.S.: 0.006%; 25°C; pulp: 4.52 g; monomer:
9 g; peroxide: 1.5 g; pH = 5)

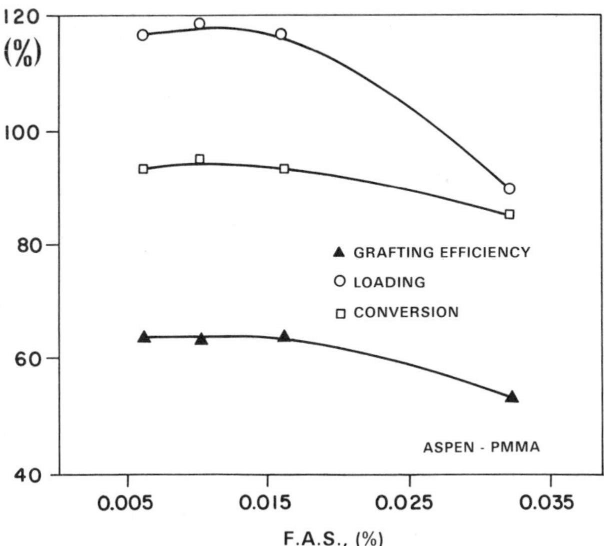

Fig. 5   Influence of monomer (MMA) concentration on grafting parameters (NaOH: 0.75N; time: 19 hrs; F.A.S.: 0.006%; 25°C;
pulp: 4.52 g; peroxide: 1.5 g; pH = 5.0)

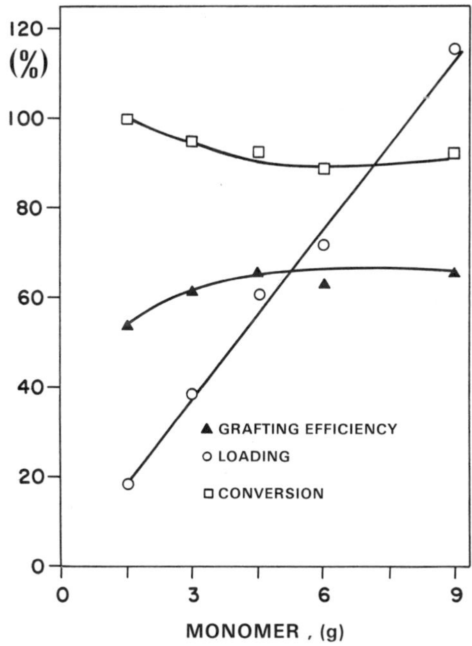

Fig. 6   Influence of ferrous ion concentration on grafting parameters
(NaOH: 0.75N; time: 19 hrs; 25°C; pulp: 4.52 g; monomer:
9 g; peroxide: 1.5 g; pH = 5.0)

The results in Figures 1 to 7 seem to indicate that the optimum conditions of aspen CTMP pulp grafting with MMA give a level of grafting
efficiency of roughly 75% and conversion of 80%. These results are compared in Table 1.

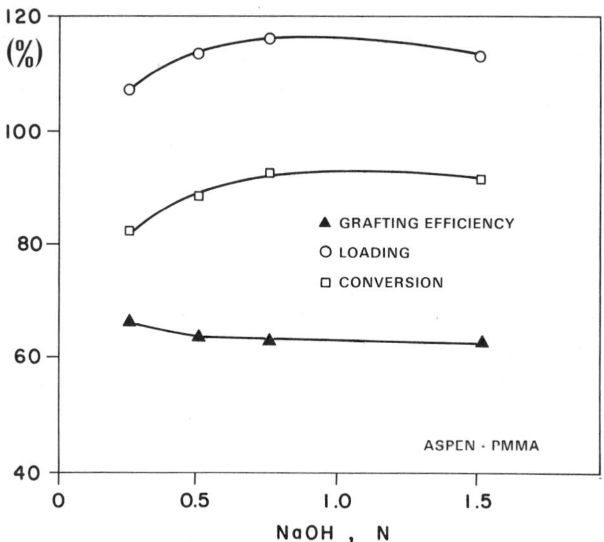

Fig. 7   Influence of sodium hydroxide concentration on grafting
parameters (time: 19 hrs; 25°C; pulp: 4.5 g; monomer:
9.0 g; peroxide: 1.5 g; F.A.S.: 0.006%; pH = 5)

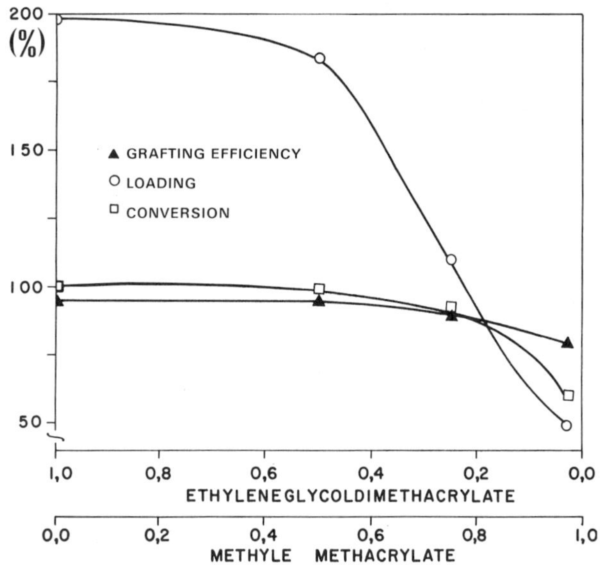

Fig. 8   Influence of EGDMA on grafting parameters. (NaOH: 0.75N;
F.A.S.: 0.006%; 25°C; time: 19 hrs; H_2O_2 = 1.5 g; pulp: 4.52 g;
monomers (EDGMA + MMA) = 9 g; pH = 5.0)

Table 1 Optimum grafting conditions for very high-yield (90%+) hardwood CTMP pulps

Pulp	Birch[9]	Birch[9]	*Aspen[8]	Birch[11] / Aspen	Aspen
Pulp/Monomer	0.5	0.5	6	0.5	0.75
H_2O_2	7 g/1	7 g/1	8 g/1	2–3 g/1	3 g/1
Initial pH	5	5	5	5.5	5–5.5
Monomer	STY	ACN	ACN	AcAm	MMA
$(NH_4)_2$ Fe SO_4	150ml 0.004%	150ml 0.004%	150ml 0.004%	150ml 0.004%	2000ml 0.006%
Time	18 hres	18 hrs	18 hrs	2 hrs	1 hr
Temperature	25°C	25°C	25°C	25°C birch / 35°C aspen	25°C
NaOH	0.25N	0.75N	0.75N	0.75N	0.75N
Conversion	96%	97%	100%	35%	80
Grafting efficiency	90%	80%	80%	80% birch / 85% aspen	75%
*CMP pulp	150ml 0.004%	150ml 0.004%	150ml 0.004%	150ml 0.004%	2000ml 0.006%

It is clear that the xanthate method of grafting is quite efficient
for all the monomers studied, except for conversion in the case of acrylamide grafting.

In one of our former studies [10], it was shown that grafted parameters could be improved by grafting while another monomer was present.
The present research focused on the effect of the addition of either
EGDMA or STY to MMA on grafting efficiency and polymer conversion. Results of this study are presented in Figures 8 and 9

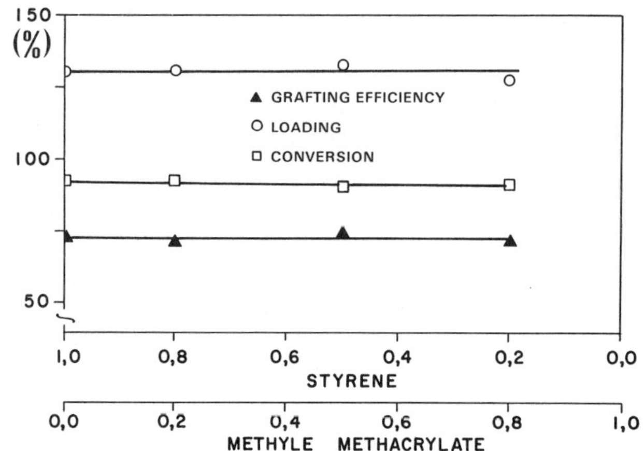

Fig. 9   Influence of styrene on grafting parameters. (NaOH: 0.75N; F.A.S.:
0.006%; 25°C; time: 19 hrs; H_2O_2 = 1.5 g; pulp: 4.52 g;
monomers (STY + MMA): 9 g; pH = 5.0)

It is obvious that only the presence of monomers having electronac-
ceptor groups (EGMA), as well as groups which can lead to crosslinking,
results in positive effects as far as grafting parameters are concerned.

When using a 50%/50% monomer mixture of MMA with EGDMA (Fig. 8), the
degree of conversion increased from 80% to 97%, and grafting efficiency
from 75% to 95%. On must note that the smaller values are found in the
case of MMA grafting. The positive effect of EGDMA was already well established at its 33% addition level which gave a 90.5% grafting efficiency and a 92.7% degree of conversion.

Effect of grafting on paper sheet properties

Kokta et al. [13] have shown that paper sheet properties could be significantly improved when using hardwood fibers grafted with acrylamide,
styrene and/or acrylonitrile. In addition, Daneault et al. [12] have found
slight deterioration of all paper properties due to grafting with PMMA.

CONCLUSION

The xanthate method of grafting, initiated by partially xanthated
aspen chemicothermomechanical pulp in the presence of hydrogen and ferrous
ion is quite efficient for methyl methacrylate grafting. The grafting parameters could be further enhanced by the addition of ethyleneglycoldimethacrylate to produce a 95% grafting efficiency and a 97% + conversion.

ACKNOWLEDGEMENT

The authors express their gratitude to the F.C.A.C. Fund (Québec)
and Natural Sciences and Engineering Research Council of Canada for their
financial support.

REFERENCES

1. N. Jensen, «Technology 2001», CPPA, Technical Section, Atlantic City
    (September, 1974).

2. R.W. Faessinger, J.C. Conte, U.S. Pat. 3,359,224 (Dec. 19, 1967): U.S.
    Pat. 3,330,787 (July 11, 1967).

3. B. Ranby, Pulp and Paper International, 52 (September, 1969).

4. V. Hornof, B.V. Kokta, J.L. Valade, Journal of Applied Polymer Scien-
    ce, 19, pp. 1543-1554 (1976).

5. V. Hornof, B.V. Kokta, J.L. Valade, Journal of Applied Polymer Scien-
    ce, 20, pp. 1543-1554 (1976).

6. V. Hornof, B.V. Kokta, J.L. Valade, Journal of Applied Polymer Scien-
    ce, Vol. 21, pp. 477-485 (1977).

7. P. Fuertes, B.V. Kokta, Pulp and Paper of Canada, Transaction, TR53-
    TR59, (September 1981).

8. B.V. Kokta, R. Chen, C. Daneault, ACS Symposium Series 187, pp. 269-
    283 (1982).

9. B.V. Kokta, C. Daneault, J.L. Valade, J. TAPPI, Proceedings, Vol. 1,
    pp. 58-63 (1983).

10. B.V. Kokta, J.L. Valade, TAPPI, Vol. 55, No. 3, pp. 366-369 (1972).

11. B.V. Kokta, C. Daneault, Int. Chem. Congress, Proceedings, 09P41,
    29 pp., Hawai (December, 1984).

12. C. Daneault, B.V. Kokta, J.L. Valade, Inter. Poplar Commission Symp.
    Proceedings, 22 pp., Ottawa (October, 1984).

13. B.V. Kokta, C. Daneault, 1983 Pulping Conference, Proceedings, Book 2,
    pp. 429-442, Houston (October, 1983).

OIL ABSORBENCY OF GRAFT COPOLYMERS FROM SOFTWOOD PULP

George F. Fanta, Robert C. Burr, and William M. Doane

Northern Regional Research Center
Agricultural Research Service
U.S. Department of Agriculture
Peoria, Illinois 61604

INTRODUCTION

Considerable research has been devoted to developing oil absorbents
for clean-up of oil spills and removal of emulsified oil from waste water.
Agricultural products and residues, such as kapok fiber,[1] cotton,[2,3] rice
hulls,[4] corncob meal,[5] bagasse fibers,[6] and peat moss[7] have been used for
this application, and these materials have the advantage of being inexpensive and readily available. Moreover, cellulosic products exist in fibrous
form and can be easily formed into mats, pads, and nonwoven sheets. Mats
and column packings for oil removal have also been prepared by mixing cellulosic fibers with synthetic fibers, e.g., polyethylene[8] or polypropylene-
polyethylene-nylon blend.[9] Cellulosics have also been mixed with inorganic materials, such as clay,[10] magnesium hydroxide, and alumina[11] to give
oil-absorbent compositions.

The hydrophobicity of cellulosic materials (and thus their affinity
for oil) has been increased by reaction with organic isocyanates[12] and with
fatty acid derivatives, such as anhydrides[13] and acid chlorides.[14] Also,
an oil-absorbing cellulose-polystyrene composite (no doubt containing some
grafted polystyrene) was prepared by immersing a cellulose sponge into
styrene-benzoyl peroxide and then heating the mixture to 100°C.[15]

Coating cellulosics with hydrophobic compounds, such as paraffin
wax,[16-18] insoluble fatty acid salts,[19] or low melting polymers, such as
polyolefins[20,21] is another technique used to increase the affinity of
fibers for oil; some polymers (e.g., ethylene-vinyl acetate copolymer)
also have been deposited onto fibers from aqueous emulsions.[22,23]

Although graft polymerizations onto cellulosics have been studied
with a wide variety of monomers and initiating systems,[24,25] little
research has been done on synthesizing graft copolymers to maximize their
oil-absorbing properties (e.g. by varying composition and amount of grafted
polymer). Grafted branches should obviously be hydrophobic, to improve the
affinity of the graft copolymer for oil; however, the weight percentage of
synthetic polymer in the final product (% add-on) should be kept as low as
possible, both for economic reasons and to allow some swelling and separation of individual fibers in water. The latter factor is especially important in a graft copolymer for removing emulsified oil from waste water.

In this research, we prepared low add-on graft copolymers from wood pulp and either acrylic acid or 2-acrylamido-2-methylpropanesulfonic acid (AASO$_3$H); we then made these graft copolymers hydrophobic through an ion exchange reaction with hexadecyltrimethylammonium bromide (CTAB). Reactions of this general type have been used with sodium cellulose sulfate[26] and with sodium alginate[27] to prepare thickeners for organic solvents. Hexadecyltrimethylammonium salts of wood pulp graft copolymers absorbed oil efficiently from an aqueous emulsion and had oil-absorbing properties superior to methyl acrylate and butyl acrylate graft copolymers with a comparable % add-on. Possible reasons for observed variations in oil-absorbing ability will be suggested with the hope that our work (which is in some areas preliminary) will stimulate further research on this important class of polymers.

EXPERIMENTAL

Materials

Bleached softwood pulp (Alberta Hi-Brite) was from St. Regis Paper Co., and analyses showed 0.06% lignin and 85.8% α-cellulose, the remainder being pentosans. Acrylic acid (Eastman, inhibited with 200 ppm p-methoxyphenol) and AASO$_3$H (Lubrizol Corp., Reaction Grade, Special Process) were used as received. Methyl acrylate (Polysciences) was distilled at atmospheric pressure through a 14-in. Vigreux column, while butyl acrylate (Eastman) was extracted at room temperature with 5% sodium hydroxide and then washed with water to remove inhibitor. Ceric ammonium nitrate was certified ACS Grade from Fisher Scientific Co. CTAB was Eastman technical grade, and the ethoxylated fatty amine was Varonic K205LC from Ashland Chemicals. One mole of amine contains 5 moles of ethylene oxide, and the neutralization equivalent is 445:

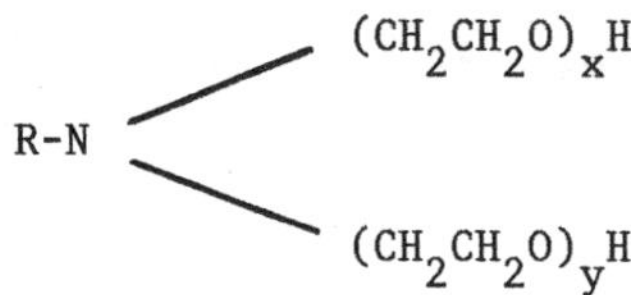

Ethoxylated Fatty Amine (x + y = 5)

Graft Polymerization

Acidic Monomers; Cobalt-60 Initiation. Thirty grams (dry basis) of softwood pulp was dispersed in water, filtered, and pressed as dry as possible. A solution of 20.0 g of acrylic acid (or AASO$_3$H) in 120 ml of water was mixed with the pulp, and excess liquid was again removed by filtration. The wet pulp, now containing about 6 g of monomer, was transferred to an 8-oz screw-cap bottle, and oxygen was displaced by evacuating four times to about 50 mm followed by repressuring with nitrogen. The reaction mass was then irradiated with cobalt-60 to 0.5 Mrad and allowed to stand for 2 h at ambient temperature. The cobalt-60 source was a Gammacell 200 unit from Atomic Energy of Canada, Ltd., and dose rate at the center of the chamber was 0.28±0.01 Mrad/h. The graft copolymer was washed several times with water and stored as a wet filter cake.

Weight percent poly(acrylic acid) or poly(AASO$_3$H) in the graft copolymer (% add-on) was determined by dispersing 1 g (dry basis) of polymer in 150 ml of 1$\underline{N}$ NaCl solution[28] and then titrating with 0.1$\underline{N}$ NaOH solution to a phenolphthalein end point. Poly(acrylic acid) grafts, containing about 20% residual carbohydrate were separated by reacting graft copolymer with sodium periodate and then degrading the oxidized polysaccharide portion with methanolic sodium methoxide.[29]

Acrylate Esters; Ceric Initiation. A stirred slurry of 75.0 g (dry basis) of softwood pulp in three 1 of water was sparged with a slow stream of nitrogen for 1 h at 25°C. Ten grams of methyl acrylate (or butyl acrylate) was added, followed after five min by a solution of 5.0 g ceric ammonium nitrate in 45 ml 1N nitric acid. The mixture was stirred for two h at 25°C; the polymer was then removed by filtration and resuspended in water, and pH was adjusted to 7 with sodium hydroxide solution. The polymer was washed with water, and the wet filter cake was extracted several times with acetone to remove 1-2 g of homopolymer. Extracted polymer was then freed of acetone by water washing and stored as a wet filter cake.

To determine % add-on and molecular weight of grafted poly(alkyl acrylate), polysaccharide was removed from the graft copolymer with sodium periodate-sodium methoxide.[29] The residue (which still contained some polysaccharide) was extracted with acetone to separate pure acrylate polymer. The poly(alkyl acrylate) content of acetone-insolubles was then estimated by comparing infrared spectra with spectra of known mixtures of poly(alkyl acrylate) and starch, and these figures were also used in % add-on calculations.

$\bar{M}_w$ and $\bar{M}_n$ of poly(alkyl acrylates) were determined by gel permeation chromatography; a 10^6 μ-Styragel column was used in addition to the columns described earlier.[30] Polystyrene standards were used for primary standardization of columns, and a standard poly(methyl acrylate) sample was run to establish the Q factor as 41.1 This same Q factor was used for poly(butyl acrylate).

Softwood pulp-g-poly(methyl acrylate) was converted to softwood pulp-g-poly(Na acrylate) by thoroughly mixing 6.0 g (dry basis) of graft copolymer with 10 ml of 1N sodium hydroxide and 10 ml of water. The resulting mixture was heated in a loosely stoppered flask in a 100°C oven for 6 h. The saponified polymer was washed with water until the slurry pH was 8.9, and was then stored as the wet filter cake. A portion of the graft copolymer was converted to carboxylic acid by stirring with 0.5N HCl, filtering, and washing with water until the slurry pH was 5.4. Titration showed 5.4% poly(acrylic acid).

Reaction of Graft Copolymers with CTAB

Five grams (dry basis) of softwood pulp-g-poly(acrylic acid) or softwood pulp-g-poly(AASO$_3$H) was suspended in 200 ml of water in a Waring Blendor and the pH was adjusted to 8 with 1N sodium hydroxide. Five grams of CTAB was added, and the mixture was stirred for 5 min. The product was separated by filtration and washed three times with 50:50 ethanol/water and three times with water. The polymer was then continuously extracted (Soxhlet) for 2 days with 95% ethanol. The extracted polymer was washed with water and stored as the wet filter cake.

Reaction of Graft Copolymer with Ethoxylated Fatty Amine

Five grams (dry basis) of softwood pulp-g-poly(acrylic acid) was suspended in 400 ml of water and titrated to pH 6.0 with a solution of 25 g of ethoxylated fatty amine in 75 ml of water. The mixture was allowed to stand for about 3 h, with periodic addition of amine solution to maintain the pH at 6.0. The product was filtered, washed three times with 50:50 ethanol/water and three times with water, and then stored as the wet filter cake. Soxhlet extraction was not carried out.

Five grams of graft copolymer in 200 ml of water was titrated to pH 7.0, and the product was worked up as described in the preceding paragraph.

<u>Tests for Oil Removal</u>

<u>Screening Test</u>. An emulsion was prepared by adding 200 ml of water, 20 drops (0.5-0.6 g) of mineral oil, and 0.1 g of dodecylbenzene sodium sulfonate to a Waring Blendor and stirring at high speed for 2 min. One gram (dry basis) of graft copolymer was shaken with 20 ml of oil emulsion in a 2-oz screw-cap bottle, and the mixture was let stand for 10 min. Supernatant was separated from fibrous solid by screening through a small Büchner funnel containing no filter paper. Percent transmission of the supernatant (% T) at 650 nm was then determined in a 1-cm ultraviolet cell.

<u>Absorption Capacity</u>. Ten milliliters of oil emulsion and 0.50 g (dry basis) of graft copolymer were placed in a 2-oz screw-cap bottle and shaken periodically for 5 min. Supernatant was separated, and % T was determined. Supernatant from the ultraviolet cell was returned to the bottle, and an additional 5 ml of oil emulsion was added. The mixture was shaken, and % T was again determined. This procedure was repeated with 5-ml increments of emulsion until % T dropped to about 10 or less.

RESULTS AND DISCUSSION

Graft polymerizations of acrylic acid and $AASO_3H$ onto softwood pulp (Table 1) were carried out by first allowing the pulp to absorb a water solution of monomer and next removing excess monomer solution by filtration. The resulting monomer-wet pulp was then irradiated with cobalt-60 to initiate graft polymerization. This semidry procedure resembles a technique used by us earlier to graft water-soluble monomers onto starch.[31] Repeat experiments 2 and 3 were run under conditions given in the Experimental, whereas 4 was run with half the amount of acrylic acid. Conditions for 5 were the same as for 2 and 3, except that $AASO_3H$ was used instead of acrylic acid.

After polymerization, homopolymer and unreacted monomer were removed by water washing, and the graft copolymer was then stored as a wet filter cake. In experiment 4, a 20% conversion of monomer to homopolymer was observed. Synthetic polymer content was determined by titration. For practical purposes, we found that the add-on should not exceed about 6%, since larger amounts of grafted poly(acrylic acid) led to excessive swelling of fibers and precluded their easy separation by filtration from a water dispersion. Poly(acrylic acid) grafts, isolated from the product of experiment 4 by degrading the polysaccharide portion with sodium periodate followed by methanolic sodium methoxide,[29] had an intrinsic viscosity of only 0.12 in 1$\underline{M}$ sodium chloride at 25°C, indicating a low molecular weight for grafted branches.

The acidic graft copolymers were next converted to their sodium salts and subjected to ion exchange reactions with CTAB (Fig. 1). Excess CTAB was removed by washing reaction products first with ethanol-water and then with water, and finally extracting them continuously with 95% ethanol in a Soxhlet apparatus. Extent of reaction with CTAB was determined from Kjeldahl nitrogen analyses before and after reaction. For polymer 3, Table 1, these nitrogen values were 0.016% and 0.43%, respectively; calculations based on these analyses indicated that 37% of the carboxyl groups in the graft copolymer had reacted as indicated in Figure 1.

Graft copolymers were screened for their ability to absorb emulsified oil by shaking them with mineral oil emulsions and then determining the clarity of supernatants, as measured by percent transmission (% T) at

Table 1. Graft Polymerizations onto Softwood Pulp. Screening Tests for Oil Removal.

| | | | | Oil removal test, % T of filtrate[b] | |
Reaction Number	Monomer	Initiator	% Add-on[a]	As is	CTAB[c] derivative
1	None	Cobalt-60	0	--	0.2
2	Acrylic acid	Cobalt-60	6.5	--	97
3	Acrylic acid	Cobalt-60	6.2	0.2	94
4	Acrylic acid	Cobalt-60	3.4	--	93
5	$AASO_3H$[d]	Cobalt-60	4.1	--	93
6	Methyl acrylate	Ce^{+4}	8	0.3	92
7	Butyl acrylate	Ce^{+4}	9	39	--

[a] Weight % synthetic polymer in the graft copolymer.

[b] % Transmission at 650 nm of oil emulsion after treatment with graft copolymer. See Screening Test in Experimental.

[c] Hexadecyltrimethylammonium bromide.

[d] 2-Acrylamido-2-methylpropanesulfonic acid.

650 nm. On the basis of this single-point screening test, CTAB-treated graft copolymers were equally effective at clarifying oil emulsions (Table 1) and gave % T values in excess of 90. The dependence of oil absorption on reaction with CTAB is seen in data presented for polymer 3. Before CTAB treatment, the poly(acrylic acid) graft copolymer gave only 0.2% T in our screening test, whereas % T was 94 in the test with CTAB-treated polymer. The dependence of oil absorption on the grafting of acidic polymer is shown in reaction 1 (Table 1), where cobalt-60 irradiation of softwood pulp and subsequent treatment with CTAB were

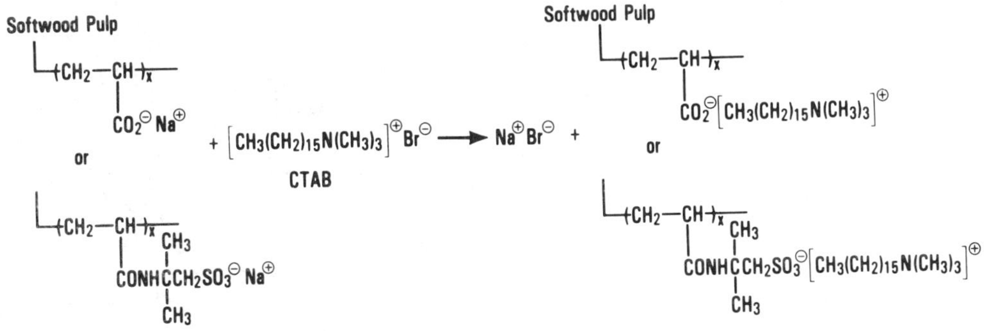

Fig. 1. Preparation of hexadecyltrimethylammonium bromide (CTAB) derivatives of wood pulp-g-poly(acrylic acid) and wood pulp-g-poly(2-acrylamido-2-methylpropanesulfonic acid).

carried out in the same way as for 2 and 3, except that acetic acid was substituted for acrylic acid during irradiation to avoid polymer formation. The resulting product produced little clarification of our test emulsion (0.2% T).

Although CTAB-treated graft copolymers behaved similarly in the one-point screening test, it is to be expected that differences will exist among these polymers in terms of total oil-absorbing capacity. These capacity differences were shown by adding oil emulsion in 5-ml increments and then measuring % T of the supernatant after each addition. Curves A through D of Figure 2 show test results with CTAB derivatives of graft copolymers 2, 4, and 5 of Table 1. Although these rather crude tests gave a broad scatter of points and only fair reproducibility (compare curves B and C for polymer 4), the general shapes of the curves can still be deduced. If we base our calculations of oil-absorption capacity on the assumption of complete oil removal from 40 ml of emulsion by 0.5 g of graft copolymer with 4% add-on, it is apparent that each gram of graft copolymer absorbs about 0.2 g of oil, whereas the capacity of grafted branches alone is on the order of 5 g/g. It is significant that the poly(acrylic acid) graft copolymer with 3.4% add-on had a capacity for oil at least as large as that of the polymer with 6.5% add-on. Also, poly(acrylic acid) graft copolymers appear to have a larger oil capacity than copolymers with poly(AASO$_3$H) grafts.

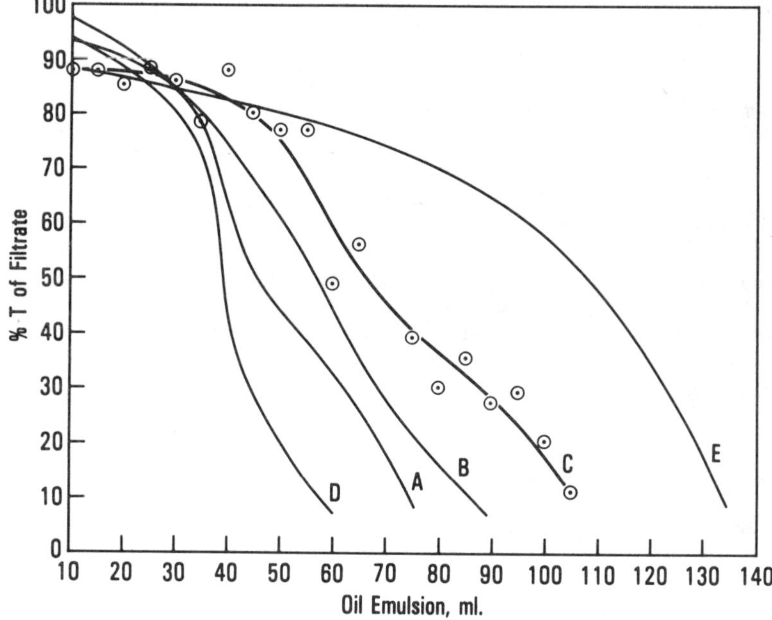

Fig. 2. Oil-absorption capacity of CTAB derivatives: (A) Wood pulp-g-poly(acrylic acid) with 6.5% add-on (polymer 2, Table 1), (B) wood pulp-g-poly(acrylic acid) with 3.4% add-on (polymer 4, Table 1), (C) repeat determination with polymer 4, Table 1 (individual points are shown to indicate point scatter), (D) wood pulp-g-poly(2-acryl-amido-2-methylpropanesulfonic acid) with 4.1% add-on (polymer 5, Table 1), and (E) saponified wood pulp-g-poly(methyl acrylate) (polymer 6, Table 1).

Because direct titration of acidic graft copolymers with fatty amine
is an obvious method for preparing polymers similar to our CTAB derivatives, this technique was tried with a poly(acrylic acid) graft copolymer
having 6.0% add-on and prepared in a manner identical to products 2 and 3
of Table 1. This method, however, suffers from some drawbacks. In
addition to the problem of incomplete reaction between weak acid and weak
base, there are difficulties associated with dispersing a high-molecular-
weight, water-insoluble amine in an aqueous solvent system. Although the
problem of poor dispersibility was avoided by choosing an ethoxylated
fatty amine (see Experimental) that forms an emulsion when added to water,
titration of the graft copolymer with this amine to pH 6.0 gave a product
with poor oil-absorbent properties (3.4% T in our screening test). Titration to pH 7.0 improved oil absorbency (47% T); however, properties were
still inferior to the corresponding CTAB derivative.

In reactions 6 and 7 of Table 1, we attempted to prepare oil absorbents by grafting hydrophobic acrylate esters onto softwood pulp with ceric
ammonium nitrate initiation. The ratio of monomer to softwood pulp in
these polymerizations was chosen to give graft copolymers with roughly the
same % add-on as the poly(acrylic acid) and poly($AASO_3H$) graft copolymers
prepared earlier. Percent add-on was calculated from loss in weight
resulting from degrading the polysaccharide portion of the polymer with
sodium periodate-sodium methoxide. Earlier work[29] showed that this procedure does not affect molecular weight of grafted poly(methyl acrylate).
Molecular weights of isolated poly(methyl acrylate) ($\bar{M}_w$ 55,000; $\bar{M}_n$ 20,000)
and poly(butyl acrylate) ($\bar{M}_w$ 321,000; $\bar{M}_n$ 95,000) were determined by gel
permeation chromatography. As expected (because of the longer alkyl ester
substituents), the poly(butyl acrylate) graft copolymer was more effective
in removing emulsified oil than its poly(methyl acrylate) counterpart
(39% T vs. 0.3% T); however, its performance was still inferior to CTAB
derivatives.

Since the CTAB derivative of wood pulp-g-poly(acrylic acid) should be
less hydrophobic than wood pulp-g-poly(butyl acrylate) based on the two
grafted polymer structures, it is interesting and rather unexpected that
the CTAB derivative has the best oil-absorbing properties. This observation suggests that hydrophobicity is not the only factor in determining
affinity of a fibrous polymer toward emulsified oil. Perhaps the ionic
nature of both CTAB derivative and unreacted sodium carboxylate allows
individual fibers to swell and separate in water, and this balance between
hydrophobicity and hydrophilicity could give improved contact between emulsified oil droplets and hydrophobic CTAB substituents.

Since poly(alkyl acrylate) graft copolymers can be saponified with
sodium hydroxide solution to form copolymers with poly(sodium acrylate)
grafts,[32] this route represents an alternative to the direct cobalt-60
initiated grafting of acrylic acid. To determine whether CTAB derivatives
of saponified acrylate ester copolymers have oil-absorbing properties
comparable to CTAB derivatives prepared from poly(acrylic acid) copolymers,
polymer 6 was heated with sodium hydroxide solution and the saponified
polymer was allowed to react with CTAB. The resulting product gave 92% T
in our screening test (Table 1), and its absorption capacity (Fig. 1E) was
superior to the other graft copolymers. Based on both performance and
simplicity of operation, this two-step procedure with ceric-initiated
grafting would probably be the method of choice for preparing these graft
copolymers.

ACKNOWLEDGMENT

We are indebted to C. L. Swanson for molecular weight determination
by gel permeation chromatography.

The mention of firm names or trade products does not imply that they are endorsed or recommended by the U.S. Department of Agriculture over other firms or similar products not mentioned.

REFERENCES

1. Y. Kobayaski, _Mizu Shori Gijutsu_, _19_, 517 (1978).
2. R. F. Johnson, T. Manjreker, and J. E. Halligan, _Environ. Sci. Technol._, _7_, 439 (1973).
3. G. F. Meenaghan, J. E. Halligan, A. A. Ball, and J. F. Leary, _Proc. Annu. Offshore Technol. Conf._, _8_, 945 (1976).
4. L. E. Bertram, U.S. Pat. 3,902,998, September 2, 1975.
5. J. VanderHooven and D. I. B. VanderHooven, U.S. Pat. 3,617,564, November 2, 1971.
6. K. O. P. Fischer, Eur. Pat. Appl. 2,070, May 30, 1979.
7. F. Hennezel and B. Coupal, _CIM_ (_Can. Inst. Mining Met._) _Bull._, _65_(717), 51 (1972).
8. D. E. Wiegand, F. H. Riedel, and O. R. Videen, U.S. Pat. 4,070,287, January 24, 1978.
9. J. Ohkita, H. Segawa, K. Saito, and M. Nakamura, Jpn. Kokai 74:84,979, August 15, 1974.
10. T. Suzuki, T. Ito, and T. Onuma, Jpn. Kokai 76:92,791, August 14, 1976.
11. H. Sato, F. Kotani, and H. Uchida, Jpn. Kokai 77:101,860, August 26, 1977.
12. A. Holst, M. Kostrzewa, H. Lask, and G. Buchberger, Ger. Offen. 2,358,808, June 5, 1975.
13. F. J. Ball, U.S. Pat. 3,770,575, November 6, 1973.
14. J. Teng and M. C. Stubits, U.S. Pat. 3,874,849, April 1, 1975.
15. J. W. Marx, U.S. Pat. 3,677,982, July 18, 1972.
16. K. S. Peterson and G. R. Palkie, U.S. Pat. 3,630,891, December 28, 1971.
17. Y. Matsuda, S. Tomita, K. Terajima, K. Abe, R. Ito, Y. Sumi, and Y. Takeuchi, Jpn. Kokai 77:76,285, June 27, 1977.
18. G. O. Orth, Jr., Ger. Offen. 2,301,176, July 26, 1973.
19. S. Aoso, Jpn. Kokai 74:64,577, June 22, 1974.
20. Y. Kunitomo, H. Ono, T. Saida, K. Nakarai, and T. Wagatsuma, Ger. Offen. 2,621,961, April 21, 1977.
21. T. Saida, A. Kunitomo, H. Ono, K. Nakarai, and T. Azuma, Jpn. Kokai Tokkyo Koho 78:04,790, January 17, 1978.
22. H. Sato and F. Itani, Jpn. Kokai 77:89,244, July 26, 1977.
23. H. Sato and F. Oriya, Jpn. Kokai 77:90,486, July 29, 1977.
24. J. C. Arthur, Jr., in _Advances in Macromolecular Chemistry_, W. M. Pasika, Ed., Academic, New York, 1970, Vol. II, p. 1.
25. J. C. Arthur, Jr., in _Graft Copolymerization of Lignocellulosic Fibers_, D. N.-S. Hon, Ed., ACS Symp. Ser. 187, 1982, p. 21.
26. R. G. Schweiger, U.S. Pat. 3,637,520, January 25, 1972.
27. National Lead Co., Brit. Pat. 876,603, September 6, 1961.
28. M. L. Miller, _Encyl. Polym. Sci. Technol._, _1_, 217 (1964).
29. G. F. Fanta, R. C. Burr, and W. M. Doane, _J. Appl. Polym. Sci._, _27_, 4239 (1982).
30. G. F. Fanta, C. L. Swanson, R. C. Burr, and W. M. Doane, _J. Appl. Polym. Sci._, _28_, 3003 (1983).
31. R. C. Burr, G. F. Fanta, W. M. Doane, and C. R. Russell, _J. Appl. Polym. Sci._, _20_, 3201 (1976).
32. L. A. Gugliemelli, M. O. Weaver, and C. R. Russell, U.S. Pat. 3,377,302, April 9, 1968.

CHEMICALLY MODIFIED CARBOHYDRATES AS HIGHLY EFFICIENT

REGIO- AND STEREOSELECTIVE CATALYSTS FOR HYDROGENATION

Ernst Bayer, Wilhelm Schumann, and Kurt E. Geckeler

Institute of Organic Chemistry
University of Tuebingen
D-7400 Tuebingen, FRG

INTRODUCTION

Polymers from natural sources have gained increasing interest for a variety of reasons. One of the main advantages is the high biological compatibility of these macromolecules in comparison to synthetic polymers. We introduced different polysaccharides as polymeric supports for catalysts in order to obtain polymer catalysts which are applicable in homogeneous phase.

In general, homogeneous hydrogenation catalysts have the advantage of a higher substrate selectivity, reproducibility, and reactivity when using even lower metal concentrations.[1] Substrate selectivity can be changed by variation of ligands at the catalyst metal; for example, asymmetric hydrogenation can be accomplished by use of chiral ligands with yields up to 99 % enantiomeric purity.[2]

The main disadvantage of homogeneous hydrogenation is the difficult separation of metal catalyst and hydrogenated product. This disadvantage can be eliminated by fixation of the homogeneous hydrogenation catalyst to a soluble polymer. Thus, separation of polymer catalyst and hydrogenated product can be carried out by membrane filtration while retaining the polymer reagent to be used for another reaction cycle. Both continuous and discontinuous procedures of membrane filtration can be used in this approach[3].

In this context it should be explained that for hydrogenation in food technology, synthetic polymers are not permitted in many countries. Therefore, natural polymers, such as cellulose, starch, collagen and gelatin, play a key role in this new methodological approach for hydrogenation of unsaturated fatty acids.

EXPERIMENTAL PART

Materials

Carboxymethyl cellulose (CMC, A 400 P) was purchased from Henkel, Düsseldorf, starch from Merck, Darmstadt, poly(vinyl-pyrrolidone) (K 90) from Fluka, Neu-Ulm, collagen (1000) from Heyl, Berlin, and gelatin from Dr. Oetker, Bielefeld. Palladium chloride, ruthenium chloride (with crystal water) was from Degussa, Hanau, and nickel chloride from Merck, Darmstadt.

Membrane filtration was carried out using systems which are described elsewhere.[4] Membranes with a cut-off level of 10 000 were manufactured in our laboratory from aromatic polysulfone polymers.[5]

Preparation of the Polymer Palladium Catalyst

To an aqueous solution of carboxymethyl cellulose (100 ml; 0.1 %) a solution of K_2PdCl_4 (0.1 ml; 1 %) was added under stirring. In order to activate the catalyst, the reaction vessel was then evacuated and flushed with hydrogen three times. After stirring for two hours at room temperature, the polymer catalyst was ready for use.

Preparation of the Polymer Nickel Catalyst

Forty-one mg of $NiCl_2 \cdot 6\ H_2O$ (= 10 mg Ni) were added to 100 ml of a CMC solution (0.4 %) under stirring at room temperature. After evacuating and flushing with hydrogen (three times), 5 mg of sodium borohydride were added and the solution stirred for two hours at room temperature.

Hydrogenation Procedure

The solution of polymer catalyst (100 ml; 0.1 %) was warmed up to $50^{\circ}C$ under stirring and then 50 g of preheated ($50^{\circ}C$) oil were added. The emulsion was hydrogenated until the hydrogen uptake corresponded to the desired hydrogenation degree which was controlled by monitoring the iodine value. The heating and stirring was stopped and the emulsion was allowed to separate. The oily phase was isolated using a separation funnel and analyzed by gas chromatography after saponifi-cation with alcoholic potassium hydroxide and subsequent esterification with diazomethane. The aqueous phase was recovered and used for another hydrogenation cycle after reactivation with hydrogen.

Preparation of CMC-Ru catalyst and Hydrogenation of Glucose and Fructose

To a solution of $RuCl_3 \cdot H_2O$ (20 mg) in 10 ml water, which was heated to $85^{\circ}C$, a sodium hydroxide solution (10 %) was added until pH 7 was reached. In order to obtain an easy-to-handle precipitate, the mixture was allowed to stand at the same temperature for ten minutes. Then, the precipitate was filtered off, washed three times with warm water ($60^{\circ}C$) and then suspended in a solution of 1 g CMC in 100 ml water

at room temperature. After replacing the air by hydrogen the
suspension is heated to $70^{\circ}C$ and stirred for 24 hours.
Cooling to room temperature yielded a homogeneous solution
of the polymer ruthenium catalyst.

Three go of the monosaccharide dissolved in 15 ml water
were added to the polymer catalyst solution in an autoclave and
hydrogenated overnight at 50 bar and room temperature. The
solution was then transferred into an ultrafiltration system
in order to separate the polymer catalyst from the low-
molecular sugar. Analysis of the filtrate by HPLC showed
quantitative conversion.

RESULTS AND DISCUSSION

Several natural polymers, which seem to be suitable for
coordinative metal binding according to their hydrophilic
nature, were investigated as soluble polymeric supports for
homogeneous catalysis. To this end, starch, carboxymethyl
cellulose, collagen, and gelatin were applied in their water-
soluble form as non-crosslinked biopolymers in combination with
ultrafiltration.

Prerequisite for the attachment of hydrogenation catalysts
to polymeric supports is a certain number of functional groups
which allow a covalent or ionogenic binding of the metal. Such
groups are CN, OH, C=O, COOH, NH_2, NR_2. Also, a strong
linearity of the polymer chain is required in order to provide
a high solubility and an appropriate accessibility to the
reaction centers thus yielding a higher selectivity. A good
solubility is also required during membrane filtration for
prevention of precipitation which could delay or inhibit the
separation process.

Cellulose was carboxymethylated in alkaline solution in
order to provide water-solubility and good complexing properties.[6,7] Moreover, the high molecular mass of cellulose
facilitates the separation of the polymer from low-molecular
compounds. Starch needs no chemical modification for this
approach because it is already water-soluble. However, the
solubility is reduced in comparison to CMC.[8,9]

Collagen is a protein which consists of units of tropocollagen,[10,11] which are 3 helical amino acid chains of 280 nm
length and 1.4 nm width. The collagen fiber can be subdivided
in collagen fibrilles which are parallel aligned tropocollagen
molecules. Collagen is slightly soluble in water (1 %)
yielding high-viscous solutions.[12] Gelatin is a degradation
product of collagen and represents a good complexing agent
based on the terminal groups or side-chains of amino acids.

Suitable metals for homogeneous hydrogenation are nickel,
cobalt, palladium, platinum, iridium, rhodium, ruthenium and
osmium. Nickel and palladium linked to the appropriate ligands
can serve for selective hydrogenation of C-C triple bonds to
double bonds as well as complete hydrogenation of unsaturated
systems. Whereas platinum favors the cis C-C double bond,
rhodium and ruthenium are especially suitable for the hydrogenation of aromatic compounds to the corresponding cycloalkanes
and sugars to sugar alcohols yielding practically only one type

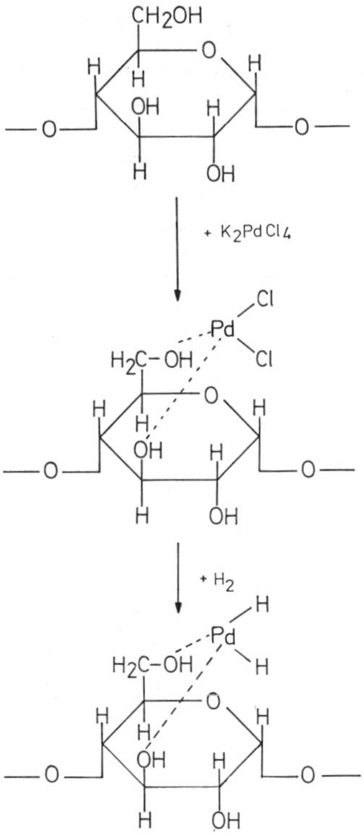

Fig. 1. Binding of the active palladium center to the polymer
by reaction of potassium palladium tetrachloride with
starch yielding the chlorine containing polymer derivative. The polymer catalyst is obtained by subsequent
activation with hydrogen. [13]

of isomer. Also aromatic ketones are hydrogenated to cyclo-
alkane ketones and phenols to cyclohexanone in 85 % yield.

As a first example of the preparation of a polymer-bound
catalyst, starch was dissolved in water and then allowed to react
with potassium palladium tetrachloride yielding a polymer-bound
palladium complex (Fig. 1). In the next step, sodium carbonate
was added in order to facilitate the reduction of Pd(II) to
Pd(0) with hydrogen. This subsequent activation step took
about 15 minutes, however, usually the hydrogen atmosphere was
maintained for about two hours to make sure that the system was
equilibrated completely. In the same manner, the carboxymethyl
cellulose palladium catalyst was prepared from CMC and
K_2PdCl_4 (Fig. 2). In this case, the metal atom is not only
coordinated to the oxygen of the hydroxyl groups but also the
carboxymethyl groups contribute to the binding. Therefore, a
polymer chelate was formed and finally the chlorine atoms
were removed by activation with hydrogen.

In order to demonstrate the high selectivity and stability
of the soluble polymer catalyst compared to similar heterogeneous systems natural materials were chosen as substrate for
hydrogenation. Basis materials for fat hydrogenation are soybean and rape-seed oil which have been investigated with
several polymer catalyst varying polymer backbone and metal
atom (Table 1).

Noteworthy is that many countries require natural products
for modification of food. Therefore, catalysts based on natural
polymers fulfill these requirements. Some data on these
experiments are summarized in Fig. 3. In the case of rape-seed
oil (a), hydrogen uptake correlated linearly with the content
of the fatty acids. The content of stearic acid ($C_{18=0}$) and
linolic acid ($C_{18=2}$) remained constant whereas the content
of oleic acid ($C_{18=1}$) showed a slight increase.

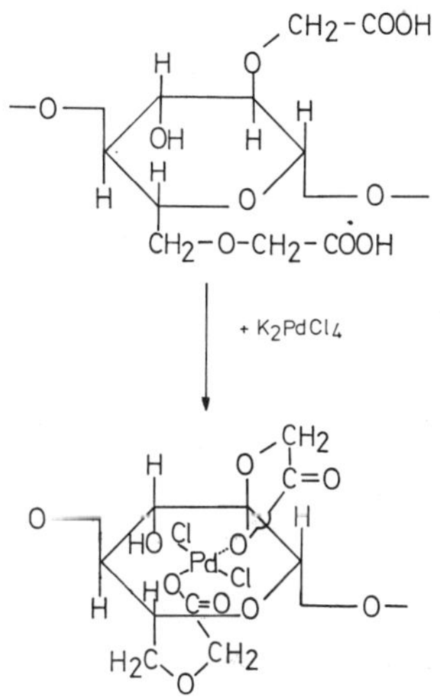

Fig. 2. Preparation of the CMC palladium catalyst by reaction
        of carboxymethyl cellulose with potassium palladium
        tetrachloride. In this compound the metal atom is
        attached to the polymer as a chelate by the carboxylic
        groups of the acetic acid. Activation is achieved by
        exposure to hydrogen.

In contrast, the concentration of linolenic acid ($C_{18=3}$)
decreased to nearly zero which is the main goal of the hydrogenation process in food technology. In addition, the content
of trans isomers is very low in comparison to conventional
heterogeneous catalysis (Table 1). The same positive effects
were observed for the hydrogenation of soybean oil (Fig. 3b).
The lower selectivity for oleic and linolic acid is here
different from rape-seed oil. However, the desired elimination
of linolenic acid and the low formation of trans isomers were
attained.

Table 1. Summarized Data of the Hydrogenation of Soybean Oil with Different Polymeric Catalysts at 50°C and Normal Pressure. The Values are Referred to Reactions of 50 g Soybean Oil with a Solution of 200 mg Polymer Including 1 mg Palladium or 10 mg Nickel Respectively.

Type of Oil	Polymer	Metal	Fatty Acid Composition (%)						Iodine Value	H_2 Uptake (ml/min)
			$C_{16=0}$	$C_{18=0}$	$C_{18=1}$	$C_{18=2}$	$C_{18=3}$	Trans		
Soybean Oil	Before Hydrogenation		10.3	3.4	26.0	51.9	8.4	--	130	--
	CMC	Pd	10.3	4.0	43.7	41.0	0.7	5.7	103	17
	CMC	Pd	10.3	19.7	58.0	12.1	–	14.0	72	14
	CMC	Ni	10.3	4.2	41.9	42.8	0.8	5.6	100	12
	Starch	Pd	10.3	4.5	40.8	43.5	1.0	5.0	101	12
	Collagen	Pd	10.3	5.8	43.8	38.0	2.1	6.1	102	8
Rape-Seed Oil	Before Hydrogenation		6.0	1.1	62.9	20.4	9.6	–	119	–
	CMC	Pd	6.0	2.2	73.4	18.4	1.0	4.2	99	16
	CMC	Pd	6.0	3.3	80.7	11.0	–	6.0	91	16
	Starch	Pd	6.0	2.8	72.8	18.0	1.4	4.8	97	10
	Collagen	Pd	6.0	3.0	75.9	13.0	2.1	5.1	98	8

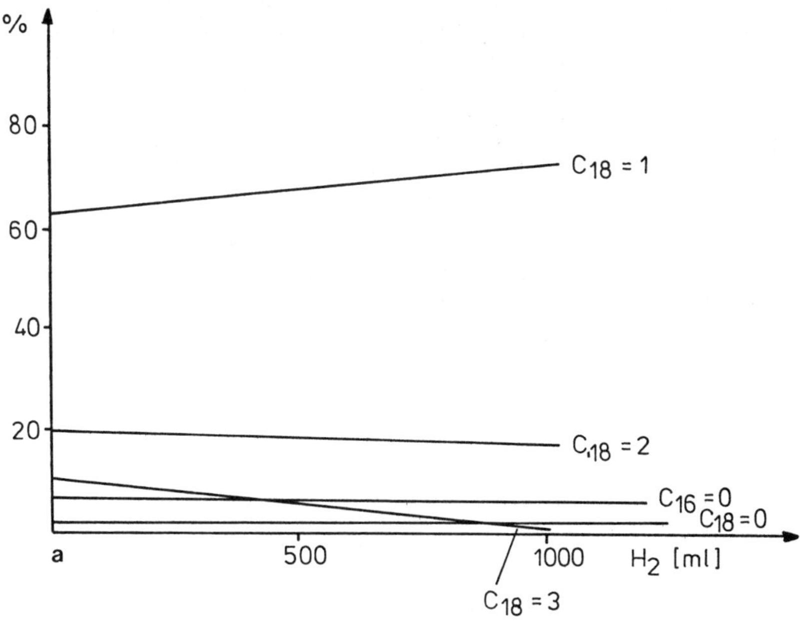

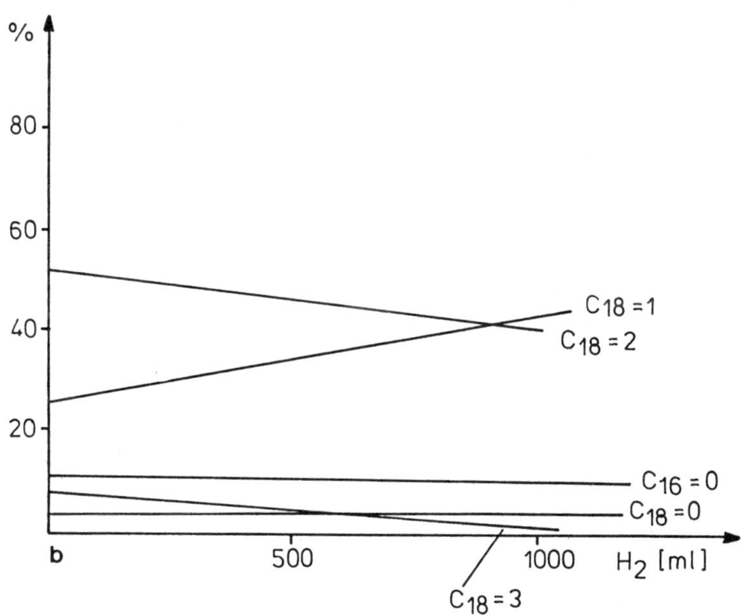

Fig. 3. Relation between the product composition of rape-seed oil (a) and soybean oil (b) after hydrogenation with CMC-Pd. The diagram shows the fatty acid composition as a function of the hydrogen uptake.

Another example of selectivity is shown by the hydrogenation
of glucose using a carboxymethyl cellulose-ruthenium catalyst
(Fig. 4). This is important because sorbitol is the starting
material for the synthesis of ascorbic acid and is also used
as a substitute for sugar.[14] Xylose can be hydrogenated to
xylitol in the same manner. For comparable rhodium, palladium,
and platinum catalysts decomposition of the sugars was observed
and lowered the turn-over considerably. The ruthenium catalyst,
however, did not show such effects and gave a complete turn-
over. Noteworthy is the selectivity in this case yielding 99 %
of sorbitol and 1 % of mannitol whereas usual hydrogenation
gives 75 % and 25 % of each hexose.[15] The same type of selectivity was observed for fructose which yielded by usual hydrogenation about 50 % of each product. Using the polymer catalyst
we have obtained 60 % of sorbitol and 40 % of mannitol
(Table 2).

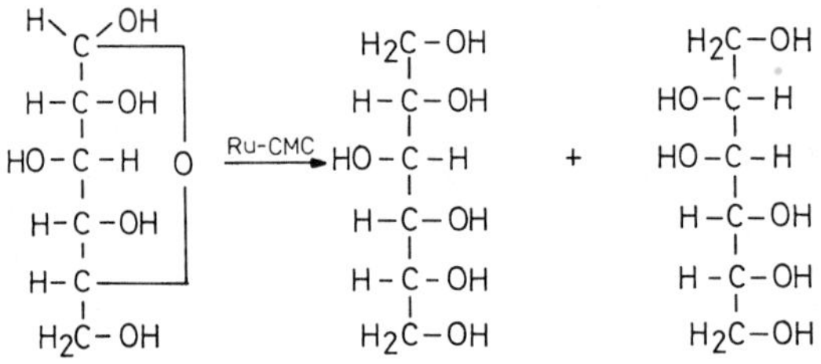

Fig. 4. Hydrogenation of glucose using the CMC-ruthenium
catalyst. The hydrogenated product contains 99 %
of sorbitol and 1 % of mannitol.

In order to study the effect of the polymer backbone on
the hydrogenation rate we have compared several types of
natural polymers with synthetic polymers, e.g. poly(vinylpyrrolidone), poly(vinyl alcohol), and poly(acrylnitrile).[16]
As an example for a synthetic catalyst, the formation of
poly(vinylpyrrolidone)-palladium is depicted in Fig. 5. The
metal atom is here coordinated to the lactam oxygen atoms.[17]

Table 2. Hydrogenation of Glucose and Fructose with a
Ruthenium-CMC Catalyst in Aqueous Solution at
Room Temperature and 50 bar. Values are
Referred to a Reaction of 3 g Sugar and 13 mg
Ruthenium Bound to 1 g CMC.

			Hydrogenated Product	
No.	Sugar	pH	Sorbitol	Mannitol
1	Fructose	7	60	40
2	Fructose	8	61	39
3	Glucose	7	99	1
4	Glucose	8	99	1

Table 3. Comparison of a Natural (CMC) and a Synthetic (PVP)
Basis Polymer for the Hydrogenation of Soybean Oil
with Palladium. (PVP = Polyvinylpyrrolidone).

Experimental Data		Polymer	
		CMC	PVP
Fatty Acid Composition			
$C_{16} = 0$	10.3	10.3	10.3
$C_{18} = 0$	3.4	4.0	3.9
$C_{18} = 1$	26.0	43.7	41.0
$C_{18} = 2$	51.9	41.0	43.6
$C_{18} = 3$	8.4	0.7	1.2
Trans	–	5.7	5.4
Iodine Value	130	109	102
H_2 Uptake (ml/min)	–	14	4

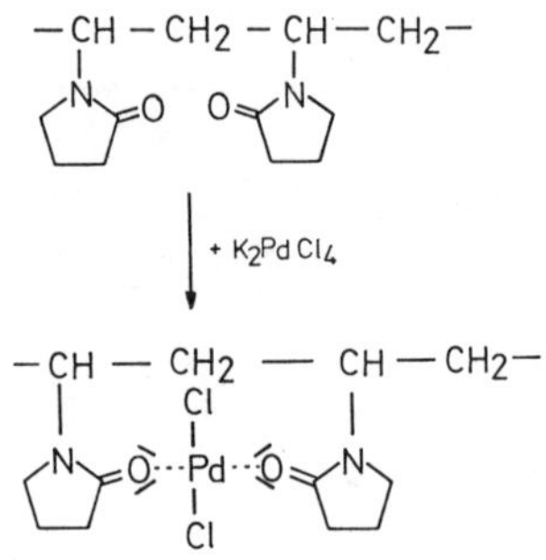

Fig. 5. Formation of the poly(vinylpyrrolidone)-palladium
catalyst by reaction of the complex salt with the
polymer. In this macromolecule the metal is oxygen
coordinated.

The comparison between CMC and PVP in terms of the fatty
acid composition of the hydrogenated product shows that
practically no difference was found (Table 3).

Investigating the hydrogen uptake as a function of time
during hydrogenation of soybean oil with palladium as metal
we have found that the three natural polymers studied exceeded
poly(vinylpyrrolidone) significantly (Fig. 6). Among the
natural polymers CMC showed the highest rate which can be
explained by the chelating of the additional carboxymethyl
groups existing in this polysaccharide.

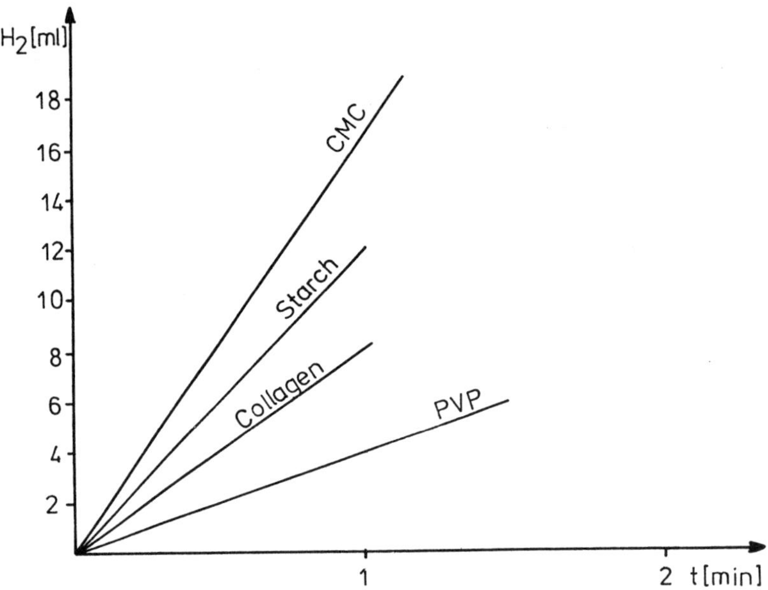

Fig. 6. Comparison of different polymers in their form as
palladium-activated catalysts. Dependence of the
hydrogen uptake as a function of time for CMC, starch,
collagen and PVP.

REFERENCES

1.   J. A. Osborn, F. H. Jardine, J. P. Young, and G. Wilkinson,
     J. Chem. Soc. A 1711 (1966).
2.   D. Valentine and J. W. Scott, Synthesis 329 (1978).
3.   E. Bayer and V. Schurig, Chem. Techn. 6:212 (1976).
4.   E. Bayer, H. Eberhard, and K. Geckeler, Angew. Makromol.
     Chem. 97:217 (1981).
5.   W. Schumann, Dissertation, University of Tuebingen,
     Tuebingen (1983).
6.   E. Gruber, Th. Krause, and J. Schurz, "Ullmanns
     Encyklopädie der technischen Chemie", Vol. 9, (1975).
7.   R. N. Hader, W. F. Waldeck, and D. Smith, Ind. Eng.
     Chem. 44:2803 (1952).
8.   R. J. McIlroy, "The Polysaccharides", E. Arnold, ed.,
     London (1948).
9.   R. L. Whistler and C. L. Smarth, "Polysaccharide
     Chemistry", Academic Press, New York (1953).
10. K. Kühn, Naturwiss. 54:101 (1967).
11. E. J. Miller, "Connective Tissue", Elsevier, North
     Holland (1982).
12. A. Veis, "The Macromolecular Chemistry of Gelatin",
     Academic Press, New York, London (1964).
13. Y. Nakao and S. Fujishigo, Bull. Chem. Soc. Japan
     53:1267 (1980).
14. G. Gräfe, Stärke 21:183 (1969).
15. E. Haidegger, Stärke 29:430 (1977).
16. W. Doll, Diplomarbeit, University of Tuebingen,
     Tuebingen (1978).
17. G. Biedermann, W. Graf and F. Steininger, Chem. Ztg.
     100:141 (1976).

DEXTRAN HEMOPOLYMERS AS MODEL SYSTEMS FOR BIOLOGICALLY
ACTIVE PROTEINS

Ernst Bayer, Martin Keck, and Kurt E. Geckeler

Institute of Organic Chemistry
University of Tuebingen
D-7400 Tuebingen, FRG

INTRODUCTION

A great number of model compounds for biologically
active metal proteins has been synthesized during the last
decade. Most papers focused on the development of systems[1-4]
which are able to form complexes with molecular oxygen.
These investigations contributed considerably to the understanding of mechanisms of biologically relevant reactions,
e.g. the reversible oxygenation of hemoglobin and myoglobin
and the activation of molecular oxygen during hydroxylation
in biological media and cells. The active centers of the
proteins, which catalyze these reactions, are iron porphyrin
systems.

When designing models for oxygen complexes the following
criteria have to be fulfilled:

1) The stereochemical arrangement around the
   prosthetic group of the model should correspond
   to that of the natural molecule.

2) Water-solubility is necessary; however, solubilization can also be provided by water-soluble high-
   molecular compounds.

3) Bimolecular autoxidation, which takes place when
   using unprotected iron(II) complexes, must be
   inhibited.

The development of model compounds for the oxygen
carriers hemoglobin and myoglobin requires additional
criteria:

1) At the central iron atom a heterocyclic amine
   should take over the function of the proximal
   histidine as a fifth ligand.

2) The active center must be protected by a hydrophobic
   microenvironment in order to avoid oxidation
   catalyzed by protons:

$$B-\overset{\textstyle |}{\underset{\textstyle |}{Fe}}^{II}-O_2 \; + \; H^+ \; \longrightarrow \; B-\overset{\textstyle |}{\underset{\textstyle |}{Fe}}^{III} \; + \; HO_2\cdot$$

B = basic axial ligand

3) Inhibition of the autoxidation according to the dimeric[5]
   mechanism can be accomplished by sterical separation:

$$B-\overset{\textstyle |}{\underset{\textstyle |}{Fe}}^{II}-O_2 \; + \; B-\overset{\textstyle |}{\underset{\textstyle |}{Fe}}^{II} \; \longrightarrow \; \overset{\textstyle |}{\underset{\textstyle |}{Fe}}^{III}-O-O-\overset{\textstyle |}{\underset{\textstyle |}{Fe}}^{III} \; + \; 2\,B$$

$$\overset{\textstyle |}{\underset{\textstyle |}{Fe}}^{III}-O-O-\overset{\textstyle |}{\underset{\textstyle |}{Fe}}^{III} \; \longrightarrow \; 2\,\overset{\textstyle |}{\underset{\textstyle |}{Fe}}^{IV}=O$$

$$\overset{\textstyle |}{\underset{\textstyle |}{Fe}}^{IV}=O \; + \; B-\overset{\textstyle |}{\underset{\textstyle |}{Fe}}^{II} \; \longrightarrow \; \overset{\textstyle |}{\underset{\textstyle |}{Fe}}^{III}-O-\overset{\textstyle |}{\underset{\textstyle |}{Fe}}^{III} \; + \; B$$

This mechanism has been investigated in detail and is still
the subject of many studies. The postulated species μ-peroxo
dimer,[6,7] ferryl intermediate,[8,9] and μ-oxo dimer[10]
could be proved for the oxidation of porphyrin systems
without axial ligand B.

Several concepts for the stabilization of oxygen
complexes can be applied. Sterical hindrance by voluminous
substituents can inhibit autoxidation of iron(II)
complexes[11,12] and iron(II) porphyrin systems.[2-5,13-15]
Another way to achieve this is immobilization on solid
supports, e.g. polystyrene[16] or silica gel.[17] In this
case, the active centers remain separated by matrix isolation.

Another concept, which has been developed recently, is
the application of soluble polymers as macromolecular supports
for the synthesis of reversibly oxygenable model systems.
Porphyrins have been bound by coordination to synthetic
polymers, e.g. poly(oxyethylene-styrene-N-vinyl-imidazole)[18]
or poly(1-vinyl-2-methyl imidazole).[19] The undesired tendency
of porphyrins to aggregation can be prevented by embedding
them in a polymer coil. In order to attain maximum protection
in this prospect, heme derivatives were bound covalently to
functionalized polymers in our laboratory. Supports included
poly(oxyethylene), poly(vinylpyrrolidone), and poly(urethane)s[20]. Other examples have been published by Tsuchida et
al., using poly(vinylpyrrolidone)[21] and polystyrene[22].

The polymer chain imitates two main tasks of the synthetic hemopolymers which are fulfilled by the protein component in naturally occuring proteins: solubilization and protection of the prosthetic group. In addition, an imidazole derivative, which is bound covalently to the heme periphery, replaces the proximal histidine.

In order to mimic the natural molecules and to meet the physiological and toxicological requirements for such synthetic derivatives we preferred the synthesis of a hemopolymer, which consisted of naturally occuring components. As a natural polymer we introduced the polysaccharide dextran which is nontoxic and biodegradable. It consists of almost linear chains of D-glucopyranose which are linked α-1,6-glucosidically. The extent of branching, based on α-1,3 and α-1,4 linkage, depends on the type of bacteria used as source for the dextran.

MATERIALS AND METHODS

The molecular mass of native dextran is in the range of $40-50 \cdot 10^6$ g·mol^{-1} which can be reduced to $40-60 \cdot 10^3$ g·mol^{-1} by partial hydrolysis with mineral acids or enzymatically. This type of dextran is physiologically compatible. Dextran 32 pharm. (from Serva, Heidelberg) with a nominal molecular mass of 32000 - 48000 g·mol^{-1} was found to have 45000 g·mol^{-1} as determined viscosimetrically and by gel filtration. The first step of amino functionalization was performed by reacting dextran with ethyl chloroformate in a mixture of dimethylsulfoxide/dioxane and in the presence of triethylamine. Then, the intermediate obtained was subjected to aminolysis with ethylenediamine.[23,24] The degree of substitution was checked by elemental analysis and quantitative UV spectroscopy using the 2,4-dinitrophenyl derivative.

N-protected amino acids were bound covalently to the modified dextran by the use of dicyclohexyl carbodiimide as the coupling reagent[25]. Photolysis of the protecting groups was carried out in quartz vessels with a mercury lamp ($\lambda >$ 320 nm).

For coupling of the porphyrin system to the carrier we used a competitive reaction of the activated propionic acid groups of the porphyrin ring[26] with both the dipeptide H-Gly-His-OMe and the amino functions of the dextran spacer. After each synthetic step the products were isolated by precipitation into ethanol. The final hemopolymer was purified by chromatography on Sephadex G 75.

The hemopolymer synthesized contained as central ion Fe(III). In order to obtain Fe(II), which is capable to form an oxygen complex, the reduction was performed in dimethyl sulfoxide using a sodium dithionite crown ether complex and also in aqueous media (phosphate buffer, pH 7,45) with dithionite.

Chemical modification of dextran is possible in a selective manner and under mild conditions at C-atom 2 by aminolysis of a cyclic carbonic acid diester.

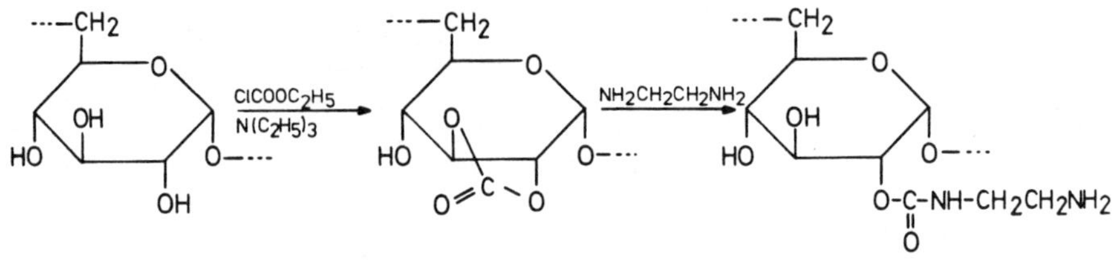

The amino-terminal spacer can be extended by N-protected amino acids which have been synthesized according to the following scheme:

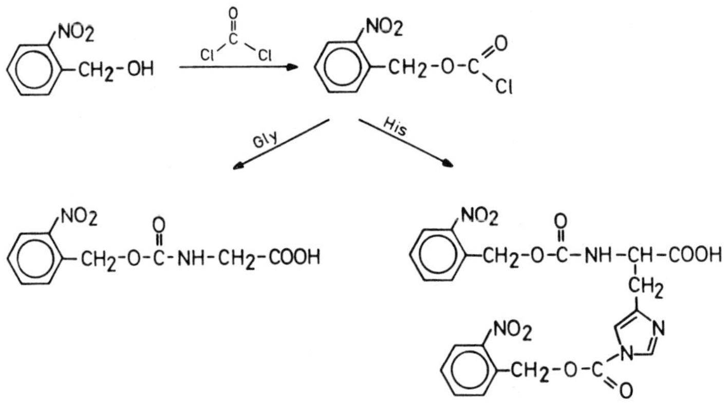

For coupling of the amino acids to the spacer of the polymer carrier the carbodiimide method was used[25]. After cleavage of the N-protecting groups the spacer group was ready for the fixation of the porphyrin component.

According to the reduced chemical stability of dextran, only very few protecting groups can be applied during synthesis. Protecting groups to be cleaved in acid media are not appropriate because dextran is acid-labile. Another limiting factor in this context is the instability of the urethane group by which the reactive side-chain of the polysaccharide is connected. We have selected the o-nitrobenzyloxycarbonyl group[27,28] for protecting because it can be cleaved under very mild conditions by photolysis. Using IR, the cleavage can be monitored spectroscopically. Photolysis was found to proceed very rapidly and quantitatively. Both chain-scission and formation of by-products were not observed under the cleavage conditions applied.

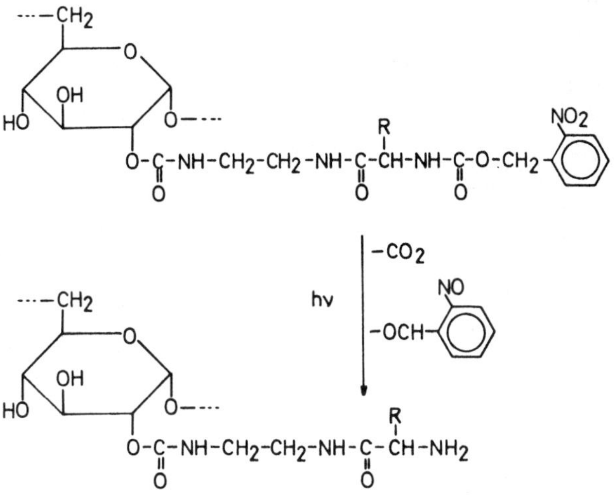

For the coordination of the proximal histidine F8 at the
central iron ion the dipeptide glycine histidine methylester
was chosen as ligand. The synthesis was carried out via the
activ ester of N-hydroxysuccinimide and benzyloxycarbonyl
glycine.

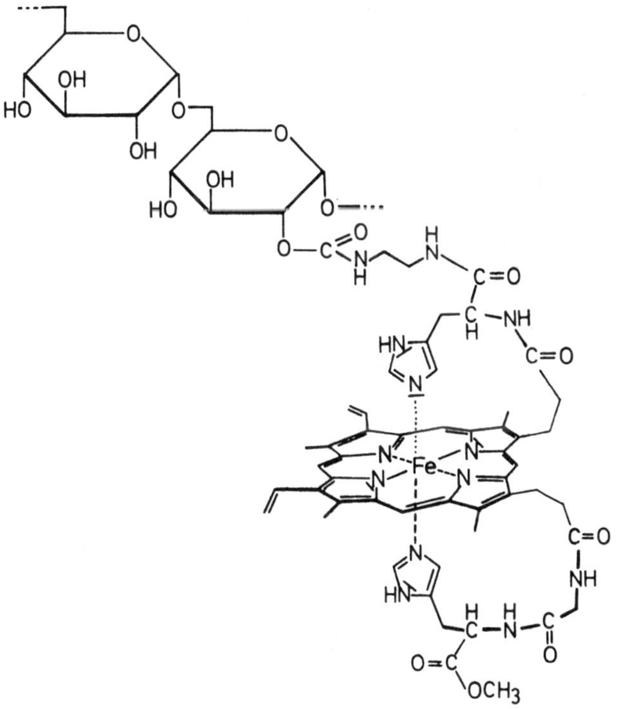

Fig. 1.   The active center of the hemopolymer
           dextran-NH-His-heme-Gly-His-OMe

The hemopolymer synthesized contains all essential
structural elements of the active center of hemoglobin and
myoglobin (Figure 1). Whereas four coordination positions
of the iron are covered by the nitrogen atoms of the
porphyrin, the fifth position is taken by the imidazol
nitrogen of the glycine histidine tail. Optimum coordination
is guaranteed by the chain length equivalent to the proximal
histidine in the natural model. The second propionic acid
group of the hemin is linked directly to the histidine of
the functionalized polymer. Thus, an optimum coordinative
fixation of the imidazol nitrogen is sterically hindered
comparable to the structural situation on the distal part
of myoglobin. According to this, the distal histidine, as
well as the polymer component, shield the coordination
position of the oxygen molecule.

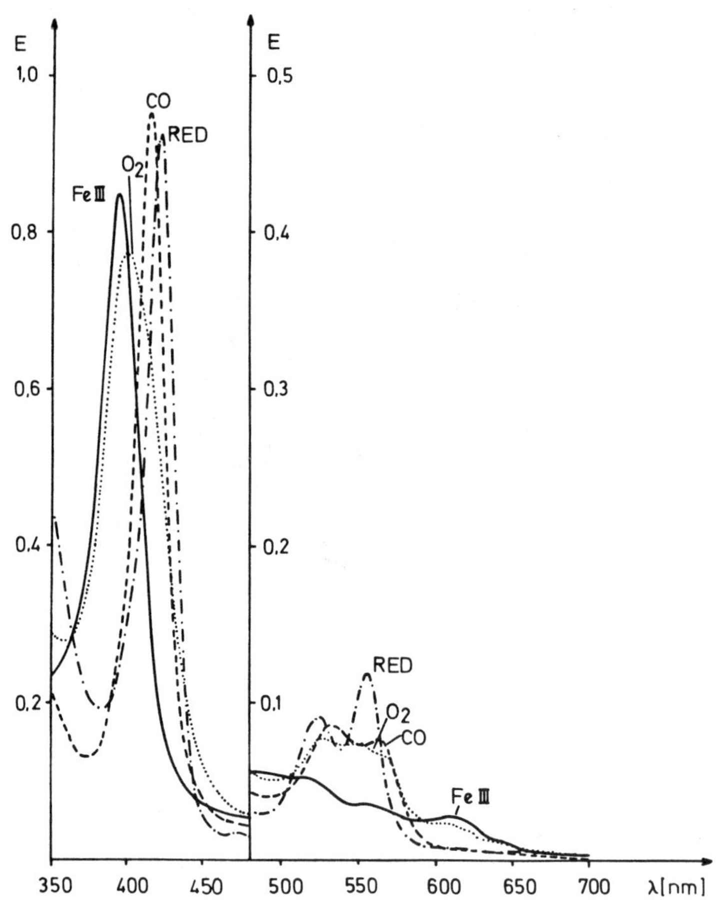

Fig. 2.   Absorption spectra of the hemopolymer during one
           oxygenation cycle in dimethyl sulfoxide
           — ·· — reduced   ---- CO-complex
           ············ oxygenated   ——— oxidized

132

The ferro-form of the hemopolymer is reversibly
oxygenable at room temperature when using the aprotic dipolar
solvent dimethyl sulfoxide (Fig. 2). Then, the coordinated
oxygen molecule is shielded sufficiently and the rate of the
oxidation reaction is lowered considerably. The half-life
period of the oxygen complex is about 35 minutes. The reduced
form can be re-obtained by evacuating or flushing with argon.
The extent of the irreversible oxidation reaction after
serveral cycles can be determined from the decrease of
extinction of the carbon monoxide complexes.

The hemopolymer is also soluble in aqueous media because
of the covalent linkage to the hydrophilic dextran. By reason
of the great number of hydroxyl groups of the polymer carrier
a proton transfer to the coordinated oxygen is possible in
aqueous solution. Therefore, the oxidation potential is
increased considerably and leads to an acceleration of the
irreversible oxidation reaction.

In another approach we have synthesized analogously the
hemopolymer dextran-NH-Gly-heme-Gly-His-OMe in which the distal
histidine is replaced by glycine. An interesting effect of
the altered stereochemical situation around the active center
was observed. From the absorption spectra of the reduced form
and the CO-complex it can be concluded that this dextran
hemopolymer could serve as a model for cytochrome oxidase.

Table 1.   Electronic Absorption Data of Cytochrome Oxidase
            and Model Systems as Reported by Various Authors.
            Values represent Absorption Maxima in nm.

	Oxidized Species		Reduced Species (by Dithionite)		Reduced CO-Complexes		Ref.
Cytochrome Oxidase[a]							
Subunit I	425	595	440	597	430	600	29
Subunit II-III	425	597	441	600	431	602	29
Cytochrome Oxidase	421	598	443	603	–	–	30
Cytochrome a	426	595	444	604	–	–	30
Cytochrome a_3	414	600	442,5	602	–	–	30
Cytochrome Oxidase	423	599	444	605	431	590	31
Heme a poly-lysine complex[b]	402	–	428	574	424	584	32
Dextran-hemopolymer	404	–	421	613	416	614	

[a] Subunit I and II-III in polyacrylamide gels, measured at $0\,^{\circ}C$

[b] At pH 10.6

REFERENCES

1. E. C. Niederhoffer, J. H. Timmons, and A. E. Martell, Chem. Rev. 84:137 (1984).
2. J. E. Baldwin and P. Perlmutter, Top. Curr. Chem. 121: 181 (1984).
3. R. D. Jones, D. A. Summerville, and F. Basolo, Chem. Rev. 79:139 (1979).
4. J. P. Collman, Acc. Chem. Res. 10:265 (1977).
5. J. W. Buchler, Angew. Chem. 90:425 (1978).
6. D. H. Chin, G. N. La Mar, and A. L. Balch, J. Am. Chem. Soc. 102:4344 (1980).
7. D. H. Chin, J. Del. Gaudio, G. N. La Mar, and A. L. Balch, J. Am. Chem. Soc. 99:5486 (1977).
8. A. L. Balch, G. N. La Mar, L. Latos-Grazynski, M. W. Renner, and V. Thanabal, J. Am. Chem. Soc. 107: 3003 (1985).
9. A. L. Balch, Y.-W. Chan, R.-J. Cheng, G. N. La Mar, L. Latos-Grazynski, and M. W. Renner, J. Am. Chem. Soc. 106:7779 (1984); and literature cited therein.
10. A. B. Hoffman, D. M. Collins, V. W. Day, E. B. Fleischer, T. S. Srivastava, and J. L. Hoard, J. Am. Chem. Soc. 94:3620 (1972).
11. J. E. Baldwin and J. Huff, J. Am. Chem. Soc. 95:5757 (1973).
12. N. Herron, J. H. Cameron, G. L. Neer, and D. H. Busch, J. Am. Chem. Soc. 105:298 (1983).
13. M. Momentean, J. Mispelter, B. Loock, and J.-M. Lhoste, J. Chem. Soc., Perkin Trans. I, 221 (1985).
14. T. G. Traylor, N. Koga, L. A. Deardurff, P. N. Swepston, and J. A. Ibers, J. Am. Chem. Soc. 106:5132 (1984).
15. J. E. Baldwin, J. H. Cameron, M. J. Crossley, I.J. Dagley, S. R. Hall, and T. Klose, J. Chem. Soc., Dalton Trans., 1739 (1984).
16. J. H. Wang, J. Am. Chem. Soc. 80:3168 (1958).
17. O. Leal, D. L. Anderson, R. G. Bowman, F. Basolo, and R. L. Burwell, J. Am. Chem. Soc. 97:5125 (1975).
18. K. Shigehara, K. Shinohara, Y. Sato, and E. Tsuchida, Macromolecules 14:1153 (1981).
19. N. Nishide, M. Sekine, and E. Tsuchida, Polymer J. 14:629 (1982).
20. E. Bayer and G. Holzbach, Angew. Chem. Int. Ed. 16:117 (1977).
21. E. Tsuchida, H. Nishide, and Y. Sato, J. Chem. Soc., Chem. Commun., 556 (1982).
22. E. Tsuchida, H. Hasegawa, and T. Kanayama, Macromolecules 11:947 (1978).
23. E. A. Kuznetsova, L. S. Shishkanova, G. E. Koroleva, E.D. Sinyagina, V. M. Shlimak, and A. E. Vasil'ev, J. Gen. Chem. USSR 48:1292 (1978).
24. E. Hasegawa, personal communication.
25. M. Mutter, R. Uhmann, and E. Bayer, Liebigs Ann. Chem. 1975:901.
26. A. van der Heijden, H. G. Peer, and A. H. A. van den Oord, J. Chem. Soc., Chem. Commun., 369 (1971).
27. A. Patchornik, B. Amit, and R. B. Woodward, J. Am. Chem. Soc. 92:6333 (1970).

28. J. A. Barltrop, P. L. Plant, and P. Schofield, J. Chem. Soc., Chem. Commun., 822 (1966).
29. D. B. Winter, W. J. Bruyninckx, F. G. Foulke, N. P. Grinich, and H. S. Mason, J. Biol. Chem. 255:11408 (1980).
30. W. H. Vanneste, Biochemistry 5:838 (1966).
31. D. C. Wharton, Cytochrome Oxidase, in: "Inorganic Biochemistry", Vol. 2, G. L. Eichhorn, ed., Elsevier Scientific Publishing Company, Amsterdam-London-New York (1973).
32. T. E. King, F.C. Yong, and S. Takemori, J. Biol. Chem. 242:819 (1967).

GRAFT COPOLYMERIZATION ONTO CELLULOSE ACETATE AND WOOD USING ANIONIC POLYMERIZATION

Ramani Narayan and Margaret Shay

Laboratory of Renewable Resources Engineering
Purdue University
West Lafayette, IN 47907

INTRODUCTION

The grafting of synthetic polymers to natural polymers from renewable resources offers the potential of preparing newer varieties of engineering materials with specific and improved properties for a wide range of applications. The properties of this molecular composite would be substantially better than the properties of either homopolymer alone. If the individual polymer segments in copolymers have opposite properties such as polar and nonpolar and/or flexible and rigid natures, many interesting properties are conceivable. For example, if an amorphous polymer like polystyrene is grafted onto cellulose, a molecular composite will be produced, where the stiff cellulose chain is a reinforcing fiber in a polystyrene matrix. The mechanical properties of this molecular composite would be substantially better than the properties of either homopolymer alone. The stiffness of the cellulose chain can be controlled by chemical modification prior to grafting, and the mechanical properties of the grafted polymer can be engineered by controlling the chemical composition and molecular weight of the graft. Thus, we can make materials with fibers of controlled stiffness covalently bound in a polymer matrix, where the matrix mechanical properties can be varied from a rubber to a hard glassy polymer. However, it is impossible to predict exactly what the mechanical properties of these polymers will be until these systems are synthesized and tested.

In view of the tremendous applications potential of cellulosic graft polymers, there has been a considerable amount of research and development effort expended on the "macromolecular engineering" of cellulose to produce a composite containing cellulose and grafted poly (vinyl) or any other synthetic polymer.

Almost all the work done to date in the preparation of cellulosic graft polymers involves radical polymerization methods. Using these methods, grafting of only a few high molecular weight molecules involving a low level of graft substitution has been obtained, i.e., only a small portion of the cellulose substrate is grafted (1-3). There is no control of the molecular weight of the grafted side chains and no knowledge of the nature of the backbone-graft linkage. A lot of homopolymer formation occurs, there is poor reproducibility and little control of the graft process, graft yields, properties and other features of the graft copolymer (4).

Thus, if the tremendous applications potential of the cellulosic graft polymers is to be realized, new and better processes for grafting onto cellulose is needed. The new processes should provide for:

1) Control of the molecular weight and molecular weight distribution of the graft side chains.

2) Elimination or at least minimization of concurrent homopolymer formation.

3) Maximizing the participation of the cellulose substrate towards grafting, i.e., shorter graft chains and high degree of substitution.

4) Knowledge of the nature of the linkage between the cellulose backbone and the synthetic graft.

5) Better reproducibility and control of the grafting process, graft yields, properties and other features of the graft polymers.

We have recently described a new method (5,6) for the preparation of cellulosic graft polymers involving anionic polymerization techniques that would overcome the drawbacks of the radical polymerization methods. The method involves 1) preparation of the *living* synthetic polymer of desired molecular weight by anionic polymerization. 2) introduction of electrophilic groups or leaving groups onto the cellulose backbone by chemical modification. 3) reaction of the *living* synthetic polymer with the modified cellulose. If the cellulose derivative is soluble in an organic solvent, then an additional advantage of the new method is that the coupling reaction between the modified cellulose and the living polymer can be carried out under homogeneous conditions.

Thus, polystyryl carbanions and polyacrylonitrile carbanions prepared by anionic polymerization were reacted with cellulose acetate in THF under homogenous conditions. The carbanions displaced the acetate groups in a SN2 type nucleophilic displacement reaction to give CA-g-PS and CA-g-PAN. Since the grafting occurs by this mechanism the backbone-graft linkage must be a carbon-carbon linkage. IR, ^{13}C NMR and thermogravimetric analysis were used to support the formation of these graft copolymers. Since the grafted synthetic polymer was prepared by anionic polymerization, the molecular weight of the graft can be controlled. The acetate groups are distributed over the length of the cellulose chain and because the grafting mechanism involved the displacement of the acetate groups by the carbanions, maximum participation of the cellulose substrate is ensured.

In this paper, we describe our further work on the preparation of cellulosic graft polymers, namely the grafting of polystyrene onto tosylated cellulose acetate, using the anionic polymerization approach developed by us. Detailed proof (IR, solubility, elemental analysis, wet chemical analysis) for the formation of these grafts is also presented. The extension of our new grafting process to prepare wood-g-polystyrene and wood-g-polyacrylonitrile directly from wood will also be described.

RESULTS AND DISCUSSION

Grafting of Polystyrene (PS) Onto Tosylated Cellulose Acetate (TCA)

Acetone-soluble cellulose acetate (DS acetyl = 2.47, DP = 108) was modified by tosylation of the unesterified hydroxyl groups using literature procedures (7). Introduction of the tosyl groups provided: (1) a better leaving group than acetate in nucleophilic displacement, and (2) capping of the free hydroxyl groups on the cellulose backbone, thus eliminating the possibility of any homopolymer formation due to quenching of the polystyryl carbanions by the hydroxyl protons. Since 40 to 60% of the free hydroxyls in the cellulose acetate are present as primary hydroxyls ($-CH_2OH$) on the C-6 carbon (8), a significant amount of the tosyl groups would be present on the C-6 carbon. Thus, the tosylation step not only introduced a better leaving group, but also put the tosyl groups on the least sterically hindered carbon

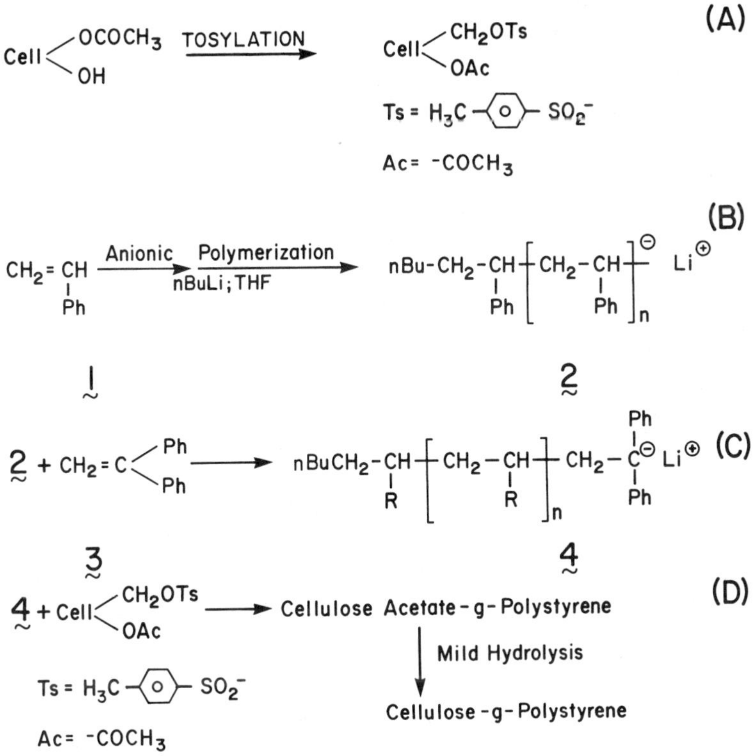

Scheme-1

(C-6) position which made it easier to introduce the bulky polystyryl chain (A, Scheme 1). The "living" polystyryl carbanion 2 was prepared by anionic polymerization using n-butyllithium as the initiator (B, Scheme 1). However, when the tosylated cellulose acetate was reacted with the polystyryl carbanions, no grafting occurred. There was a lot of homopolymer formation with concurrent degradation of the cellulose chain. It appears that the very strongly basic polystyryl carbanion abstracts the C-1 proton of the anhydroglucose unit, resulting in cleavage of the glycosidic bonds.

To circumvent this problem, the basicity of the polystyryl carbanion was reduced by reaction with 1,1-diphenylethylene (C, Scheme 1). This resulted in the formation of a polystyryl chain with a sterically hindered carbanion end group 4. Reaction of this carbanion 4 with the TCA (D, Scheme 1) proceeded smoothly and gave excellent yield of the graft product. The yield of the graft product, after extraction with toluene to remove any homopolystyrene formed, was 90% of the theoretical on a weight basis.

The amount of grafted PS was determined by UV spectroscopy (9). The UV absorption at 260 nm of the copolymer dissolved in THF was compared to polystyrene standards. It was found that the sample contained 39% PS, which corresponds to 1 grafted side chain for every 16 anhydroglucose units, i.e., 6.7 polystyryl chains ($\bar{M}_v = 3,400$) per cellulosic chain. In contrast, free radical grafting of styrene onto cellulose using radiation or radical initiation produces graft copolymers having 0.03 to 0.8 polystyryl chains per cellulosic chain with molecular weights ranging from 354,000 to 960,000 (4).

Proof of grafting was obtained by IR spectroscopy (Figure 1), solubility (Table 1), and elemental analysis (Table 2). The IR spectrum of the TCA-g-polystyrene showed the aromatic ring vibration bands (1500-1600 cm^{-1}) and the aromatic -CH stretching vibrations (above 3000 cm^{-1}) of the polystyrene, as well as the strong carbonyl band (1720 cm^{-1}) of the TCA. The TCA-g-PS was insoluble in toluene and glacial acetic acid, while PS and TCA are soluble in toluene and glacial acetic acid, respectively. This solubility behavior supported the formation of a "true" TCA-PS graft copolymer. The elemental analysis data (Table 2) showed that there was a substantial increase in the carbon and hydrogen content with a marked decrease in the oxygen and sulfur content of the grafted product as compared to the starting TCA. This supported the grafting of polystyrene onto the cellulosic substrate.

The TCA-g-PS has unreacted tosyl and acetate groups on the cellulose backbone, which rendered this graft polymer soluble in THF and acetone. However, removal of the unreacted tosyl and acetate groups by mild hydrolysis with ammonia furnished a cellulose-PS graft copolymer which is insoluble in all solvents, a typical behavior of cellulosic copolymers. The IR spectrum of the cellulose-g-PS (Figure 1) showed the absence of the carbonyl band (1720 cm^{-1}) and the presence of -OH stretching vibrations (3500 cm^{-1}), indicating that the acetate groups have been hydrolyzed to hydroxyl groups. Elemental analysis showed little sulfur, demonstrating that the tosyl groups have also been removed. The IR spectrum still retained the aromatic bands just above 3000 cm^{-1} and in the 1500-1600 cm^{-1} region, confirming the presence of the polystyrene graft chains. Since there are no acetate or tosyl groups remaining, this is a true cellulose-polystyrene graft copoly-

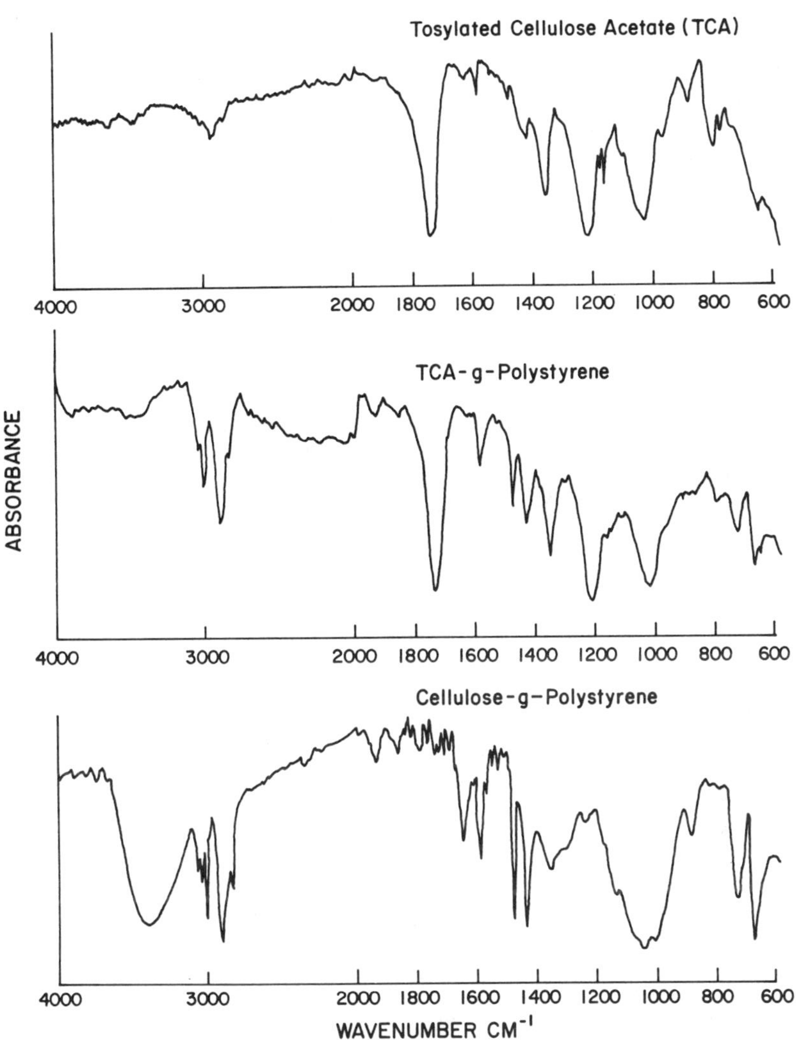

Figure 1. IR Spectra

mer. This study also overruled the possibility of the polystyrene graft being attached to the cellulose backbone via the acetate group, i.e., by attack of the polystyryl carbanion on the carbonyl carbon of the acetate group. If this had indeed occurred, then during the hydrolysis of the acetate groups all of the

Table 1. SOLUBILITY DATA

Solvent	PS	TCA	PS-g-TCA	PS-g-Cellulose
Toluene	Soluble	Insoluble	Swells	Swells
Styrene	Soluble	Insoluble	Swells	Swells
Acetic Acid	Insoluble	Soluble	Swells	Insoluble

Table 2. ELEMENTAL ANALYSIS

Element	TCA[+]	TCA-g-PS[*]	PS	C-g-PS
C	49.51%	73.56%	91.19%	79.34%
H	5.48%	7.22%	7.83%	7.21%
S	4.10%	2.30%	--	0.77%
O	40.91%	16.92%	--	12.68%

[+]DP = 108; DS of acetyl = 2.47; DS of tosyl = 0.42

[*]MW of PS = 3400; % PS = 39; 1 PS chain per 16 glucose units
or 6.7 polystyryl chains per cellulosic chain.

polystyrene would have been eliminated from the cellulose backbone. However, the IR spectrum of the cellulose graft polymer after hydrolysis showed that this was not the case, and the polystyrene grafts were still attached to the cellulose backbone. Thus, the grafting must occur by nucleophilic displacement of the acetate groups resulting in a carbon-carbon bond between the cellulose backbone and the polystyrene graft as proposed.

Potential Applications

The unique and interesting properties of graft polymers arise from their semimicroscopic heterogeneity. Each block retains all the characteristics of its respective homopolymer. Because the individual components are usually incompatible, the blocks of the graft copolymers try to avoid each other, creating well defined domains of each individual homopolymer within the graft copolymer. It is this heterogeneity of the graft copolymers that allows the individual components of the graft copolymer to retain their basic properties. For example, the grafting of 8 to 12% of polyacrylic acid onto high density polyethylene, left the melting point and crystallinity of the polyethylene backbone essentially unchanged while imparting to it increased modulus and softening point. The water absorption and permeability, were markedly increased by the grafting (10,11). The grafting reaction can thus be seen as perhaps the only method by which certain properties can be introduced permanently into a polymer without changing its chief performance characteristics. Therefore, it is possible to impart desirable properties to the cellulosic derivative without changing its main properties by grafting a synthetic polymer with the desired property onto the cellulosic derivative. In general, however, the changes brought about by grafting onto cellulose using the radical polymerization techniques have been very modest. This is probably due to the inherent problems of the radical polymerization method as described earlier, particularly the presence of only a few very long graft chains on the cellulose backbone rather than a number of short side chain grafts.

The grafting of polystyrene onto the tosylated cellulose acetate using our approach has produced a cellulosic graft copolymer with a number of short side chain polystyrene grafts. The graft copolymer is soluble in organic solvents and can be readily processed, i.e., made into films, beads, etc., for a number of membrane and chromatography applications. Films made from this graft copolymer should have better water resistance sensitivity and can be heat-sealed as compared to conventional cellophane films. Cellophane films have to be chemically treated to overcome these problems. The polystyrene grafts can impart improved mechanical strengths to cellulose acetate membranes that enjoy such specialized uses as kidney dialysis, desalination of sea water and battery separations. In fact, it has been reported (12) that radiation grafting of polystyrene onto cellulose acetate does improve mechanical properties without affecting permeability. Obviously, the molecular weight of the synthetic polymer, the degree of substitution of the graft side chains, the nature and position of the backbone-graft linkage and the amount of ungrafted polymer present will dictate the properties of the graft copolymer. Our graft polymerization approach allows us to vary and control all these features. The molecular weight of the polystyrene can be varied and controlled since it is prepared by anionic polymerization. The degree of substitution of the graft can be regulated by the ratio of the polystyryl carbanions to the leaving groups on the cellulose backbone and the nature of the linkage and position of the graft side chain can be controlled by varying the type of "reactive groups" on the cellulose backbone. Work is in progress to prepare and evaluate the properties of a series of cellulose-polystyrene graft copolymers with these different features.

Grafting onto Wood

Acetylation of wood chips with acetic anhydride in the presence of concentrated H_2SO_4 furnished a triacetate which was partially hydrolyzed to give a secondary acetate (13). IR spectrum of this product showed the characteristic carbonyl stretching at 1750 cm^{-1} confirming the introduction of acetate groups onto the wood backbone.

Using anionic polymerization techniques, polystyryl carbanion (MW = 10,000) and polyacrylonitrile carbanion (MW = 15,000) (5) were prepared. Each of these living polymers was reacted with the acetylated wood chips. The highly reactive carbanions displaced the acetate groups in an SN$_2$ reaction with the formation of the wood-polystyrene (W-PS) and wood-polyacrylonitrile (W-PAN) graft copolymers. Proof for the formation of these graft copolymers was obtained from IR and thermogravimetric analysis. The IR spectrum of the W-PS graft copolymer showed the presence of the aromatic -CH stretching vibrations above 3000 cm^{-1}. The IR spectrum W-PAN graft copolymer showed the distinctive -CN stretching vibrations at 2200 cm^{-1}.

Thermogravimetric analysis provided strong evidence for the formation of a "true" graft copolymer and discounted the formation of merely a physical blend of the two polymers. Figure 2 shows the weight loss and rate of weight loss as a function of temperature for a) acetylated wood (W); b) polystyrene; c) 1:1 physical blend of W and PS; and d) W-PS graft copolymer. The rate of weight loss curves $(\frac{d\Delta W}{dt})$ of the acetylated wood shows two decomposition maximum at 265°C and 345°C. The polystyrene homopolymer has its decomposition maximum at 425°C.

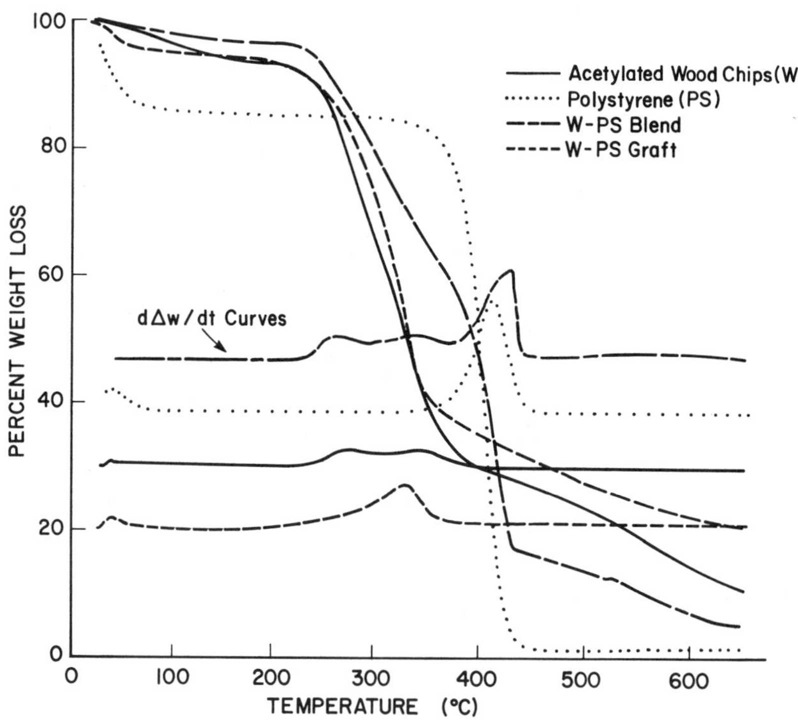

Figure 2.   TGA of W-PS

As is to be expected, the wood-PS blend shows three decomposition maxima at 265°C, 345°C and 425°C corresponding to the decomposition maximum of the individual wood and polystyrene homopolymers. However, the W-PS graft copolymer shows only one decomposition maximum at 300°C. Furthermore, 25% of residue is obtained at 650°C for the W-PS graft copolymer as compared to 5% for the blend.

Figure 3 shows the thermograms for the a) acetylated wood; b) PAN; and c) W-PAN graft copolymer. Again, the W-PAN graft copolymer shows only one

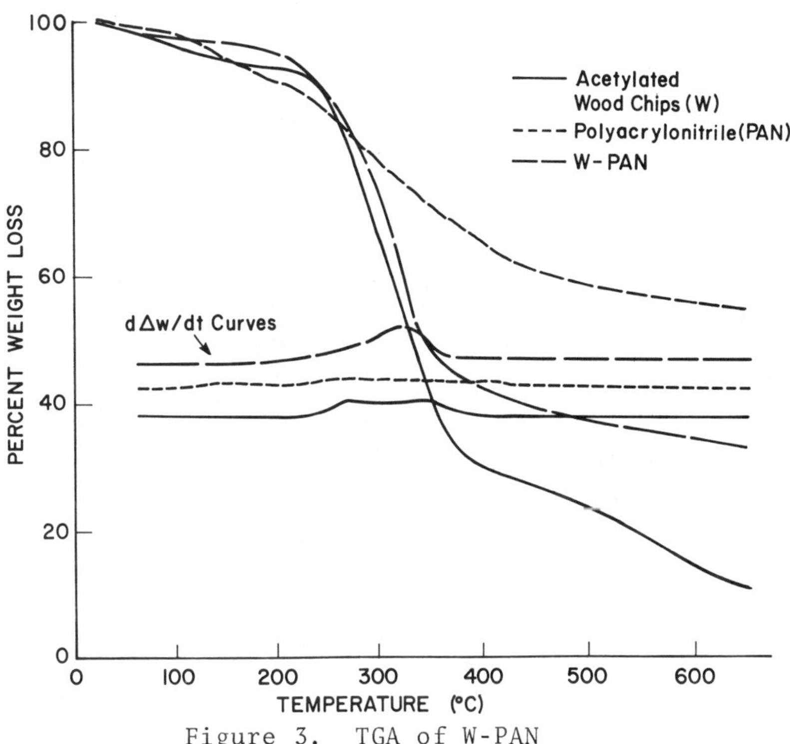

Figure 3.   TGA of W-PAN

decomposition maximum at 325°C. However, a higher amount of residue (35%) is obtained as compared to the W-PS graft copolymer. Since the PAN homopolymer leaves as much as 50% residue at 650°C, the increased residue obtained for the W-PAN graft copolymer is not surprising and can be taken as additional evidence for the formation of a W-PAN graft copolymer.

CONCLUSION

A cellulosic graft copolymer containing seven polystyryl chains ($\overline{M}_v = 3400$) per cellulosic chain (degree of polymerization =108) has been prepared in excellent yields using the new grafting approach developed by us. Infrared spectroscopy, solubility, elemental analysis and wet chemical analysis provided irrefutable proof for the formation of the graft copolymer in which the polystyryl graft chain is linked to the cellulosic backbone by a carbon-carbon bond. The TCA-g-PS is soluble in THF and can be readily processed for a variety of membrane and chromatography applications. If necessary, the processed material can then be subjected to mild hydrolysis to yield a product which is insoluble in all solvents.

Thus, this synthetic approach using anionic polymerization techniques can be used to prepare a series of cellulose-polystyrene graft copolymers wherein one can vary and control the graft chain's molecular weight, nature of the backbone-graft linkage and the degree of substitution.

We have also successfully grafted polystyrene and polyacrylonitrile onto wood itself using the new grafting process developed by us.

ACKNOWLEDGMENTS

This work was supported by a grant (85-CSRS-2-2554) from the United Stated Department of Agriculture (USDA) to the Wood Utilization Research Program (WURP) at Purdue.

REFERENCES

1. D.J. McDowall, B.S. Gupta, and V.T. Stannett, Prog. Polym. Sci., *10*, 1 (1984).

2. A Hebeish and J.T. Guthrie, *The Chemistry and Technology of Cellulosic Copolymers*, Springer-Verlag, Berlin (1981).

3. J.C. Arthur, Jr., Adv. Macromol. Chem., *2*, 1 (1970).

4. V.T. Stannett, ACS Symp. Ser., *187*, 1 (1982).

5. R. Narayan and G.T. Tsao, Amer. Chem. Soc., Polymer Preprints, *25*(2), 29 (1984).

6. R. Narayan and G.T. Tsao, "Graft Copolymers from Renewable Resource," Amer. Chem. Soc. Div., Polymer Mat. Sci. and Eng. Preprints, in press (1985).

7. F.B. Craemer and C.B. Purves, J. Amer. Chem. Soc., *61*, 3458 (1939).

8. L.J. Tanghe, L.B. Genung and J.W. Mench, *Methods in Carbohydrate Chemistry*, Vol. III, ed., R.L. Whistler, Academic Press, New York, p. 199 (1963).

9. P. Mansson and L. Westfelt, J. Polym. Sci., Polym. Chem. Ed., *19*, 1509-1575 (1981).

10. J.K. Rieke and G.M. Hart, J. Polym. Sci., *Part C, 1*, 117 (1963).

11. J.K. Rieke, G.M. Hart, and F.L. Saunders, J. Polym. Sci., *Part C, 4*, 589 (1964).

12. H.PP. Hopfenberg, V. Stannett, F. Kimura-Yeh, and P.T. Rigney, J. Appl. Polymer. Sci., Applied Polymer Symposia, *13*, 139 (1970).

13. B.L. Browning, *Methods of Wood Chemistry*, Vol. II, Interscience Publishers, New York (1967).

CHEMICAL MODIFICATION OF WOOD: REACTION WITH THIOACETIC ACID AND ITS EFFECT ON PHYSICAL AND MECHANICAL PROPERTIES AND BIOLOGICAL RESISTANCE

Satish Kumar and Kamini Kohli

Forest Research Institute
Dehra Dun, India

INTRODUCTION

Wood is a three-dimensional biopolymer complex composite made up of cellulose, hemicellulose and lignin. Being of organic origin it is readily degraded by a variety of micro-organisms (wood rotting fungi, soft rot), insects, termites and marine borers attacking wood for food or shelter and other agencies such as fire, UV light, etc. Because of the hygroscopic nature of the wood components wood adsorbs/desorbs moisture with changing relative humidity of the atmosphere. Changing moisture conditions cause alternate swelling and shrinkage resulting in physical degrade leading to mechanical failure. High moisture contents are also conducive to higher and easier degrade by biological agencies. Numerous types of treatments have, therefore, been developed to improve its dimensional stability, mechanical strength, fire endurance and durability against micro-organisms.

Although several broad spectrum wood preservatives exist and have been effectively used since the last century, their use is viewed with great concern by the environmentalists as these chemicals are toxic to other living systems including humans. The Product Control Board in Sweden has banned the use of pentachlorophenol, a most widely used preservative the world over (Richardson, 1977). Similarly arsenic containing preservatives have been restricted in three European countries (Willeitner, 1976). The policy on the use of various chemicals, which pose pollution hazards, is constantly reviewed by various agencies in several countries and although U.S. Occupational Safety and Health Administration (OSHA) has cleared the use of pentavalent arsenic, used in copper-chrome-arsenic compositions, the clearance may be revoked any time, as pentavalent arsenic is manufactured from trivalent arsenic, which has high toxicity hazards. The next on the casuality list may be chromium salts, which are important components in nonleachable type preservatives as the same are known carcinogenic compounds. Moreover, such conventional wood preservatives do not improve dimensional stability or mechanical strength of wood. Rather some of these compositions have been observed to adversely affect the same (Kumar and Jain, 1978). Inadequate treatments resulting from poor penetration of such chemicals can lead to premature failure as conditions of alternate drying and wetting or cyclic humidity changes cause checks on wood surface due to dimensional movement and expose untreated portions to hazards of biodeterioration and subsequent mechanical failure. Since protection of wood from biodeterioration is a must to extend our timber resources, several alternate methods of preserving wood are being investigated,

which are not based on toxicity. Such methods are based on chemical modification reactions, which alter the very nature of wood, make it unrecognisable
or unpalatable to the biodegrading organism as a source of food. The nontoxic approach to preservation of wood is an intriguing futuristic research
concept which is considered to have a great commercial potential as such
treated wood not only becomes resistant to most biodegrading agencies but
also improves in dimensional stability, density and mechanical strength.

Chemicals used in such treatments react with molecules in the wood
and become part of the wood structure. These need not be toxic to the organism,
because their main function is to form stable bonds with wood components
to render the same unpalatable/indigestible or unrecognisable as a food source.
The most abundant reactive chemical sites in wood are the hydroxyl groups
present on all the major wood components. These hydroxyl groups act as
sites for adsorption of water and cause swelling of wood in contact with
water or in high humidity. Most microorganisms also forge a link with wood
substrate through these hydroxyl groups as the enzymes secreted by such
organisms being specific in nature require conformational sites in a lock and
key type arrangement for proper enzyme-substrate reaction (Lehninger, 1970).
Such reactions metabolise the wood into simpler compounds, which are subsequently consumed by the degrading organisms. Protection accorded by certain
wood extractives, present usually in durable woods as well as wood preservatives,
is based on the theory that these toxic compounds are also solubilised in
the enzymatic secretions and the organisms are immobilised due to their
toxicity. Protection through wood modification is achieved through substitution
of the active hydroxyl groups, thus making these conformational sites unavailable
for the microorganisms.

Brief Review of Wood Modification Reactions

Rowell (1975), in his review on chemical modification of wood listed
several reagents, which have been used to modify wood. By definition, chemical
modification is achieved through formation of nonpolar covalent linkages
of the type carbon-oxygen-carbon. Whereas formation of ethers, acetals
and esters results in chemical modification of wood, treatments which do
not form covalent bonds such as heat treatment, some coatings, polymer
inclusions filling only the voids or bulking the cell wall are excluded from
this definition. Chemical modification reactions have been classified by the
type of bond they produce. Ethers, acetals and esters being stable are of
major importance and have been studied in considerable details.

Ether type bonds are formed by methylation reactions with dimethyl
sulfate (Rudkin, 1950) or methyl iodide (Narayanamurti and Handa, 1953);
alkylation with alkyl chlorides (Risi and Arseneau, 1957); β-propiolactone
(Goldstein, 1960); and epoxides (Rowell and Gutzmer, 1976). Among the ether
yielding reactions, only epoxides have been found to be the most appropriate
and promising as other chemicals are either toxic or the reactions degrade
wood mechanically. Epoxides are, however, very expensive and relatively
higher levels of treatment are required as chemicals undergo polymerisation
reactions making determination of degree of substitution impossible from
simple weght gains (Rowell and Gutzmer, 1975).

Acetals are formed by reaction with formaldehyde and other aldehydes.
Acetal bonds are highly effective in preventing fungal attack even at very
low weight percent gains (2%) (Weaver, et al,1960; Stamm, 1959; Stamm and
Baechler, 1960), and at weight percent gains around 4%, the antishrink efficiency
(ASE) achieved is 4 times the ASE, obtained through acetylation (Tarkow
and Stamm, 1953; Stamm, 1959). A number of other aldehydes such as acet-
aldehyde, benzaldehyde and dialdehydes have been tried. Acetal bonds have
been found to be unstable and reactions with aldehydes degrade wood mechanically.

There are numerous reagents, which can be used to esterify wood but acetylation and isocyanate bonding have been widely studied. Esters can also be formed by phthaloylation (Risi and Arseneau, 1958) and reactions with carboxylic acids (Arni, et al, 1961a; Nakagami, et al, 1974). The most studied. of all the chemicals modification reactions for wood has been acetylation. The first patent on wood acetylation was obtained in 1930 using acetylchloride (Suida, 1930). Subsequently ketene gas was used to acetyl wood (Tarkow, 1945). Acetic anhydride is, however, the most common reagent used for acetylating wood (Ridgway and Wallington 1946; Stamm and Tarkow, 1947). The reaction is acid or base catalysed and many catalysts such as urea-ammonium sulphate (Clermont and Bender, 1957), boron triflouride (Risi and Arseneau, 1957), sodium and potassium acetates (Tarkow, 1959) dimethyl formamide (Risi and Arseneau, 1957; Clermont and Bender, 1957; Baird, 1969), magnesium perchlorate (Arni, et al, 1961b; Ozolina and Svalbe, 1972; Truksne and Svalbe, 1977), trifluoracetic acid (Arni, et al, 1961b) and even X rays (Svalbe and Ozolina, 1970) have been tried. The best acetylation conditions have been reported in vapor phase treatment with uncatalysed acetic anhydride in xylene solutions at 100-130°C (Goldstein, et al, 1961). Reaction with acetic anhydride proceeds as follows:

$$\text{Wood-OH} + (CH_3CO)_2O = \text{Wood-O-}\underset{O}{\overset{}{C}}\text{-CH} + CH_3COOH$$

This modification reaction meets all the requirements such as high resistance to decay fungi and termites (Goldstein, et al, 1961; Koppers, 1961; Peterson and Thomas, 1978; Ozolina and Svalbe, 1966; Tarkow, et al, 1950); good dimensional stability (Koppers, 1961; Krylova, 1970; Risi and Arseneau, 1957; Sidorenko, et al, 1973; Tarkow et al, 1950); loss in permeability (Kumar et al, 1979) and moisture absorption and over all water repellency (Rugevitsa, et al, 1977; Turksne and Svalbe 1977; Shiraishi, et al, 1972). Reactions with uncatalysed acetic anhydride have negligible effect on wood color (Goldstein and Weaver, 1963) and the mechanical properties have been found to improve in general. Modulus of rupture, compressive strength and hardness increase while there is only little change in impact and stiffness (Koppers, 1961; Dreher, et al, 1964). Shear (Dreher, et al, 1964) and modulus of elasticity (Narayanamurti and Handa, 1953) have been reported to decrease slightly. Above all, acetyl bonds are quite permanent and neither the reagent nor the treated wood is toxic to humans. Despite these, the treatment process could not be commercially exploited as there are several drawbacks in reaction with acetic anhydride. The major disadvantage is that nearly 50% of the reagent is lost in the form of acetic acid. Being entraped inside, the wood smells of acetic acid. Its continuous presence may corrode metal fasteners, hydrolyse wood components impairing wood strength and reverse the reaction as ester bonds are liable to acid hydrolysis (Rowell, 1975). Although processes have been developed to recover unreacted reagents and by-products in U.S.A. as well as U.S.S.R., high technology costs have been the limiting factor for their commercial exploitation (Koppers, 1961; Otlesnov and Nikitina, 1977).

Search for new acetylating reagents, which do not adversely affect wood properties, resulted in discovery of thioacetic acid, which was found to react well with wood (Singh, et al, 1979). The reaction proceeds liberating only H_2S gas, which is recycled for production of thioacetic acid, making the process a closed circuit one (Kumar and Agarwal, 1982).

$$\text{Wood-OH} + CH_3CO\ SH = \text{Wood-O-}\underset{O}{\overset{}{C}}\text{-CH}_3 + H_2S$$

Vapor phase reactions with uncatalysed thioacetic acid resulted in weight percent gains (WPG) between 7-8. Among the several catalysts tried, monochloroacetic acid and trichloroacetic acid produced the best results increasing weight gains to nearly 20 percent resulting in an antishrink efficiency (ASE) of over 60 percent in chir pine (Table 1).

Table 1 Effect of Various Catalysts/Swelling Agents on Acetylation in chir pine (Singh and Kumar, 1981).
(4 hour vapor phase treatment with thioacetic acid)

Pretreatment	WPG	Acetyl Content %	ASE %
Untreated control	—	1.27	—
No catalyst	7.08	6.75	25.8
Potassium acetate	6.83	6.30	30.6
Dimethyl formamide	8.10	7.46	28.1
Ammonium chloride	8.32	8.20	36.4
Urea-ammonium sulphate	9.13	7.91	28.3
Pyridine	10.39	9.33	29.4
Sulphur dioxide	11.20	-	-
Trichloroacetic acid	18.92	13.05	60.7
Monochloroacetic acid	22.39	15.30	66.6

EXPERIMENTAL

The experimental set up used for vapor phase acetylation was the same as reported earlier (Kumar and Agarwal, 1982). While chir pine (Pinus roxburghii) and mango (Magnifera indica) were used for most of the studies, for strength properties two species viz. fir (Abies pindrow) and gurjan (Dipterocarpus indicus) were also included. Treatment was carried out on oven dry samples between 93-100°C and time varied from 1 to 6 hours. Samples were redried in an oven with forced air circulation at 100 ± 2°C after acetylation. In some cases, acetylation was followed by 1 hour extraction with xylene vapors to recover unreacted reagent entraped in wood voids. WPG was obtained from difference in oven dry weights before and after acetylation. Acetyl content was determined as per standard chemical analysis procedure (Dore, 1920) Shrinkage (swollen to oven dry) was calculated as percent of swollen dimensions in tangential direction. Antishrink efficiency (ASE) imparted by acetylation was calculated as below (Tarkow, et al, 1950).

$$ASE(\%) = \frac{\text{Percentage swelling in untreated sample} - \text{Percent swelling in acetylated sample}}{\text{Percent swelling in untreated sample}} \times 100$$

Reduction in water absorptivity (RWA), a measure of water repellency was obtained by the following equation:

$$RWA(\%) = \frac{\text{Percent water absorbed by untreated control} - \text{Percent water absorbed by acetylated sample}}{\text{Percent water absorbed by untreated control.}} \times 100$$

Three strength tests viz static bending, compression and toughness/brittleness (Charpy) were carried out to evaluate the influence of treatment. Samples of the size 20 x 1.25 x 1.25 cm and 15.0 x 1.25 x 1.25 cm were used for acetylation. A static bending test was carried out on 17.5 cm span under a constant rate of loading so as to obtain deflection of 0.6 mm/min as per Indian Standard Specification 1708-1969 (Anon, 1969). Modulus of rupture was computed from maximum load applied. For compression tests, (sample size 5.0 x 1.25 x 1.25 cm) a constant loading rate so as to cause deformation of 0.5 mm/min/cm length of sample was applied till failure. Brittle ness tests (samples size 12.5 x 1.25 x 1.25 cm) were also carried out as per IS specification 1708-1969 (Anon, 1969) using the swinging pendulum machine. Average values of at least 10 samples were taken for each test.

Decay resistance tests were carried out on chir pine (Pinus roxburghii) using Poria monticola, a brown rot fungus (Kumar and Agarwal, 1983). Evaluation of fungal attack was based on weight loss after the stipulated periods as per Indian Standard Specification (Anon, 1968). Termite tests were conducted on mango blocks (25 x 12.5 x 12.5 mm) as per method developed by Sen-Sarma (1975) using Microcerotermes beesoni Snyder, a sub-terranean termite. The test is based on forced feeding and evaluation was done on the basis of weight loss after exposure for 60 days (Agarwal, et al, 1985).

All tests against decay fungi and termites were carried out on leached samples.

RESULTS AND DISCUSSION

Effect of Acetylation on Physical Properties of Wood

Acetylation being a single site reaction with no crosslinking or polymerisation, one acetyl group is required for every hydroxyl group substituted. WPG has thus been found to be directly related to the acetyl groups added (Goldstein, et al, 1961; Kumar and Agarwal, 1982; Peterson and Thomas, 1978). Most physical properties such as density, bulkling etc. increase with increase in acetyl content. Shrinkage, swelling, hygroscopicity and water repellency improve with increasing WPG (Table 2).

Table 2 Effect of Acetyl Level on Physical Properties of wood (Mango; sample size 7.5 x 1.25 x 1.25 cm)

Property	Treatment period (hr)					
	Control	1	2	3	4	5
WPG	0.0	3.73	8.14	8.70	8.04	8.40
Acetyl content %	4.44	8.58	11.33	11.33	12.48	12.90
Density increase %	0.0	0.20	3.50	3.90	2.70	2.20
Shrinkage % (tangential)	12.53	9.02	5.46	5.47	5.54	5.34
ASE %	0.0	27.90	58.00	58.00	57.40	60.05
RWA %	0.0	10.8	39.4	54.1	29.1	23.2

A comparison of acetylation with thioacetic acid and acetic anhydride with respect to antishrink efficiency has been depicted in Fig. 1. Whereas an ASE of about 70% in spruce is achieved at 25 to 30% acetyl content with acetic anhydride (Tarkow, et al, 1950), a similar ASE is achieved at 15 to 17% acetyl content by acetylating with thioacetic acid (Kumar and Agarwal, 1982). It has been reported that at an acetyl level around 13.5% most of the lignin and hemicelluloses are acetylated while 90% of cellulose remains unreacted during acetylating with acetic anhydride (Truksne, 1977). At this level the ASE is 35% (Tarkow et al, 1950). It appears that thioacetic acid because of its swelling nature, penetrates to the cellulose structure and accords higher dimensional stability even at lower acetyl levels. Poor penetration of acetic anhydride is also evident from electron microscopic studies on acetylated pine wherein early wood tracheids were found to have a very low level of acetylation. A gradient also existed across the cell wall thickness from S_3 to S_1 layer (Peterson and Thomas, 1978).

Mechanical Properties of Acetylated Wood

Most strength properties have been found related to specific gravity and moisture content (Wangaard, 1950). Both these factors should play an important part in acetylated wood as the bulking action of the chemically modifying reagent increases the density and lowers the equilibrium moisture

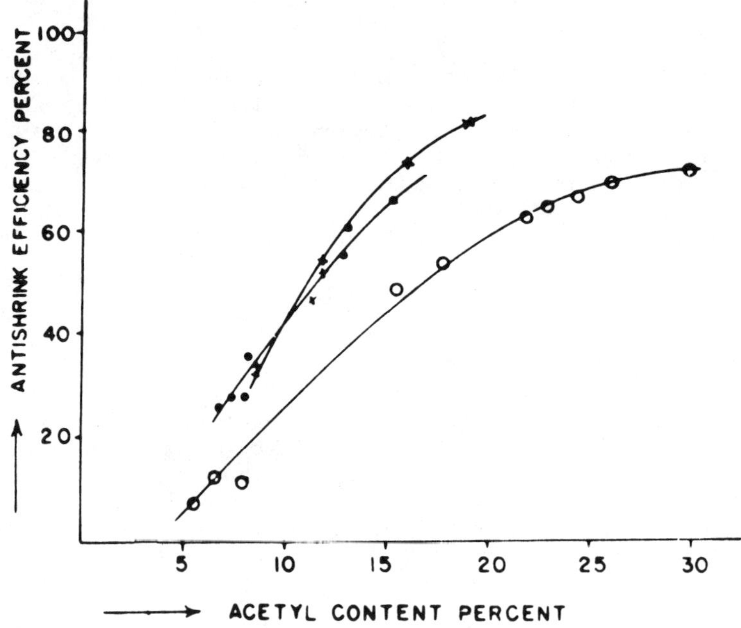

Fig.1 Relationship between acetyl content and
antishrink efficiency.Key:ospruce,chir,
*Mango

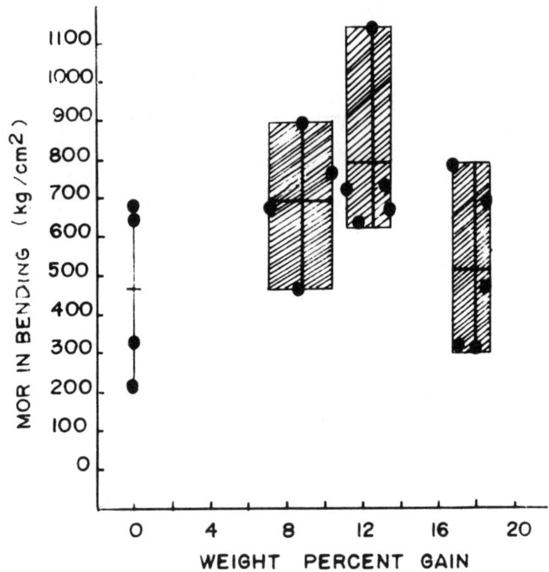

Fig.2 Scatter diagram showing variation over
strength with **WPG** in acetylated Mango

content. However, such reaction occuring in acidic or basic conditions can
be deleterious to mechanical strength. Release of acidic by-products such
as in case of reaction with acetic anhydride can impair toughness due- to
scission of cellulose chains. The scanty information based on a few studies
indicate that strength properties generally improve on acetylation (Tarkow
et al, 1950; Dreher, et al, 1964). Large increases were observed in compressive
strength and hardness apparantly due to increase in density and decrease
in equilibrium moisture content. There was slight losses in static bending
and shear.

Thioacetic acid is a known solvent for lignin, thus reaction with this
is likely to affect mechanical behavior adversely.   Strength data alongwith
data on acetylation time, WPG and specific gravity have been reported in
Table 3. MOR in bending for acetylated wood was higher than for untreated
wood and showed an increasing trend with increasing WPG (upto 3-4 hour
treatment time) in case of fir, mango and gurjan.   Results in case of chir
are inconsistent. MOR falls initially and then starts rising, reaching a maximum
at 3 hours and after which starts falling again. Decrease in strength beyond
3 hours was observed in all the wood species studied. There is a similar
change in specific gravity also. The increase in specific gravity is not proportional
to the weight added as there is an accompanied increase in volume of wood
also. Acetylated wood becomes permanently swollen due to bulking and there
are less fibres per unit area compared with untreated wood. Whereas the
increase in specific gravity is supposed to increase the mechanical strength,
the less number of fibres per unit area has the tendency to lower the strength
values. This is probably the reason for initial fall in strength in case of
chir.

The overall influence of acetylation level (WPG) may also be seen in
Fig. 2 wherein pooled data for acetylated mango samples has been represented
in a scatter form. The data represent acetylation obtained at 2, 3 1/2 and
4 hours period and the shaded blocks show the scattered pattern at various
WPGs. It is evident that strength (MOR in bending) continues to rise with
increasing WPG but a reversal in strength occurs when treatment period exceeds
3 1/2 hours, despite the fact that WPG continues to rise. The net strength is,
however, still higher than untreated wood. A similar pattern is observed
in other strength properties. Compression parallel to grain as well as toughness
(Charpy test) also showed a maximum at three hours treatment period in
all the species studied. Fig. 3 and 4 show the overall trend of all mechanical
properties as affected by increasing WPG for two softwoods and two hardwoods
respectively. The figures represent pooled data of treatments upto 3 hours
comprising (i) vapor phase treatment with thioacetic acid (ii) vapor phase
acetylation followed by extraction with xylene for one hour (iii) vapor
phase acetylation in presence of SO_2 gas. As may be seen all the strength
properties investigated tend to improve at higher WPG. The effects are
more pronounced in case of compression and toughness.   Bending results
are inconsistent   nevertheless the trend is on the increasing side.
The results are in general in confirmity with the earlier reports on acetylation
with acetic anhydride (Dreher et al, 1964; Koppers, 1961; Tarkow, et al,
1950). Dreher, et al, (1964) reported an increase in MOR for softwoods and
a decrease for hardwoods due to acetylation but no such differences have
been observed in the present investigation.

Resistance of Wood Acetylated with Thioacetic Acid to Biological Degradation

Resistance against decay fungi: Resistance of acetylated wood to various
wood destroying organisms is well established as illustrated by numerous
studies (Goldstein, et al, 1961; Tarkow, et al, 1950; Peterson and Thomas
1978). Acetylated veneers lasted 17.5 years in graveyard tests compared
with untreated controls destroyed in 2.5 years (Davidson, 1977). Such resistance
is reported at WPG above 15. Decay resitance has been reported at WPG

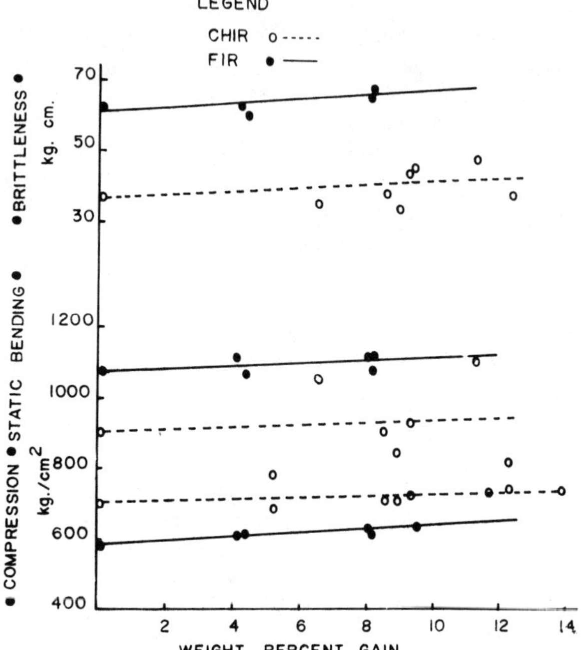

Fig.4 Strength properties as influenced by **WPG** in hardwoods

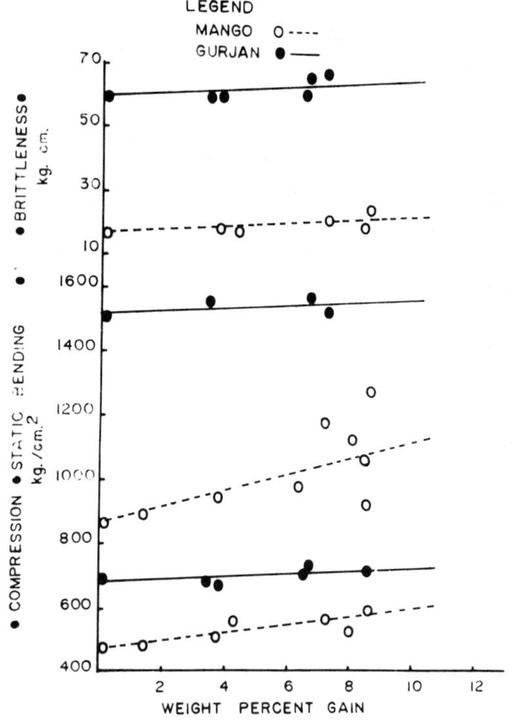

Fig.3 Strength properties as influenced by **WPG** in softwoods

around 7.5 also (Peterson and Thomas, 1978). Singh (1982) reported a threshold value of about 11.35 acetyl content (equivalent to WPG of 14) corresponding to weight loss of 7.6 percent with <u>Poria monticola</u>. A reanalysis of the data corresponding to weight loss of 5 percent as per accepted norms put the limiting value at about 12.35 % corresponding to 18.2 WPG (Fig.5). At such levels of acetylation, it is reported that nearly 86.4% of the lignin and 21.6% of the hemicelluloses present in wood are acetylated (Truksne, 1977). This indicates that the pattern of substitution of hydroxyl group with thioacetic acid is similar to that of acetic anhydride. It was, however, expected that thioacetic acid known for its affinity towards lignin should react with the same at a more preferential pattern and thus induce resistance to lignin degradation at much lower WPG. Although substitution in cellulose at such acetyl level is only 9.3% (Truksne, 1977) the resistance against brown rot fungus probably results from several factors viz.

(i) Blocking of lignin hydroxyls
(ii) Partial blocking of hemicellulose and cellulose hydroxyls.
(iii) Decrease in availability of cell wall moisture due to bulking effect.

Table-3 Effect of acetyl level and acetylation period on strength properties of wood (Agarwal et al, 1985a)

	Time (hr)	WPG	Specific gravity	MOR in Static Bending (kg/cm^2)	Max Compressive Stress parallel to grain (kg/cm^2)	Toughness (Work absorbed) (kg-cm)
Chir	Control	0.0	0.551	922	707	37.6
	1	5.2	0.541	789	695	30.8
	2	8.9	0.534	847	714	35.3
	3	9.3	0.609	934	730	46.5
	4	12.3	0.541	828	747	38.7
	6	13.9	0.538	642	746	40.1
Fir	Control	0.0	0.417	1077	577	62.9
	1	4.1	0.441	1117	609	63.4
	3	8.1	0.478	1135	618	69.2
	6	8.0	0.421	1080	594	42.0
Mango	Control	0.0	0.549	871	478	17.6
	1	4.3	0.551	1169	559	17.9
	2	7.2	0.581	1187	570	21.3
	3	8.6	0.572	1285	594	26.3
	4	8.4	0.541	1071	493	19.4
	6	8.5	0.561	927	462	20.8
Gurjan	Control	0.0	0.632	1509	707	61.3
	1	3.4	0.617	1558	689	60.9
	3	6.6	0.657	1568	730	65.5
	6	8.6	0.617	1383	724	46.5

Resistance to white rot fungi is expected at much lower levels because of the preferential acetylation of lignin.

<u>Resistance against termites:</u> The incidence of termite attack at different WPG is depicted in Fig. 6, which indicates the threshold value at about 12.8 WPG, at which level the weight loss is around 10%. Earlier studies have also indicated that acetylated wood does not become totally unpalatable

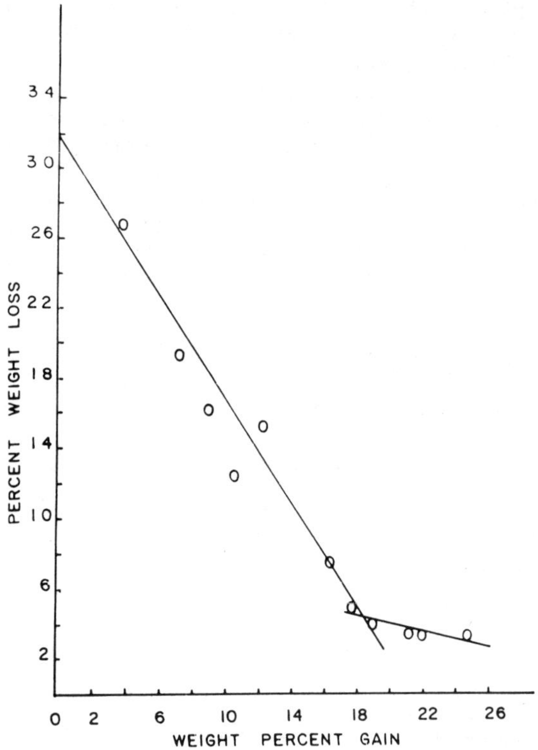

Fig. 5 Decay resistance of acetylated wood

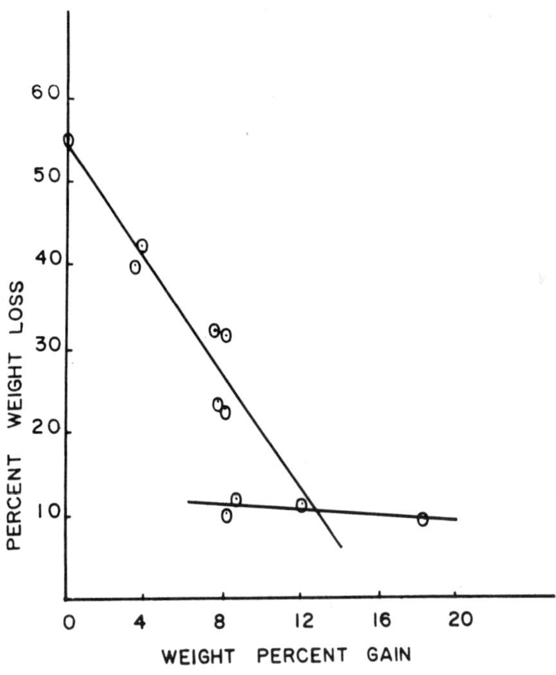

Fig. 6 Termite resistance of acetylated wood

SAMPLE NO.	109	36	233A	211B
TREATMENT TIME (hr)	0	3	4	6
WPG	0	8.70	8.04	8.40

Fig. 7 Termite attack on mango samples acetylated for different periods.

to termites even at a higher WPG of 18 (Goldstein, et al, 1961). The present results rate acetylated wood as resistant as the weight losses are comparable to teak inner heart wood (Sen-Sarma, 1975). Optimum resistance against termites is achieved at 3 hours acetylation period although WPG is around 8 only. Termites basically metabolise carbohydrates but certain termites possessing bacteria in their guts can degrade wood lignin also (French and Bland, 1975). So resistance at such low levels of acetylation probably resulted from blocking of lignin hydroxyls. Extending the reaction period beyond 3 hours results in higher degradation with termites (Fig. 7). This shows that prolonged action of thioacetic acid produces some biologically active sites to enable metabolisation by termites.

This study indicates that physical, mechnical as wellas biological resistance properties of modified wood are influenced by the method of acetylation. Both the reacting reagent and the reaction period seem to influence the behavior of acetylated wood. Almost all physical and mechanical properties deteriorate when treatment period is extended beyond 3-4 hours. The response to biological resistance is much higher even at low WPG in case of acetylation with thioacetic acid than obtained in conventional acetylation with acetic anhydride. With meagre data available so far, it is not possible to fully explain this trend. The initial substitution is reported to be on the easily accessible hydroxyl groups in lignin and hemicelluloses (Kirk, 1973; Rowell, 1982) and amorphous regions in cellulose chains (Truksne, 1977). The dissolving action of thioacetic acid on lignin especially on prolonged contact may be responsible for loosening some lignin - hemicellulose-cellulose bonds and thus weakening the structural regidity and biological resistance. Possibilities of cell wall rupture as well as disturbance in the crystallanity of cellulose chains can also not be ruled out. Such damage to cell wall has been observed in case of treatment with alkylene oxides, but the damage occures when the wood is super-swollen especially at high weight gains (Rowell, et al, 1976). These factors, therefore, need detailed studies using larger samples, modern tools like electron microscope and sophisticated testing machines.

REFERENCES

Agarwal, S.C., Kohli, K.,Shukla, N.K., and Kumar,S., 1985a, A preliminary note on the mechanical properties of wood acetylated with thioacetic acid (In Press).

Agarwal, S.C., Misra, S.C., and Kumar,S., 1985b, Termite resistance of wood acetylated with thioacetic acid (In Press).

Anon., 1968, Method for laboratory testing of wood preservatives against fungi, I.S. 4873. Indian Standards Institution, New Delhi.

Anon., 1969, Methods of testing small clear specimens of timber I.S., 1708, Indian Standards Institution, New Delhi.

Arni, P.C., Gray, J.D., and Scougall, R.K.,1961a, Chemical modification of wood I. Use of trifluoracetic anhydride in the esterification of wood by carboxylic acid, J.Appl.Chem., 11:157.

Arni, P.C., Gray, J.D., and Scougall, R.K., 1961b, Chemical Modification of wood II Use of trifluoracetic acid as catalyst for the acetylation of wood, J.Appl. Chem.,11:163.

Baechler, R.H., 1960, Fungus resistant wood prepared by cyanoethylation, U.S. Patent,2959 496.

Baird, B.R., 1969, Dimensional stabilisation of wood by vapor phase chemical treatments, Wood and Fiber, 1:54.

Clermont, L.P., and Bender, F. 1957, Effect of swelling agents and catalysts on acetylation of wood, Forest Prod. J., 7:167.

Davidson, H.L., 1977, Comparison of wood preservatives in stake tests, USDA FS FPL 02, Madison.

Dore, W.A., 1920, Determination of acetyl content J.Ind.Eng.Chem., 12:472.

Dreher, W.A., Goldstein, I.S., and Cramer, G.R., 1964, Mechanical properties of acetylated wood, Forest Prod. J., 14:66.

French, R.J., and Bland, O.E., 1975, Lignin degradation in the termites Coptotermes lacteus and Nasutitermes extiosus, Mat. Und. Org., 10:281.

Goldstein, I.S., 1960, Improving fungus resistance and dimensional stability of wood by treatment with β-Propiolactone, U.S. Patent, 2 931 741.

Goldstein, I.S., Jeroski, E.B., Lund, A.E. Nielson, J.F., and Weaver, J.W., 1961, Acetylation of wood in lumber thickness, Forest Prod. J., 11:363.

Goldstein, I.S., and Weaver, J.W., 1963, U.S.Patent, 3 094 431.

Kirk T.K., 1973, Chemistry and Biochemistry of decay, In "Wood Deterioration and its Prevention by Preserva tive Treatment" W.A. Cote, ed., Syracuse Wood Sceince Series-5, Syracuse University Press, Syracuse.

Koppers' Acetylated wood, 1961, Dimensionally stabilised wood, New Materials Technical Information No. (RDW-400) E-106.

Kumar, S., and Jain, V.K., 1978, Effects of wood preservatives on physical properties of wood II Effects of different salt loadings of copper-chrome-arsenic composition, Wood and Fiber, 9:262.

Kumar, S., Singh, S.P., and Sharma, M., 1979, Effect of acetylation on permeability of wood, J.Timb. Develop. Assoc. (India), 25(3):5.

Kumar, S., and Agarwal, S.C., 1982, Chemical Modification of Wood with Thioacetic Acid, In "Graft Copolymerisation of Lignocellulosic Fibres", David N.S. Hon, ed., ACS Symposium Series 187 Americal Chemical Society, Washington. D.C.

Kumar, S., and Agarwal, S.C., 1983, Biological degradation resistanace of wood acetylated with thioacetic acid, Intnl. Rearch Group on Wood Preservation, 14th Annual Meeting, Australia, Doc. No IRG/WP/3223.

Krylova, A.N., 1970, Modification of Wood, LSKhA, 130:59.

Lehninger, A.L., 1970, 'Biochemistry', Worth Publishers Inc., New York.

Nakagami, T., Amimoto, H., and Yokota, T., 1974, Esterification of wood with unsaturated carboxylic acids. I Preparation of several wood esters by the TFAA method, Bull. Kyoto Univ. For., 46:217.

Narayanamurti, D., and Handa, B.K., 1953, Acetylated woods, Das Papier, 7:87.

Otlesnov, Y., and Nikitina, N., 1977, Trial operation of a commercial installation for modification of wood by acetylation, LSKhA, 130:50.

Ozolina, I., and Svalbe, K., 1966, Acetylation of wood and biological testing of acetylated samples, Latvijas PSR Zinaatnu Akademija Vestis, 9:56.

Ozolina, I., and Svalbe, K., 1972, Acetylation of wood by an anhydrone catalyst, LSKhA, 65:47.

Peterson, M.D. and Thomas, R.J., 1978, Protection of wood from decay fungi by acetylation- An ultrastructural and chemical study, Wood and Fiber, 10:149.

Richardson, B.A., 1977, Developments in wood preservation, Intl. Research Group on Wood Preservation Doc No. IRG/WP/393.

Ridgway, W.B., and Wallington, H.T., 1946, Esterification of wood, British Patent, 579 255.

Risi, J., and Arseneau,D.F., 1957, Dimensional stabilisation of wood. Part I Acetylation, Forest Prod. J., 7:210.

Risi, J., and Arseneau, D.F., 1958, Dimensional stabilisation of wood Part II Phthaloylation, Forest Prod. J., 8:252.

Rowell, R.M., 1975, Chemical modification of wood: Advantages and disadvantages, Proc. Amer. Wood Preserv. Assoc., 71:41.

Rowell, R.M., 1982, Distribution of acetyl groups in southern pine reacted with acetic anhydride, Wood Sci., 15:172.

Rowell, R.M., and Gutzmer, D.I., 1975, Chemical modification of wood: Reactions of alkylene oxides with southern yellow pine, Wood Sci., 7:240.

Rowell, R.M., and Gutzmer, D.I., 1976, Treatment of wood with butylene oxide, U.S. Patent, 3 985 921.

Rowell, R.M., and Gutzmer, D.I., Sachs, I.B., and Kinney, P.E., 1976, Effect of alkylene oxide treatments on dimensional stability of wood, Wood Sci., 9:51.

Rudkin, A.W., 1950, The role of the hydroxyl group in the glueing of wood, Australian J. Appl. Sci., 1:270.

Rugevitsa, A., Iurevitsa, S., and Svalbe, K., 1977, Water repellent properties of wood of hardwoods modified by acetylation, LSKhA, 130:45.

Sen-Sarma, P.K., 1975, Wood Destroying Termites of India, PL-480 Project No. A7-FS-58, Forest Research Institute, Dehradun.

Shiraishi, N., Yokota, T., Kimura, T., and Sumizawa, K., 1972, The interaction of wood with organic solvents III. The crystal structure change of wood and the reactivity for acetylation, J.Japan Wood. Res. Soc., 18:215.

Sidorenko, A.K., Kuz'min, N.F., and Reshtnikov, E.K., 1973, Investigation of the dimensional stabilisation of modified wood by acetylation, Lesnoi Zhur. 16:157.

Singh, S.P., 1982, "Studies on Chemical Modification of Wood to improve Durability and Physical Properties" Ph.D. Thesis, Garhwal University, Srinagar, India.

Singh, S.P., Dev, I., and Kumar, S., 1979, Chemical modification of wood: Acetylation and antishrink efficiency of wood with thioacetic acid, J.Timb, Develop Assoc.(India), 27(3):30.

Stamm, A.J. 1959, Dimensional stabilisation of wood by thermal reactions and formaldehyde cross linking, Tappi, 42:39.

Stamm, A.J. and Baechler, R.H., 1960, Decay resistance and dimensional stability of four modified woods, Forest Prod. J., 10:22.

Stamm, A.J., and Tarkow, H., 1947, Acetylation of wood and boards,U.S. Patent, 2 417 995.

Suida, H., 1930, Acetylated Wood, Austrian Patent, 122 499.

Svalbe, K., and Ozolina, H., 1970, "Modification of wood by acetylation", in Plast. Modify. Drev., 145.

Tarkow, H., 1945, Acetylation of wood with ketene , Forest Products Laboratory (Madison) Office Report.

Tarkow, H., 1959, A new approach to the acetylation of wood , Forest Products Laboratory (Madison), Office Report.

Tarkow, H., and Stamm, A.J. 1953, Effect of formaldehyde treatments upon the dimensional stabilisation of wood, J.Forest Prod. Res.Soc. 3:33.

Tarkow, H., Stamm, A.J., and Erickson, E.C.O., 1950, Acetylated wood,
    USDA FS FPL(Madison), 1593.
Truksne,D., 1977, Investigation of the content of bonded acetyl group in
    the main constituents of acetylated wood, LSKhA, 130:32.
Truksne, D., and Svalbe, K., 1977, Water repellent properties and dimen-
    sional stability of acetylated pine wood in relation to the degree
    and method of acetylation, LSKhA, 130:26.
Wangaard, F.F., 1950, "The Mechanical Properties of Wood", John
    Wiley & Sons, Inc., New York.
Weaver, J.W., Nielson, J.F., and Goldstein, I.S., 1960, Dimensional
    stabilisation of wood with aldehydes and related compounds, Forest
    Prod. J., 10:306.
Willeitner, H., 1976, World survey on the status of pollution control
    in the field of wood preservation, Intl. Research Group on Wood
    Preservation, Doc. No. IRG/WP/369.

SECTION IV - OILS

BEHAVIOR OF WATER IN ORIENTAL LACQUERS

J. Kumanotani*, K. Inoue**, M. Achiwa***, and L.W. Chen****

*   Ehime University, 3 Bunkyocho Matsuyama, Japan 790
**   Ube Kosan, Ube City, Japan 750
*** Chyuo University, 27-13-1 Kasugacho Tokyo, Japan 112
**** National Taiwan University, Taipei, Formosa

INTRODUCTION

In the orient, various kinds of lacquer trees are grown, Thus vernicifera
(China, Korea and Japan), Rhus succedanea (Vietnam and Formosa) and Melanor-
hoea usitate (Burma and Tailand). Of these, the tree, Rhus vernicifera gives
the best quality sap as a material for coatings which has been used in China
for thousands of years. The lacquer wares characteristic of Korean and
Japanese culture are also based on the same sap as in China.

The major constituents of the sap has been classified according to their
solubility in ethanol and water[1]; urushiol, the alcohol-soluble part, gummy
substances and nitrogenous substances which are water soluble and insoluble
parts, respectively, of the alcohol insoluble part of the sap (acetone is
used more favorably instead of alcohol in laboratories, and the acetone
insoluble part of the sap is defined as acetone powder). Recently, it is
found that the gummy substances involve mono, oligo and polysaccharides,
laccase (an enzyme) and stellacyanine as well, and nitrogenous substances
belonging to glycoproteins as mentioned below. Moreover, the urushiol
congener is mixed with 10-20% (by wt) of oligo-urushiol.

The constituents of the sap are different due to the kinds of lacquer
trees. A typical example is shown in Table 1 for the sap, Rhus vernicifera
(China and Japan) and Rhus succedanea (Vietnam), indicating the latter sap
contains larger amount of nitrogenous substances gummy substances and
water than the former, and laccol is involved as a major phenolic component
instead of urushiol[2].

Usually the sap is used after being converted into lacquer: the sap
is stirred below 45ºC for 4-6 hr. in a special vessel with a stirrer, and
the end point of stirring is decided from changing color and viscosity of
the sap under treatment. The lacquer thus made or sap itself is coated,
and dried in a humid chamber with RH 70-80% for one day at room temperature,
making an elegant film.

The polysaccharides are found to be composed of two fractions having
respective molecular weights M_w 84,000 and 27,000 by aqueous-phase gel-
permeation chromatography. The fractions contain D-galactose (65 mol%),
4-0-Methyl-D-glucuronic acid (24%), D-Glucuronic acid (3 mol%), L-Arabinose
(4 mol%) and L-Rhamnose (3 mol%). Smith degradation of the carboxyl-reduced

Table 1. Constituents of Sap

	Rhus vernicifera	Rhus succedanea
Urushiol or Laccol	60–65 %	52 %
Gummy Substances	5–7	17
Glycoproteins	2–5	2
Laccase et al.	1	1
Water	20–25	30

polysaccharides give products of halved molecular weight, and these consist
of a β-(1→3)-linked galactopyranan main chain and side chain made up of
galactopyranose residues. Peripheral groups, such as α-D-Galp-, α-D-Galp-
(1→6)-β-D-Galp-, 4-O-methyl-β-D-GlacpA-, and 4-O-methyl-β-GlacpA-(1→6)-β-
D-Galp-, are attached to this interior core through β-(1→3)- or β-(1→6)-
linkages. The polysaccharides are existing as salts of Mg, Ca and Na
together with mono and oligosaccharides(2:8) in the sap[3](see Fig.1).

The CP/MAS spectrum of the nitrogenous substances indicated that
the nitrogenous substances are glycoproteins and showed similar IR spectral features with those of laccase isolated from the sap Thus vernicifera, and its sugar/protein ratio was found to be 1/9 (w/w). An
aqueous solution of the glycoproteins(20 mg) reduced with mercoptoethanol
(100 µl) in water(2.0 ml) added with SDS(40 mg) at 100°C for 5 min was
filtered, and submitted to SDS-polyacrylamide gel electrophoresis is
according to the Laemmli method[4] on 10% acrylamide gel using subunits of
RNA polymerase B(β' 180,000, β 14,000, X 100,000, α 42,000 and Z 39,000),
chymotrypsinogen(25,000), lysosome(14,400), insulin subunit(A 3,200,
B 2,400) as standards and comassie brilliant blue-R 250 as a dying material,
indicating that major fractions are those with molecular weights 8,000 and
17,000 together with the minor fractions with mol.wts. 25,000 and 47,000.

The separated polysaccharides are soluble in water, and the glycoproteins are not. Furthermore, a stable water in oil(urushiol) type emulsion
is made only together with gummy substances and glycoproteins, and not made
with either of them, indicating that the glycoproteins are dispersing as
giant molecules composed of subunits bonded with S-S bonding in the urushiol
(oil) phase and the gummy substances in water droplets of the W/O type sap,
and the glycoproteins seem to perticipate as an emulsion stabilizer in the
sap.

Urushiol is a mixture of 3-substituted catechol with C_{15} chains with
0-3 double bonds, polymerizing by the action of laccase, an enzyme, in the
lacquer-making process or drying process[5].

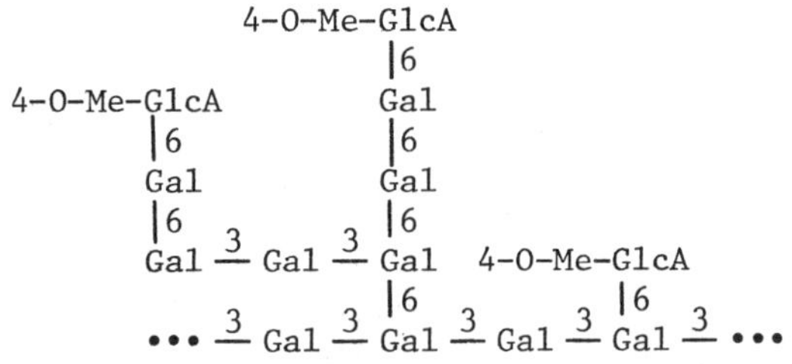

Fig. 1 Structural Feature of Polysaccharides in the Acid Form

HOW DOES WATER INTERACT WITH THE FILM ARCHITECTURE?

When thinking of the status of the water in the sap and lacquer films, not only the presence of hydrophilic constituents in the sap but also the process of making lacquer and the lacquer drying process should be taken into consideration.

The lacquer made from the sap usually contains 2-4%(by wt) water, and is dried in a humid chamber with relative humidity (RD)% of 70-80 for one day. Drying of the sap or lacquer coating requires laccase-catalyzed polymerization of urushiol where Cu^+ in laccase made after Cu^{++} being reduced by urushiol should turn over into Cu^{++} to keep activity of the laccase for polymerization of urushiol. In drying, the upper surface layer of the lacquer or sap coating facing air begins to dry into a solid film which may hinder the oxygen permeation from air into the inside of the lacquer. However, in the humid chamber, the upper surface layer when moistened with water vapor may cause easy permeation of oxygen molecules to oxidize Cu^+ to Cu^{++} in laccase, achieving complete drying of the sap or lacquer coating by the laccase-catalyzed polymerization.

The structure of the superdurable lacquer film was proposed in a previous paper[6]. The dried lacquer film is packed densely with grains with 1,000Å in size in which substantially oxidizable polymerized urushiol is involved, and protected from air-oxidation or degradation as a result of being surrounded by cell wall made of the polysaccharides which have high barrier for oxygen diffusion.

Generally speaking, when water is absorbed in the prepolymer or polymer matrix, major changes appear in the mechanical properties of the crosslinked polymers derived thereof. Furthermore, water-uptake causes easy oxygen permeation, resulting in the promotion of oxidative degradation of the polymer matrix. On the other hand, the lacquer films are well known to show excellent durability despite of involving ca 10% of highly hydrophilic gummy substances and glycoproteins as components of the film.

In exhibition in museums or show windows of stores, a cup of water or humidity controller is normally placed to conserve the lacquer wares.

When considering the presence of a relatively large amount of hydrophilic component of gummy substances and glycoproteins, the sap and lacquer film formation mechanism in a humid chamber, and a cup of water used in museum etc., attempts to describe the status of water in the formation of lacquer film which is highly durable polymeric material seem to be very attractive not only to find a method of the conservation of lacquer wares but also from a general viewpoint of demonstrating the water behavior in natural polymers, because polysaccharides are mostly abundantly existing biomass which have been used as material and will extensively expand its application in the future in combination with the synthetic polymeric materials will be expanded.

The behavior of water is described here for the sap and lacquer films as well as the acetone powder, gummy substances or glycoproteins of Rhus vernicifera and Rhus succedanea by means of DSC, dielectric measurements, water sorption, and T_2(spin-spin relaxation time) measurements by a Pulse NMR.

The results obtained indicate that water molecules in the sap and lacquer films are structurally bound very strongly in the matrix, although they may be removed from the film in a very dry atmosphere. This fact gives us an idea of conservation of artifact coated with the lacquer, and of how cellulosic materials might be utilized in combination with synthetic polymeric materials.

SAMPLES

A sample of the sap, Rhus vernicifera(Japan or China) or Rhus succedanea
(Vietnam) and the lacquer made therefrom were supplied from the Saito Co.,
Tokyo which were coated with a doctor blade(76 µm) on glass plates and dried
in a chamber with RH 70-80% at room temperature for one day, and peeled after
being dipped in water(film thickness, 40-50 µm). The water soluble and water
insoluble parts of the acetone powder were obtained conventionally.

MEASUREMENTS

Water Sorption Two methods were applied; (1) weighing varying water-
uptake of sap or lacquer films which are dipped in deionized water for two
weeks at room temperature(see Fig.2), and (2) measurements of the equilibrium
water sorption isotherm under various conditions(Fig.3 and 4).

Equilibrium water sorption was measured for the sap and lacquer films
(ca 10 mg) as well as for the acetone powder and glycoproteins(5 mg) at RH%
of 0-100 at a temperature of 40°C and 50°C with a modified Prager apparatus[7]
used for the sorption of organic vapour where a Cahn electric balance RG-HV
with a sensitivity of 0.1 µg was equipped to measure the amount of water-
uptake automatically; the sample tube is surrounded by a water jacket through
which water kept at a certain temperature was running in order to keep the
temperature of the sample more precisely. When the sorption has reached at
an equilibrium, the sample is evacuated and the desorption is started.

Diffusion equations allow us to obtain diffusion constants D_a and D_b
from the respective slopes of K_a and K_b of the linear plot of (M_t/M_∞) vs.
$t^{1/2}/h$ according to the eq.1 and 2, and an average diffusion constant D_{av} is
obtained according to the eq. $D_{av}=(D_a+D_b)/2$; a and b are symbols related to
absorption and desorption, t and h, time of the measurements and thickness
of the film, and M_t and M_∞, water uptake at time t and at the time when the
absorption reached at an equilibrium, respectively.

$$D_a=\pi D_a{}^2/16 \qquad\qquad (1)$$

$$D_b=\pi D_b{}^2/16 \qquad\qquad (2)$$

DSC DSC was run with a Model Prekin Elmer DSC-2 in the temperature
range of -40°C to 50°C for the sap and lacquer films(5-15mg) with the absorption water uptake of 17%, on Rhus succedanea and Rhus vernicifera samples.

Dielectric Measurements They were achieved in a temperature range of
-130∿+50°C over a frequency range of $10-15^5$kHz using a General Radio Model
1615-cp capacitance bridge and an ordinary thermostat[8].

Pulse NMR Spectroscopy A Brucker PC-20 at 20MHz was used for the film
or powder of acetone powder, Gummy substance or glycoprotein which was packed
in a test tube(dia, 7mm) and conditioned previousely in an atmosphere with
a certain RH% prepared from saturated aqueous salt solutions: CH_3COOK, 20%RH,
K_2CO_3 44%, NaBr 58%, NH_4Cl 80%, K_2SO_4 98%. For RH 0%, P_2O_5 was used for
drying. Samples were dried for 2 weeks. Measurements were performed at
40°C using a 90º pulse method. A pulse width of 90º pulse was 1-1.5µs. Free
induction decay after 90° pulse(FID) was recorded by a signal averager
Kawasaki Electronica KR-3160 which allows improvement of the signal to noise
ratio of FID curves by about 1000 times record. For the numerical analysis
of FID, a microcomputer SORD Mark III was connected with the signal averager
by GP-IB interface. The calculation program was BASIC. For detailed
analysis, a mini computer NEC MS 30 was also used and the program was coded
by FORTRAN. FID is expressed in terms of the Lorenzian line of eq. 3 or
Gaussian line of eq. 4.

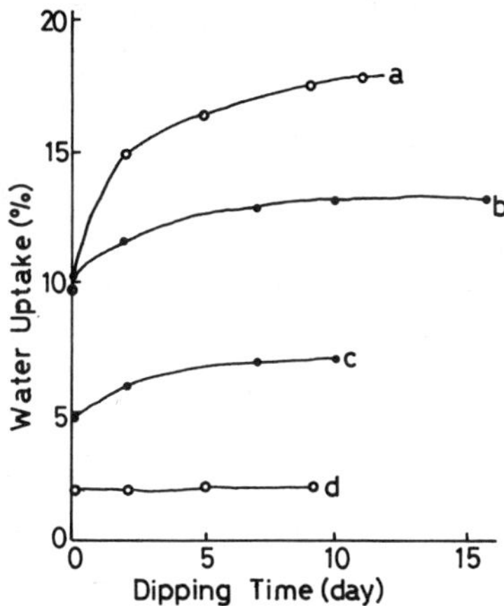

Fig.2 Water Uptake of the Films in Water at r.t.; a and b, sap and
lacquer films, aged for 7 and 15 months, respectively(Rhus
succedanea,Vietnam); c, lacquer film, aged for 15 months
and d, baked acetone-extracted urushiol at 160 °C - 6hr(Rhus
vernicifera,China).

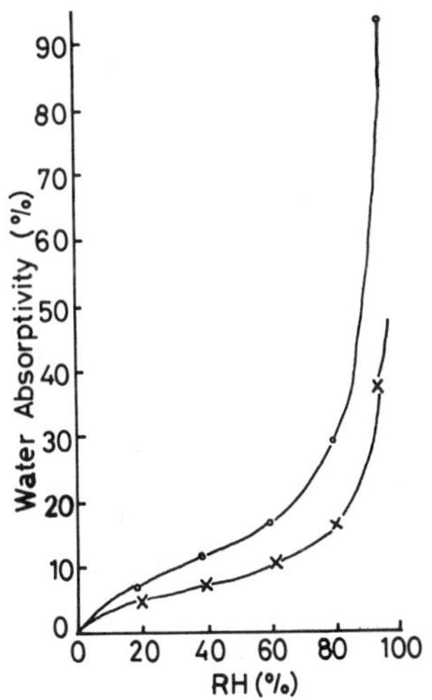

Fig. 3 Water Uptake(%) at 50°C as a Function of Relative Humidity
(RH)(%) for Polysaccharides(o) and Glycoproteins(x).

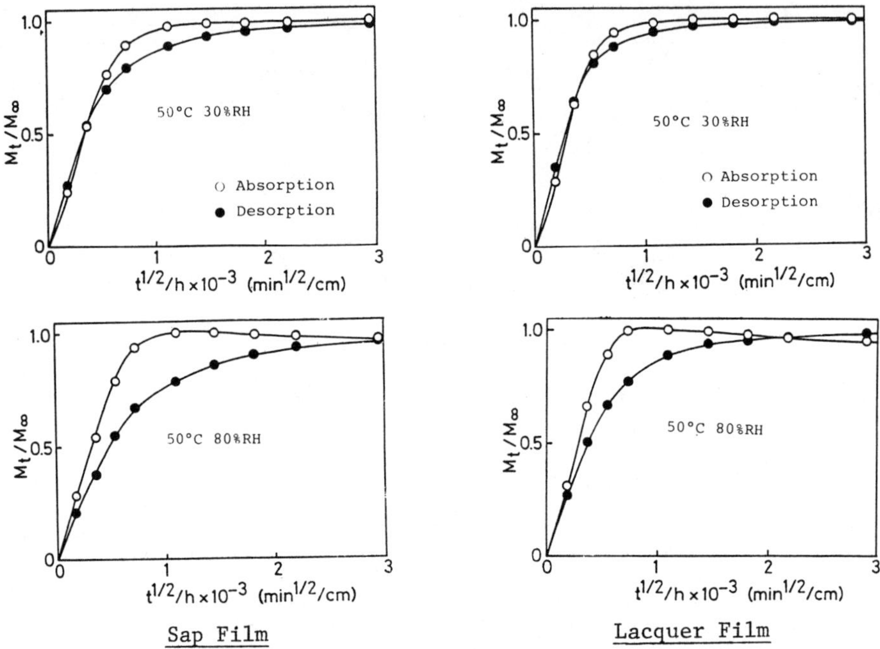

Sap Film Lacquer Film

Rhus vernicifera

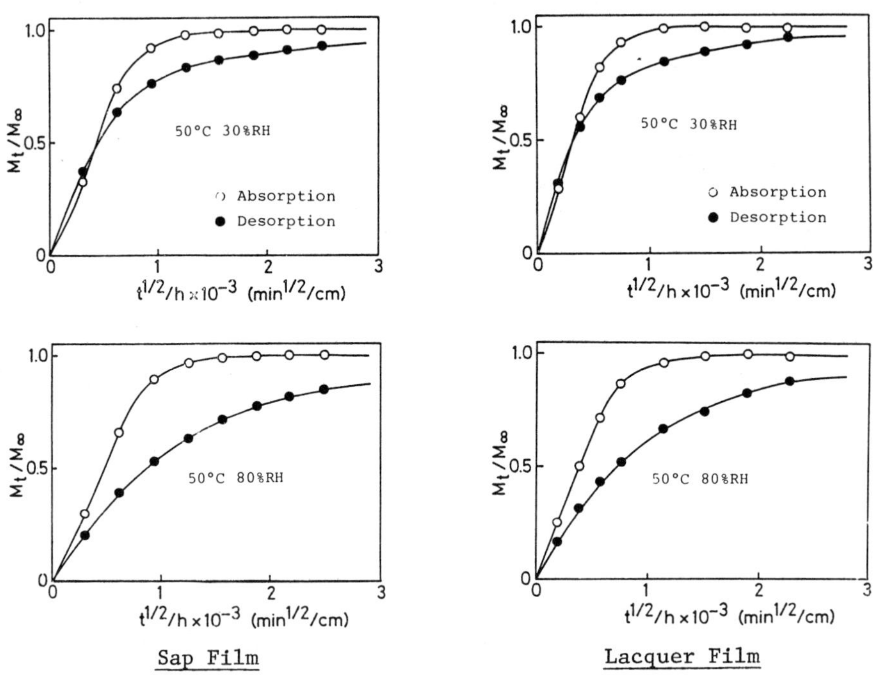

Sap Film Lacquer Film

Rhus succedanea

Fig. 4 Water Sorption Curves for Sap and Lacquer Films

$$S(t) = S \exp(-t/T_2) \qquad (3)$$

$$S(t) = S \exp(-t^2/2T_2) \qquad (4)$$

Then ln S shows a linear change vs t or t^2. From their slope K, T_2 is obtained by $T_2=-1/K$ or $T_2=(-1/2K)^{1/2}$. The effect of field inhomogenity to T_2 is neglected when T_2 is small.

DOES WATER SORPTION TAKE PLACE FOR THE SAP AND LACQUER FILMS ?

Water Absorption in Water of Films

The sap and lacquer films made conventionally from the sap, Rhus vernicifera and Rhus succedanea, were dipped in deionized water at room temperature, and the amount of water-uptake(per cent by weight for the original film) is plotted as a function of dipping time(day)(see Fig.2), indicating that water sorption is increasing and reaching an almost constant level versus time.

Larger water-uptake seems to occur for the films with larger contents of the hydrophilic components, polysaccharides and glycoproteins; with the sap film Thus succedanea, aged for 7 months after its preparation, V(S, 7M) absorbed 17% of water for 12 days, and the lacquer films 1 year after preparation Rhus vernicifera, C(L, 1 year), absorbing 7% of water.

In order to know the status of water, DSC was run for the film V(S, 7 months) absorbing 17% of water at a heating rate of 10°C/min, exhibiting no peak of ice-melting or no presence of free water in the film in the temperature range of −40∿50°C.

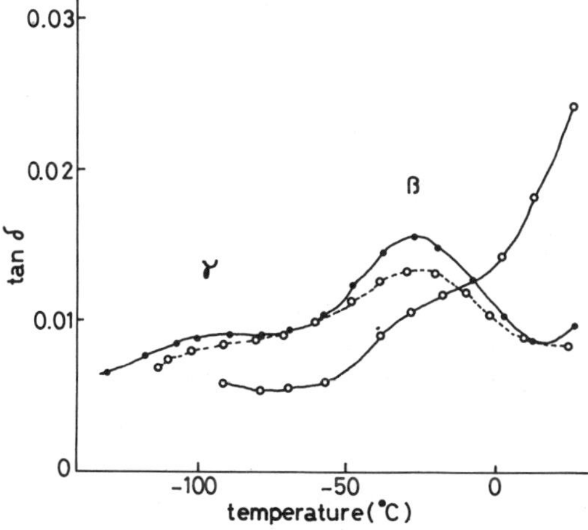

Fig. 5. Dielectrical tan δ vs. Temperature (ºC) correlation at 1 kHz; baked acetone extracted urushiol at 160°C-6hr in air, -o- and -●- after being dipped in water for 2 weeks, and --o-- lacquer film(Jpn) aged for 18 years.

However, evidence for the presence of water in the lacquer film was
given from the measured dielectrical properties. As may be seen in Fig.5,
a small sloped shoulder is seen in the temperature range $-50 \sim -10°C$ at 1 kHz
for the baked acetone-extracted urushiol at 160°C for 6 hr, which appears
more clearly and remarkably at -25°C when absorbing 2% water after being
dipped in the deionized water for 2 weeks. This peak was designated as
β-peak and ascribed to the presence of water in the film.

The β-peak is found for the sap and lacquer films aged for 18 years at
-22°C and -25°C, respectively(see Fig.5).

Water Vapor Sorption

Polysaccharides and Glycoproteins Equilibrium water sorption of the
acetone powder, gummy substances or glycoproteins(5 mg) was measured as a
function of time at 50°C and at relative humidity(RH) in the range of 20-90%,
respectively(Fig.5), and the obtained equilibrium water uptake(per cent by
weight) is summarized in Table 2. It is clear that the gummy substances
can absorb a larger amount of water than the glycoproteins, indicating that
the former is more hydrophilic than the glycoproteins, and at RH over 60%,
water sorption is rapidly increasing remarkably.

Sap and Lacquer Films Fig.3 shows water sorption isotherms at 50°C
and at RH 30 and 80% for the sap and lacquer films, Rhus vernicifera(Japan)
and Rhus succedanea(Vietnam), which are described in terms of (M_t/M_∞) vs.
$t^{1/2}/h(\times 10^{-3})(\min^{1/2}/cm)$ for the water absorption and desorption,
respectively.

The type of the sorption isotherm obtained is similar to each other,
and to those already known for major synthetic polymers.

The amount of equilibrium water sorption for the sap or lacquer film
is almost same with that obtained from those for the gummy substances and
glycoproteins involved in the films(see Table 3), indicating that water
sorption is mainly due to the glycoproteins and gummy substances. Moreover
a larger water sorption is shown for the sap film than for the lacquer film.

It is to be noted however, that, in the case of the absorption process
of the sap and lacquer films at 80% RH, the amount of water absorption
increasing with time reached at a maximum value, then decreasing gradually.
This isotherm is characteristic for the crystallizable polymers such as
nylons which may crystallize as a result of water absorption[9].

In a SEM-photograph of Fig.6 for the etched gummy substance cast from
an aqueous solution is seen fibrillar form of the polysaccharides very
clearly, which is interesting for the highly branched polysaccharides, particularly, in connection with the excellent durability of the lqcquer film
densely packed with grains, having a cell structure made of polysaccharides
cell wall and the inside polymerized urushiol(6), because the fibrillation
of the cell polysaccharides may increase the barrier of the cell wall towards
oxygen permeation, resulting in increasing durability of the film. This
fibrillation might occur mainly in the drying process of the lacquer in the
humid chamber.

Larger hysterisis of the sorption isotherms was found for the sap
films, reflecting the difference in morphology of the films as already reported, in the sap films the polysaccharides particles are dispersed in the
polymerized urishiol matrix and seem to arrest the water molecules strongly in its matrix; on the other hand the lacquer film, in which the polysaccharides are complexed with the polymerized hydrophobic urushiol release
water molecules more easily from the matrix with smaller hysteresis in the

Table 2. Equilibrium Water Sorption of Polysaccharides and Glycoproteins as a Function of Relative Humidity (RH) at 50°C.

Polysaccharides		Glycoproteins	
RH (%)	Equilibrium Water Sorption (%)	RH (%)	Equilibrium Water Sorption
18.0	7.55	19.0	4.65
39.0	11.7	40.0	7.35
60.0	16.4	60.0	10.4
79.0	29.3	80.0	16.3
94.1	85.4	94.0	37.5

Table 3. Equilibrium Water Sorption of Films, derived from Thus vernicifera and Rhus succedanea (in blanket).

Temperature (°C)	Relative Humidity(%)	Equilibrium Water Uptake(%)	
		Sap	Lacquer
30	30	1.5	1.3
	60	2.1	1.0
	80	3.1	1.1
40	30	1.2	1.0
	80	3.2	2.9
50	30	1.3(2.5)	1.1(3.0)
	80	3.5(8.1)	3.1(7.3)

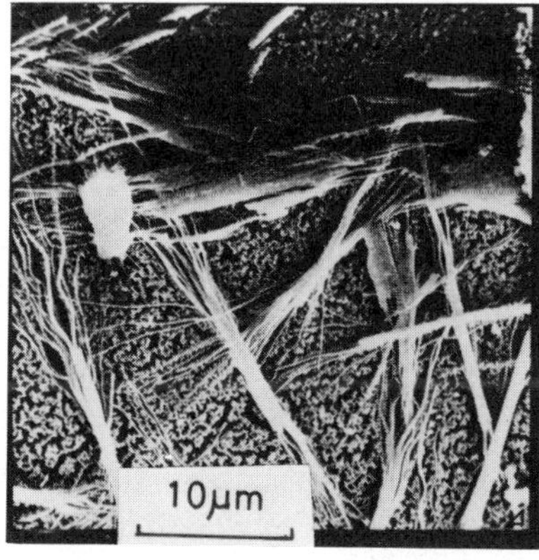

Fig. 6 A SEM photograph of fibrillated polysaccharides.

Table 4. Equilibrium Sorption Water $C\infty$(%) and Diffusion Constant D_{av}(cm^2/sec) for Films, derived from Rhus vernicifera

| | Temperature(°C) | | | | | |
| | 30 | | 40 | | 50 | |
RH(%)	C_∞(%)	$D_{av}\cdot10^{-7}$	C_∞(%)	$D_{av}\cdot10^{-7}$	C_∞(%)	$D_{av}\cdot10^{-7}$
[SAP]						
30	1.5	0.78	1.2	1.55	1.3	3.13
60	2.1	0.96				
80	3.1	1.12	3.2	2.29	3.5	4.01
[LACQUER]						
30	1.3	1.16	1.0	2.52	1.1	5.27
60	1.0	1.44				
80	1.1	1.67	2.9	3.34	3.1	6.25

Table 5. Equilibrium Sorption Water, $C\infty$(%) and Diffusion constant D_{av}(cm^2/sec) at 50°C for Films, derived from Rhus succedanea

| | [SAP] | | [Lacquer] | |
RH (%)	C_∞ (%)	$D_{av}\cdot10^{-7}$ (cm^2/sec)	C_∞ (%)	$D_{av}\cdot10^{-7}$ (cm^2/sec)
30	2.5	1.25	3.0	1.47
80	8.1	3.02	7.3	3.31

measured sorption isotherm. This result becomes more clear from the obtained diffusion constant for the lacquer and sap films based on the water sorption isotherms (Tables 4 and 5): larger diffusion constant values are given for the lacquer films than for the sap films, indicating a quick response toward the change of environmental humid condition occurs for the larger films than for the sap films. These are excellent properties of the lacquer films keeping less amount of water in the film from a view point of durability.

T_2(SPIN-SPIN RELAXATION TIME) MEASUREMENTS TO KNOW WATER STATUS IN THE SAP AND LACQUER FILMS

The obtained FID curves based on the Lorenzian(for the short T_2 component) and Gaissian(for the long T_2 component) equations, 1 and 2 clearly show a linear change of ln S vs. t and $t^{1/2}$ as shown in Fig.7 and 8, respectively. From the slope k, T_2(spin-spin relaxation time, μs) is obtained and T_2 values are indicated in respective figures together with RH% of the moistening conditions of the measured sample.

As may be seen in Fig.7, the acetone powder dried over P_2O_5 for 2 weeks exhibited two relaxation processes in the FID curves, with respective T_2 of 12.1 μs and 45.1 μs, and the glycoproteins treated under the same conditions show T_2 at 12.2 and 53.3 μs. From the level of the values of T_2, the shorter T_2 component seems to appear from the water molecules arrested very strongly in the matrix, and the longer one, from water molecules rather loosely bonded.

Longer T_2 time for the acetone powder than that for the glycoproteins is in
agreement with the result that the acetone powder or its major component,
polysaccharides are more hydrophilic than glycoproteins as already pointed
out from the equiliblium water sorption results shown in Table 2.

With incleasing RH% of moistening conditions of the acetone powder,
the longer T_2 components became longer; for the acetone powder, T_2 at RH
44% which may absorb an equiliblium water of 11.7% is 120.0 μs, becoming
longer to 230.0 μs at RH 80%, with an equiliblium water 29.0%, for the
glycoproteins, T_2, 84.5 μs(RH 44%, 7% water-uptake) to T_2, 150.0 μs(RH 80%,
16% water-uptake). These results exhibit the increased amount of the water
loosely bonded in the matrix with an increase of the amount of the water
absorbed in the matrix.

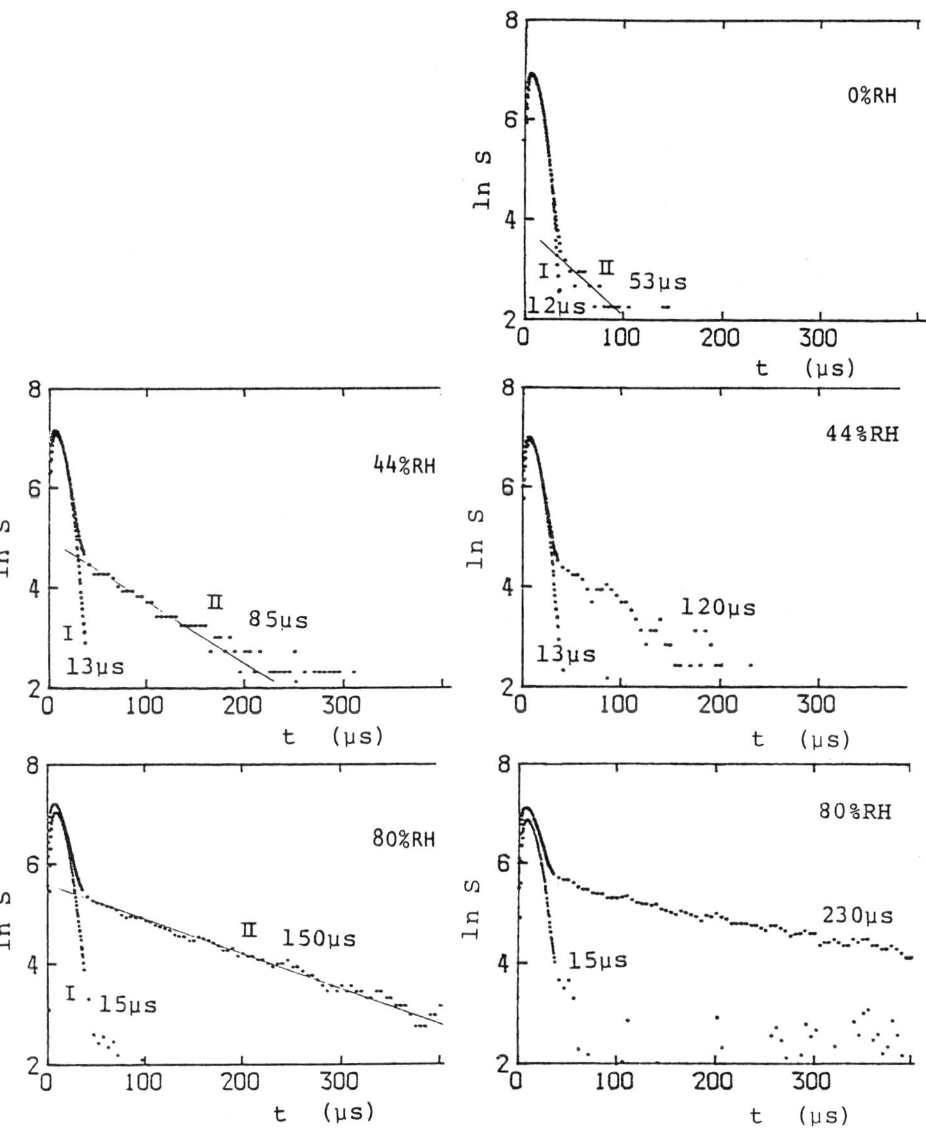

Fig. 7 FID Curves for Glycoproteins(left) and Acetone Powder(right)
        at40°C, Rhus vernicifera,Japan.

It is tremendously surprising that for the sap and lacquer films,
the mode of T_2 is remarkably different from those observed in the acetone
powder and glycoproteins.

For the sap films Rhus vernicifera(Japan) at RH 0%(see Fig. 8), the
longer time component found in the glycoproteins or in the gummy substances
could not be observed clearly, and at RH 80%, only a small fraction of the
long time component with T_2 of 30.0 μs appears.

For the lacquer films it was different to find clearly the long time
signal at 44% RH or even at 80% RH.

The same trend in T_2 is found for the sap and lacquer films derived from
Rhus succedanea (Tietnam) involving a larger amount of the acetone powder in
them.

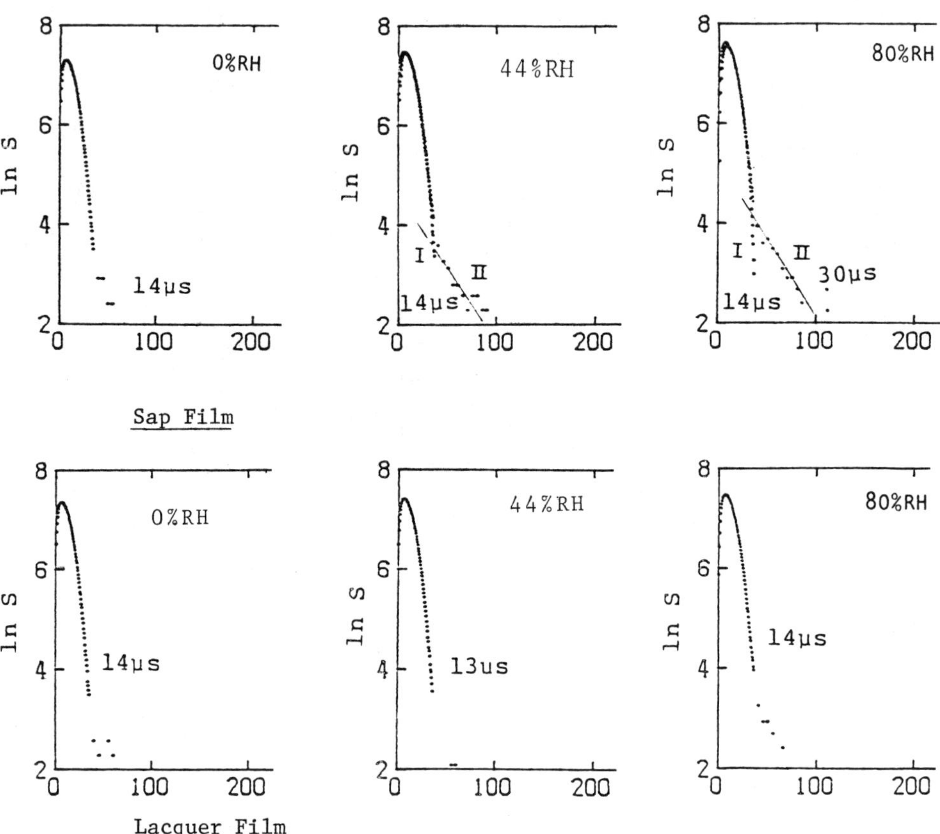

Fig.8·1 FID Curves for the Sap and Lacquer Films(40°C) of
Rhus vernicifera(Japan)

From the results of T_2 measurements it is clear that the hydrophilic
components of the polysaccharides and glycoproteins may absorb water greatly
at high RH such as 80%, however, when confined in the hydrophilic polymerized urushiol, their water-sorption behavior changed remarkably, particularly, in the lacquer film.

T_2 and DSC measurements gives us important information that water
molecules are confined in the matrix as structural components of the films
judging from the observed very short T_2 level in FID curves. Really when
the film is evacuated in the isotherm measurements, it curls, and then
recovers its oridinal shape when being moistened.

Therefore, the lacquer wares have been made as composite materials
with sap or lacquer, wood, bamboo, paper, boiled rice or wheat, clay, dried
blood, and wire etc, in order to reinforce the wares against the internal
stress brought about by evaporation of water from the lacquer coating
depending on the atmospheric moisture where the wares are kept.

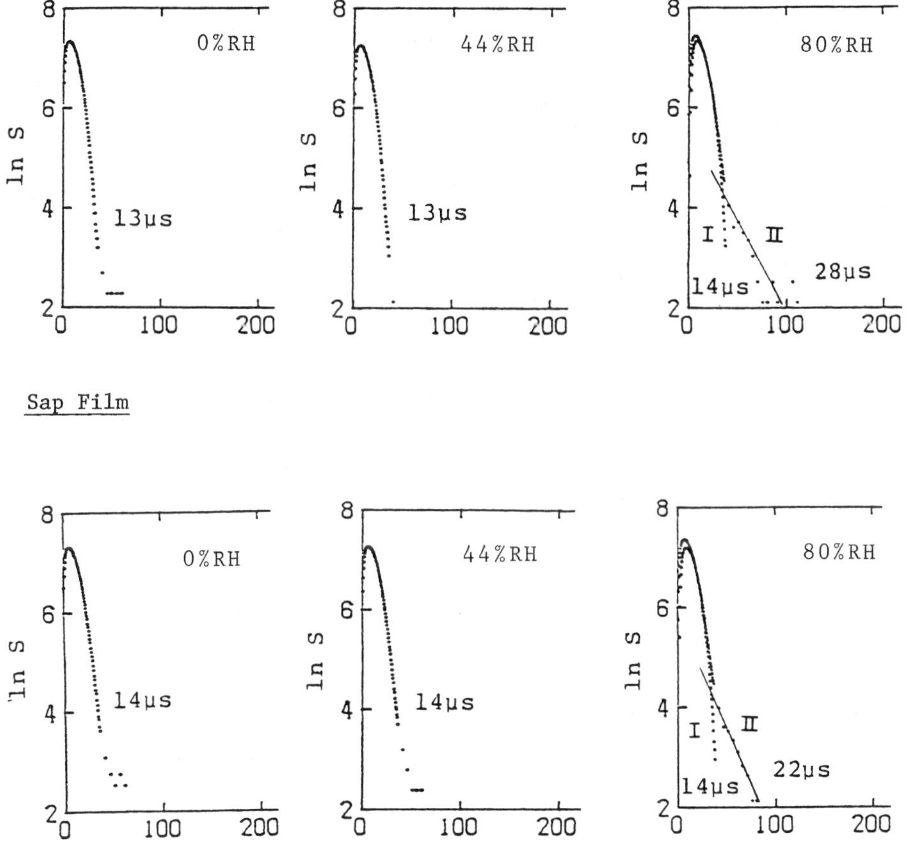

Sap Film

Lacquer Film

Fig. 8·2 FID Curves for the Sap and Lacquer Films(40°C) of
Rhus succedanea(Vietnam)

175

CONCLUSION

It should be noted that the lacquer film is composed of polysaccharides, glycoproteins and polymerizable urushiol which have different solubility parameters, and it makes a clear film in appearance, although an inhomogeneous structure of the film is found with electron microscopy.

From a series of studies on Oriental lacquers, the following processes may participate in the lacquer preparation: (1) water evaporation, followed by deposition of the polysaccharides from water droplets in the oil(urushiol) phase (2) association of the polysaccharides with the glycoproteins dispersing in the urushiol phase, or urushiol-grafted glycoproteins that are made from urushiol and glycoproteins. These phenomena seem to occur mainly due to secondary bond forces of the constituents which are enhanced partly by occurrence of the reaction between them.

The difficulty of blending different kinds of polymeric materials is well known, and the technique of grafting or blocking of polymers has been applied to improve mixing of the polymers concerned. However, modification seems to be limited to non-polar or less hydrophilic polymers such as acrylonitrile-butadiene-styrene copolymers. The study of the Oriental lacquers concludes that polar hydrophilic polymeric material as cellulose, hydroxyl, carboxyl, amide, or amino-group containing polymeric material should be used as the renewable components of the polymer matrix with a consideration of the reinforcement against internal stress caused by the water sorption.

REFERENCES

1. JIS K5950
2. G.Sawaguchi, Nippon Shikko No Kenkyu, Bijutsu Shuppan Sha(Tokyo), (1966)
3. R.Oshima and J.Kumanotani, Carbohydr. Res,:127.43(1984).
4. U.K.Lammli, Nature,227:680(1970)
5. R.Majima, Chem. Ber,:55B:172(1922),
   S.V.Sunthanker and C.R.Dawson, J. Am. Chem. Soc.,76:5070(1956),
   Y.Yamauchi, R.Oshima and J.Kumanotani, J. Chromatgr.,243:71(1982).
6. J.Kumanotani, Application of Renewable-resorce Materials, eds. C.E.
   Carraher and L.H.Sperling (Plenum Publishing Co., N.Y.) p 225 (1983).
7. S.Prager and F.A.Long, J. Am. Chem. Soc.,73:4072(1951)
8. H.Sasabe, Research Electrochemical Labratory Report No. 721(1971), p 13.
9. K.Inoue and S.Hoshino, J. Polym. Sci. Phs. Ed.,14:1513(1976) and
   17:1517(1976).

SIMULTANEOUS INTERPENETRATING NETWORKS BASED ON VERNONIA OIL POLYESTERS
AND POLYSTYRENE: II. A COMPARISON OF THE REACTIVITIES OF VERNONIA OIL
AND CASTOR OIL TOWARD THE FORMATION OF POLYESTERS

A. M. Fernandez, J. A. Manson*, and L. H. Sperling**

Polymer Science and Engineering Program, *Departments of
Chemistry and Metallurgy and Materials Engineering
**Department of Chemical Engineering
Materials Research Center #32
Lehigh University
Bethlehem, PA   18015

INTRODUCTION

Triglyceride oils such as linseed, oitica, tall, castor, and tung
have long been used in varnishes, alkyds and other coating formulations.
These days, new polymers based on such oils represent an alternative
approach to reducing the dependence on petrochemical derivatives (1-7).
Triglyceride oils can be polymerized to form interpenetrating polymer
networks, IPN's. An IPN may be defined as a combination of two polymers in
network form, at least one of which is polymerized and/or cross-linked in
the immediate presence of the other (3). The IPN's of the present study
are simultaneous interpenetrating networks, SIN's. This involves the mixing
of two monomers or prepolymers in an early stage of their reaction, followed
by the formation of both networks. The reactions proceed simultaneously,
through independent mechanisms, i.e. step growth and chain growth. Recently,
experimental oils from potential new oil seed crops were investigated. In
particular, naturally epoxidized vernonia oil, as well as chemically
epoxidized linseed, crambe, lunaria, and lesquerella oils were used (2,4-6).
Vernonia oil comes from a wild plant native to Kenya, Africa. The synthesis
and characterization of vernonia oil SIN's formed the first paper in this
series (5).

The purpose of this paper is twofold. First, the reactivities of
castor oil and vernonia oil toward the formation of polyesters will be
compared. Secondly, the synthesis of SIN's based on vernonia oil-sebacic
acid polyester and polystyrene, as well as their viscoelastic behavior and
their morphology will be presented.

EXPERIMENTAL

Materials

Elastomeric polyester homopolymers synthesized from castor oil,
structure (1), and vernonia oil, structure (2), with sebacic acid were
studied. Both castor oil and vernonia oil are triglycerides. They are
comparable in having the same number of carbon atoms, and a double bond
in identical positions. However, they bear different functional groups;

castor oil, hydroxyl groups; and vernonia oil, epoxy groups. Also, they differ in the nature and amount of diluent components. Nevertheless, on reacting with sebacic acid, polyester materials will result. Neglecting the influence of the diluents toward their reactivities, the reaction kinetics and material properties will be examined and compared.

For reaction purposes, the composition of castor oil is considered as the triglyceride of ricinoleic acid. As discussed above, the interesting reactive sites are three hydroxyl groups. This compound is approximately 90% pure in ordinary pressed castor oil.

Vernonia oil is a natural triglyceride that contains a bountiful supply of epoxy groups. It consists mainly of the triglyceride of 12,13-epoxy oleic acid. Nuclear magnetic resonance studies indicated that this oil contains 80% epoxy groups (5). Based on the experimental value of 80% epoxy content, an average functionality of 2.4 epoxy groups per molecule was considered (5).

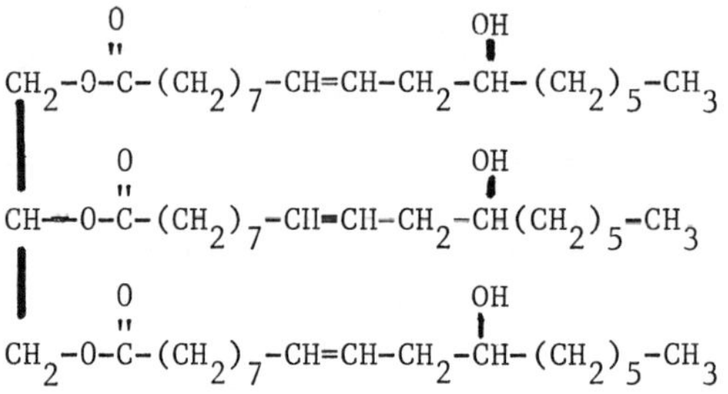

$$
\begin{array}{l}
CH_2{-}O{-}\overset{O}{\overset{\|}{C}}{-}(CH_2)_7{-}CH{=}CH{-}CH_2{-}\overset{OH}{CH}{-}(CH_2)_5{-}CH_3 \\[2ex]
CH{-}O{-}\overset{O}{\overset{\|}{C}}{-}(CH_2)_7{-}CH{=}CH{-}CH_2{-}\overset{OH}{CH}(CH_2)_5{-}CH_3 \\[2ex]
CH_2{-}O{-}\overset{O}{\overset{\|}{C}}{-}(CH_2)_7{-}CH{=}CH{-}CH_2{-}\overset{OH}{CH}{-}(CH_2)_5{-}CH_3
\end{array} \qquad (\underline{1})
$$

$$
\begin{array}{l}
CH_2{-}O{-}\overset{O}{\overset{\|}{C}}{-}(CH_2)_7{-}CH{=}CH{-}CH_2{-}\overset{O}{\overset{\triangle}{CH{-}CH}}{-}(CH_2)_4{-}CH_3 \\[2ex]
CH{-}O{-}\overset{O}{\overset{\|}{C}}{-}(CH_2)_7{-}CH{=}CH{-}CH_2{-}\overset{O}{\overset{\triangle}{CH{-}CH}}{-}(CH_2)_4{-}CH_3 \\[2ex]
CH_2{-}O{-}\overset{O}{\overset{\|}{C}}{-}(CH_2)_7{-}CH{=}CH{-}CH_2{-}\overset{O}{\overset{\triangle}{CH{-}CH}}{-}(CH_2)_4{-}CH_3
\end{array} \qquad (\underline{2})
$$

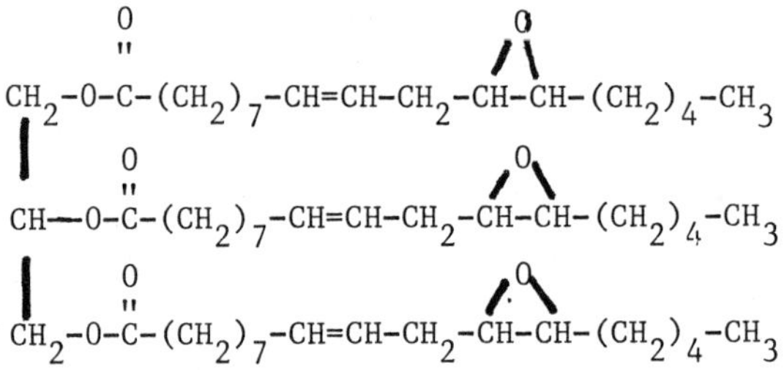

When castor oil and vernonia oil react with difunctional sebacic acid, a three dimensional esterification reaction occurs, forming a network. Hydroxyl and epoxy groups react to form cross-linked polyesters with dibasic acids, but by different mechanisms. Sebacic acid, itself commercially obtained from castor oil (and hence constituting a renewable resource), serves as an excellent crosslinker.

The reaction between a dibasic acid, such as sebacic acid, and an epoxy bearing oil can be written as eq. ($\underline{3}$).

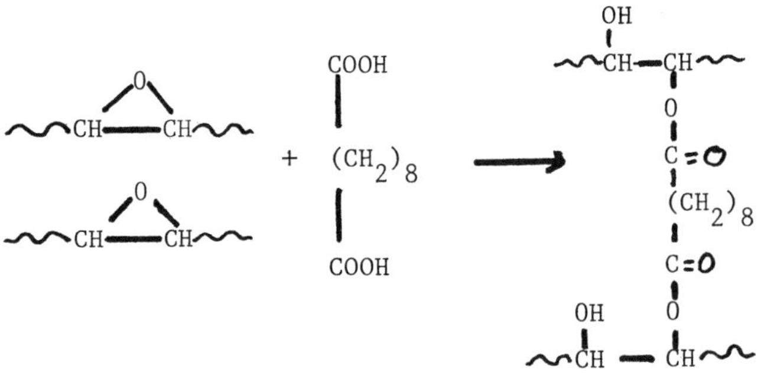

$$
\begin{array}{ccc}
\overset{O}{\overset{\triangle}{\sim\sim CH{-}CH\sim\sim}} & & \overset{OH}{\underset{}{\sim\sim CH{-}CH\sim}} \\
& COOH & O \\
& | & \| \\
+ & (CH_2)_8 & \longrightarrow \quad C{=}O \\
& | & | \\
\overset{O}{\overset{\triangle}{\sim\sim CH{-}CH\sim\sim}} & COOH & (CH_2)_8 \\
& & C{=}O \\
& & \overset{OH}{\sim\sim CH}{-}\overset{O}{CH\sim\sim}
\end{array} \qquad (\underline{3})
$$

Simultaneously with the formation of ester groups, free hydroxyl
groups are formed. Given favorable reaction conditions, these OH groups
constitute a new locus for further reactions. The reaction between castor
oil and sebacic acid forms a regular polyester; along with ester groups,
water is formed as a by-product which has to be removed from the reaction
medium. The eight methylene units in sebacic acid in eq. (3) contribute
to the low glass transition temperature, T_g, of the final products. This
aspect attains importance in the development of novel elastomers for impact-
resistant plastics, and tough leathery products.

Synthesis and Characterization Techniques

Homopolymer Polyesters. Homopolymer vernonia oil-sebacic acid polyester, poly(VO,SA), was obtained by weighing the reaction components sepa-
rately, mixing quickly, stirring, and placing the reaction vessel in a
silicone oil bath at 140°C. The reaction was carried out with continuous
magnetic stirring under a nitrogen atmosphere. Equivalent amounts of the
two substrates were reacted in order to obtain a ratio of unity between
their respective functional groups. Close to the gel point, the prepolymer
was poured into teflon-lined molds. The reaction was completed at 140-
150°C in a nitrogen atmosphere for 30 hrs. Homopolymer castor oil-sebacic
acid polyester, poly(CO,SA) was synthesized in an identical fashion.
Synthetic details are given elsewhere (7).

Vernonia Oil-Sebacic Acid Polyester and Polystyrene, Poly(VO,SA)-SIN-
(S,DVB), Simultaneous Interpenetrating Networks (SIN's). Styrene (S) was
mixed at room temperature with divinylbenzene (DVB) at a 1% concentration
level, and 0.4% of benzoyl peroxide was added as an initiator. Vernonia
oil-sebacic acid prepolymer was charged into a three necked flask, followed
by the styrene polymerization at 70°C. During this stage, the reaction
vessel was stirred under a nitrogen atmosphere. Just before the styrene-
DVB component gelled, the mixture was poured into preheated molds and the
reaction continued at 80°C in an oven for 30 hrs. After this time, the
temperature was raised to 140°C for 24 hrs. to complete the vernonia oil-
sebacic acid network formation.

Polyesterification Kinetics. The reaction progress was followed by
titrating the unreacted carboxyl groups with a standard methanolic potassium
hydroxide solution.

Viscoelastic Measurements. The 10-second shear modulus, G, was
determined with a Gehman torsion stiffness tester. For the elastomeric
polyesters, values of the average molecular weight between crosslinks,
M_c, were estimated from G by means of the theory of rubber elasticity,

$$G = \frac{\rho}{M_c} RT \ + \tag{4}$$

where ρ is the density, R the gas constant, and T the absolute temperature
(in this case, at 373°K, well above the T_g of the polyester homopolymers).
In addition, the crosslinking density, n, which represents the number of
networks chains per unit volume was evaluated by $n = \rho/M_c$.

Swelling Experiments. Vernonia oil and castor oil-based polyesters
were extracted with acetone, and the amount of extractable material determined. The swelling ratio, q, was calculated from the equilibrium swelling
degree

$$q = \frac{\text{swollen volume of the network}}{\text{dry volume of the network}} \tag{5}$$

The swelling experiments were performed at room temperature with acetone. A swelling time of 15 days was employed.

Electron Microscopy. Kato's osmium tetroxide staining technique was employed, which attacks the double bonds of the triglyceride oils. Specimens with cross sections of 0.2 x 0.2mm and lengths ranging from two to ten mm were exposed to osmium tetroxide vapor at room temperature for one week. Portions of the stained specimens were embedded in an epoxy resin, trimmed to a truncated pyramid shape, and microtomed on a Porter Blum MT-2 ultra-microtome using a diamond knife. Ultra-thin sectioning at room temperature to a thickness of 60-80 nm (600-800Å) yielded satisfactory results. Transmission electron micrographs were taken employing a Phillips 300 electron microscope.

RESULTS

At 140°C, the gel point of an equimolecular composition of the poly-(VO,SA) system was observed after about 7 hrs. of reaction (5). At 180°C, for an equimolecular composition of the poly(CO,SA) system, the gel point was observed, on the other hand, only after about 18 hrs. (7). Since water removal is difficult beyond the gel point, final network formation was carried out under vacuum at 80°C. The reader should note that water is not evolved in the vernonia oil reaction. The gel point results were considered in light of two theories. The Carothers approach relates the extent of reaction at the gel point, P_c, to the composition at the stage where the number-average degree of polymerization, X_n, becomes infinite.

For an equimolecular ratio of reactants, P_c is given by

$$P_c = 2/f_{avg} = 2\Sigma N_i / \Sigma N_i f_i \tag{6}$$

where f_{avg} represents the average functionality of the system, and N_i is the number of molecules of monomer i with functionality f_i, and the summations are over all the monomers present in the system. For the poly (VO,SA) system, (taken 2.4 epoxy groups/molecule as the functionality of vernonia oil), equation (7) yields $f_{avg}=2.30$ (7).

Flory and Stockmayer developed a statistical theory to predict the value of P_c at the point where the weight-average degree of polymerization, X_w, becomes infinite. For an equimolecular ratio of reactants,

$$P_c = 1/(f-1)^{1/2} \tag{7}$$

The quantity f represents the functionality of the branch units; e.g. of the monomer with functionality greater than two. Table 1 gives the experimental and predicted extents of reaction at the gel point for both homopolymer polyesters.

The shear modulus-temperature curves of the two homopolymer polyesters, polystyrene network, and two SIN's based on vernonia oil are shown in Figure 1. The polyester materials poly(CO,SA), (sample 1); and poly(VO,SA), (sample 2); and the polystyrene network (PS with 1% DVB), (sample 5), have sharp, well defined single glass transition temperatures near -60, -50 and 100°C respectively. Two rather broad transition temperatures are observed for the SIN's as a consequence of their two-phase morphology. Taking the parent homopolymer T_g's as reference (samples 2 and 5), an inward shift of both phase transitions is observed. The 60% polyester/40% polystyrene SIN, poly(VO,SA)-SIN-(S,DVB)) 60/40, (sample 3), show T_g's near -40 and

Table 1. Comparison of the Extents of Reaction at the Gel Point, P_c, for the Vernonia Oil and Castor Oil Based Polyesters (5,7)

Polymer Designation	Functional[a] Group Composition Ratio	Extent of Reaction, P_c		
		Exp.	Carothers	F-S
Poly(VO,SA)	1.0	0.619	0.917	0.845
Poly(CO,SA)	1.0	0.760	0.870	0.767

[a]For Vernonia oil a functionality of 2.4 was assumed. The functional composition ratio is: Epoxy/COOH=1.0. For castor oil a functionality of 2.7 was assumed. The functional group composition ratio is: OH/COOH-1.0.

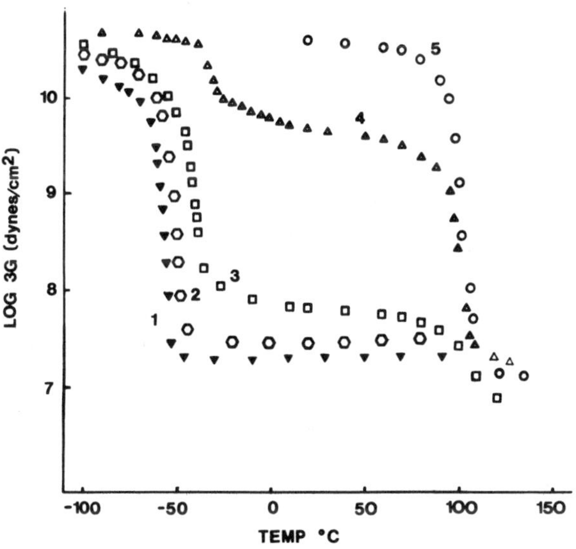

Figure 1. Viscoelastic behavior of polymerized triglyceride oils and their SIN's. Note two glass transitions in the SIN compositions. 1. Poly(CO,SA); 2. Poly(VO,SA); 3. Poly(VO,SA)-SIN-Poly(S,DVB), 60/40; 4. Poly(VO,SA)-SIN-Poly(S,DVB), 15/85; 5. PS network.

$90^{\circ}C$, while the 15% polyester/85% polystyrene SIN, poly((VO,SA)-SIN-(S,DVB)) 15/85, (sample 4), shows T_g's near -30 and $98^{\circ}C$. Above $100^{\circ}C$, samples 3 and 4 show a lower rubbery modulus, probably due to the lower crosslinking density of their continuous phase (elastomeric phase).

In between the two transitions, a lower than expected modulus is found, suggesting that the oil phase is continuous. Values of $E=10^{8}-10^{9}$ dynes/cm^{2} indicate leathery behavior. Table 2 shows the estimated values of the plastic phase transition temperature, T_g, the rubber phase T_g, as well as other parameters for the material studied, determined from the modulus-temperature curves. Both elastomer polyesters are yellow in color, transparent, soft and of low tensile strength. The SIN materials are opaque white, having tough elastomeric properties. As a result of the very low miscibility between the components, all the SIN's exhibit a two-phase morphology. Figure 2 shows electron micrographs of sample 3 and a poly{(VO,SA)-SIN-(s,DVB)} 50/50 material. Given their very similar compositions, both samples have almost the same properties. In both cases, the vernonia oil elastomeric phase, stained, is continuous. The polystyrene phase domains show a bimodal distribution. The large domains size range is $0.25-0.45\mu$, while the smaller domains are in the range of $0.10-0.15\mu$, being the more numerous of the two. Figure 3 shows the morphology of sample 4. Even at 15/85 composition, the elastomeric phase remains continuous. As before, the PS domains show a bimodal distribution. However, the difference in the domains dimensions is more pronounced. The larger domains size range is $2.0-2.8\mu$ and the smaller domains size range is $0.1-0.15\mu$, and as before are the more numerous of the two. No phase inversion takes place.

DISCUSSION

Polyesterification Kinetics

Examination of Tables 1 and 2 show that the Flory-Stockmayer theory offers a more accurate treatment of the problem. It is well known that intramolecular cyclization reactions with the production of elastically ineffective loops are "destructive" and require the polymerization to be carried out to a greater extent of reaction to reach the gel point; as a consequence, higher than predicted values of P are observed. The same result is observed when the system has a non-equivalent concentration of reactive functional groups.

The opposite situation, in which the gel point value occurs before the predicted point, has to be considered. Situations where the gel point value occurs prior to the predicted point may be due to interchain connectivity reactions occurring during the polymerization, caused when a different functionality is developed. In the case of vernonia oil-sebacic acid reaction, free hydroxyl groups are formed, which in turn, constitute a new locus of reaction.

It must be noted, however, that M_c values of the order of 3000-4000 gms/mole are shown in Table 2. Noting that the glyceryl moiety should be a trifunctional crosslink point at full reaction, values below 1000 gms/mole would be expected. Hence, many reaction sites must be "wasted" in yet unknown ways.

Morphology and Extent of Mixing Between the Phases

The electron micrographs show that the elastomeric phase remains continuous for sample 4 (15/85 composition). At this composition, the system poly(CO,SA)/P(S,DVB) shows a phase inverted morphology (7), with a continuous polystyrene phase and a cellular structure within the elastomer droplets, resembling polybutadiene/polystyrene high impact plastics.

Table 2.   Characterization Data from Modulus-Temperature and Swelling
            Experiments

Polymer Sample Designation	T_g ^oC		q	M_c (g/mol)[a]	(g/cm^3)
	Elast.	Plast.			
1	−60	---	2.60	4,150	0.980
2	−50	---	2.06	3,020	1.024
3	−40	90			
4	−30	98			
5	---	100			

[a]Determined at 100°C.

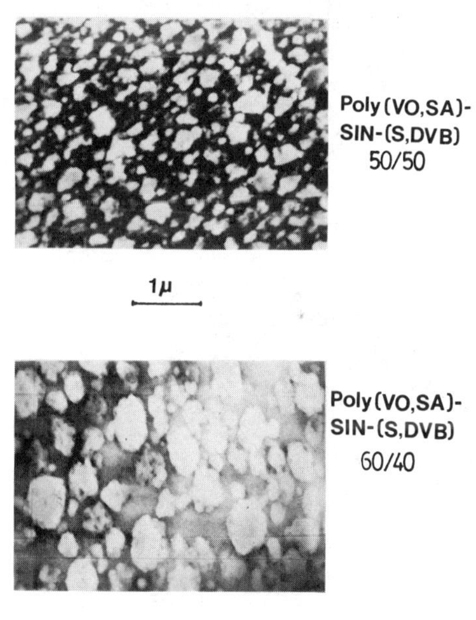

Figure 2.   Transmission electron micrographs of mid-range SIN compositions
of polymerized vernonia oil and crosslinked polystyrene.

Table 3. Economic Development Locations

Oil	Regions
Castor	India and Brazil Columbia, S.A.
Lesquerella (Bladder pods of pop weeds)	Arizona, USA
Vernonia	Zimbabwe and Kenya, Africa

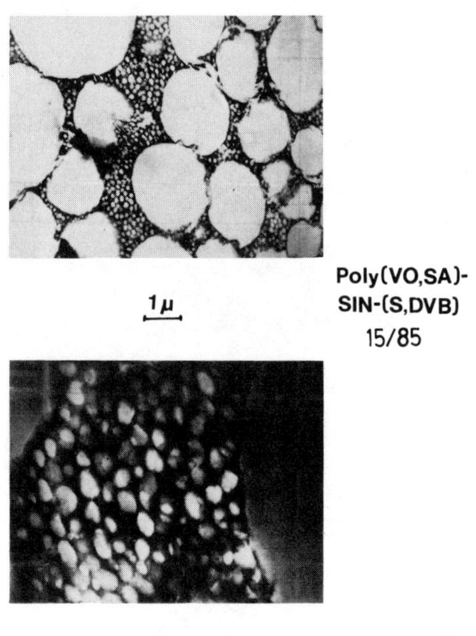

Figure 3. A 15/85 SIN composition, illustrating the bimodal distribution of domain sizes. Poly(VO,SA)-SIN-Poly(S,DVB).

The extent of mixing between the phases is qualitatively reflected by the viscoelastic behavior of the material. The SIN's do not show a considerable broadening of their T_g's values. Previous studies made on the poly(CO,SA) system indicate that, in general, the degree of molecular mixing for this kind of material ranges from 0-20% (7).

Due to the evolution of the hydroxyl groups, vernonia oil is more reactive towards sebacic acid than castor oil. To some extent, this causes an increased rubbery modulus, increased T_g, and a decreased swelling.

<u>Industrial Status</u>

The state of industrial development of vernonia oil is illustrated in Table 3. Vernonia grows in a broad expanse throughout central Africa, and actually consists of several species of plants. Vernonia plantations are now being developed in Zimbabwe. Also shown in Table 3 are industrial patterns of two other special functional group oils, Castor and Lesquerella. Castor oil is grown commercially on a world wide basis, but especially in India and Brazil. Countries such as Colombia, S.A. are highly interested in also growing this oil which is widely used in polyurethane elastomers, paints, and adhesives.

Lesquerella oil, like castor, bears hydroxyl groups, but has nine $-CH_2-$ groups in between the glycerol moiety and the hyroxyl rather than the seven borne by castor. This feature gives it a slightly lower glass transition temperature. It grows in the desert regions of western United States, particularly in Arizona, and consists of several species of wild flower. Lesquerella is known to the residents of Arizona by the names Pop Weeds or Bladder Pods. This oil could bring a new industry to the desert areas of the Western U.S. All three of these oils contain a special functionality, hydroxyl or epoxy, which makes them rare in the huge number of triglyceride oils.

CONCLUSIONS

When reacted with equimolar quantities of sebacic acid, vernonia oil forms a somewhat tighter network than castor oil. The SIN's of vernonia oil and sebacic acid with styrene and divinyl benzene exhibit two glass transitions which are shifted inwards, suggesting phase separation with some molecular mixing. The polymerized oil phase forms the continuous phase.

ACKNOWLEDGMENT

The authors wish to thank the National Science Foundation for support under Grant No. CPE-8109099.

REFERENCES

1. (a) L. H. Princen, J. Coat. Technol. $\underline{49(12)}$, 88 (1977);
    (b) J. Am. Oil Chem. Soc. $\underline{56(9)}$, 845 (1979).
2. L. H. Sperling and J. A. Manson, JAOCS $\underline{60}$, 11 (1983).
3. L. H. Sperling, "Interpenetrating Polymer Networks and Related
    Materials," Plenum, New York, 1981.
4. S. Qureshi, J. A. Manson, L. H. Sperling, and C. J. Murphy, in
    "Polymer Applications of Renewable-Resource Materials," C. E.
    Carraher, Jr. and L. H. Sperling, Eds., Plenum Press, New York,
    1983.
5. A. M. Fernandez, C. J. Murphy, M. T. DeCrosta, J. A. Manson,
    and L. H. Sperling, in: "Polymer Applications of Renewable-
    Resource Materials," C. E. Carraher, Jr. and L. H. Sperling,
    Eds., Plenum, New York, 1983.
6. L. H. Sperling, J. A. Manson, and M. A. Linne, J. Polymeric
    Materials $\underline{1(1)}$, 54 (1984).
7. N. Devia, J. A. Manson, L. H. Sperling, and A. Conde, Macro-
    molecules $\underline{12(3)}$, 360 (1979).

POLYMERS FROM RENEWABLE RESOURCES:

CROSSLINKING AND THERMAL BEHAVIOR

Sukumar Maiti[a], Sabyasachi Sinha Ray,
and Achintya K. Kundu

Polymer Division, Materials Science Centre
Indian Institute of Technology
Kharagpur 721302   India

INTRODUCTION

The scarcity and high price of crude oil have resulted
in research and development activities worldwide for the use
of renewable resources as feedstocks for polymers.[1] Recently
gum rosin from the exudate of pine trees has been developed as
a raw material for copolyimides and other polymers.[3-10] Rosin
was made to react with maleic anhydride to form a rosin-maleic
anhydride Diels-Alder adduct (RMA). RMA has been found to be
a suitable substitute for trimellitic anhydride (TMA), a key
chemical for high temperature resistant polymers such as poly-
esterimides, polamideimides and other copolyimides.[2-5] Schuller
and Lawrence have also reported some polyamideimides using RMA
as the raw material.[6-8] Synthesis of polyamideimides from
maleopimaric acid and diamines has been reported by Aldrich.[9]
Penczek and coworkers have reported the synthesis of polyester-
imides by reacting maleopimaric acid with amino alcohols.[10]

In previous reports[2-5] polyamideimides have been prepared
by the formation of an intermediate imidodicarboxylic acid.
Polyamideimides were prepared by reacting the monoacid chloride
derivative of RMA with diamines. Besides maleic anhydride, other
dienophile such as acrylic acid has been used to form the Diels-
Alder adduct which is used for the synthesis of a polyamide and
a polyester by reacting the diacid chloride derivative of the
adduct with a diamine or a diol.

Polymers from rosin, unlike the polymers derived from TMA,
offer an interesting possibility of subsequent crosslinking of
the polymer formed. The residual unsaturation in the hydrophen-
anthrene ring of the rosin moiety has been utilized for crosslinking. As expected the crosslinked polymers from rosin exhibit
improved thermal and chemical resistance. Since the unsaturation
site in the rosin moiety present in such polymers is sterically

[a]Present address: Plant Polymer Research, USDA Northern
Regional Research Center, Peoria, IL   61604

hindered the onset of crosslinking reaction requires rather
higher temperatures.

EXPERIMENTAL

Preparation of Monoacid Chloride of Maleopimaric Acid (RMA-Cl)

RMA was dissolved in excess thionyl chloride and refluxed
for 6 h. Excess thionyl chloride was removed by azeotropic
distillation with benzene. The crude product was purified by
crystallization from chloroform.[6] The yield of the product
was 92%, color brown, m.p. 170°C.

Preparation of Rosin-Acrylic Acid Adduct (RAA)

Ten g rosin and 2.4 g acrylic acid containing 0.5% hydroquinone (to inhibit polymerization of acrylic acid) were heated
under a nitrogen atmosphere at 140°C for 2 h, 160°C for 2 h and
finally at 175°C for 1 h. After the reaction was over the
solid mass was cooled and washed with water to remove unreacted
acrylic acid. The adduct was finally purified by dissolving in
diethyl ether solution and reprecipitated by petroleum either.[12]
The yield of the product was 70%, color white, m.p. 220°C.

Preparation of Diacid Chloride of Rosin-Acrylic Acid Adduct (RAA-Cl)

RAA was dissolved in escess thionyl chloride and refluxed
for 10 h. After the reaction was over excess thionyl chloride
was removed by azeotropic distillation with benzene. The crude
product was crystallized from chloroform.[12] The yield of the
product was 90%, color brown, m.p. 150°C.

Synthesis of Polyamideimides

Polyamideimides were synthesized by reacting monoacid chlorides of rosin maleic anhydride adduct (RMA-Cl) with different
diamines in a 1:1 mole ratio in the presence of a triethylamine
acid acceptor by heating at 30°C for 2 h, 60°C for 2 h and
135°C for 4 h using a mixture of dimethyl formamide (DMF) and
N-methyl 2-pyrrolidone (NMP) 3:1) containing 4% LiCl.[11] After
the reaction is complete the polymer was separated by precipitation in ice/water. The product was prified by repeated precipitation from DMF solution by methanol. The diamines used were
ethylenediamine, hexamethylenediamine, p-phenylenediamine,
diaminodiphenylmethane, diaminodiphenyl-sulfone and p,p'-
(bis-amino cyclohexyl) methane. The yields of the products
were found to be 70-80%.

Synthesis of Polyamide from Diacid Chloride of Rosin-Acrylic Acid Adduct

Polyamide was synthesized by reacting the diacide chloride
of rosin acrylic acid adduct (RAA-cl) with hexamethylenediamine
(RAA-Cl:diamine = 1:1 mole ratio) in the presence of a triethylamine acid acceptor by heating under a nitrogen atmosphere at

30°C for 2 h, 60°C for 2 h and 100°C for 4 h in a reaction medium of 4% LiCl in DMF : NMP = 3:1. After the end of the reaction the reaction mixture was poured in ice/water and the polymer was separated by filtration. The polyamide was purified by repeated precipitation from DMF solution by methanol.[12] The yield of the product was 70%.

Synthesis of Polyester

Polyester was synthesized by reacting RAA-Cl with hexane-diol in the mole ratio of 1:1 in the presence of triethylamine as the acid acceptor under the same reaction condition as used in the case of polyamide.[12] The polyester was purified following the same procedure as described in the case of polyamide.[12] The yield of the product was 68%.

Crosslinking of the Polymers

The polymer containing 1% initiator (w/w) was coated on glass plates and kept in a closed container fitted with nitrogen inlet and outlet tubes. The glass plates were heated at the appropriate temperature in the presence of nitrogen. The percentage of crosslinking was determined by taking out the coated glass plates from the crosslinking chamber and separating the insoluble fraction from the soluble one by the sol-gel analysis technique as described below.

Analysis of Sol-Gel Fraction

The crosslinked polymer was subjected to the soxhlet extraction with NMP as the solvent for 96 h. The insoluble product was dried to constant weight.

Swelling Behavior

Crosslinked polymer (0.1 g) was equilibrated with NMP for 48 h at a constant temperature of 30°C. The sample was separated by filtration, blotted by filter paper and subsequently washed quickly with diethylether. The weight of the swollen sample was determined immediately. The swollen sample was dried to constant weight under vacuum at 100°C.

RESULTS AND DISCUSSION

The structures of the polymers are shown in Fig. 1. Physical characteristics of the polyamideimides, the polyamide and the polyester are shown in Table 1.

Molecular weight of a few samples of the polyamideimide from rosin has been determined by vapor pressure osometry. Number average molecular weight of the polymer ranges from 13,000 to 16,500 corresponding to the inherent viscosity values of 0.31 dl/g to 0.37 dl/g (Table 1).

Table 1. Physical Characteristics of the Polyamideimides, Polyamide and Polyester from Rosin[a]

Polymer	Reactants		Inherent viscosity (dl/g)[b]	IR absorption band for olefinic double bond appeared at (cm^{-1})
PAI-1	RMA-Cl	Ethylenediamine	0.33	1625
PAI-2	RMA-Cl	Hexamethylene-diamine	0.35	1625
PAI-3	RMA-Cl	p-Phenylenediamine	0.31	1630
PAI-4	RMA-Cl	Diaminodiphenyl-methane	0.31	1625
PAI-5	RMA-Cl	Diaminodiphenyl-sulfone	0.28	1620
PAI-6	RMA-Cl	p,p'-Bix(amino-cyclohexyl)methane	0.37	1620
PA-1	RAA-Cl	Hexamethylene-diamine	0.30	1610
PE-1	RAA-Cl	Hexanediol	0.28	1610

[a]Data from Refs. 11 and 12.
[b]Measured with 0.5 wt.% solution in DMF at 30°C.

Various routes for the synthesis of the polyamideimides from rosin may be schematically shown in Fig. 2. In the earlier process the polyamideimide was synthesized by three steps: first, a diacide containing an imido group was prepared by the reaction between RMA and p-aminobenzoic acid; next, this diacid was converted to diacid chloride by refluxing with thionyl chloride; and finally the diacid chloride was condensed with a diamine to obtain the polyamideimide.[3,5] The molecular weight of the polymers as evident from the inherent viscosity values was low. This may be due to the stoichiometric defect of the reactants and the presence of impurities in the diacid-chloride because of incomplete reaction.

These drawbacks have been overcome in the present study by synthesizing the rosin imidoamino acid (RIAA) by reacting RMA with a diamine. Since RIAA can polymerize by itself--the amino group can react with its carboxyl group to from an amide linkage-- the stoichiometric imbalance can thus be avoided. However, the polymer obtained by melt-polycondensation is insoluble.[13] The

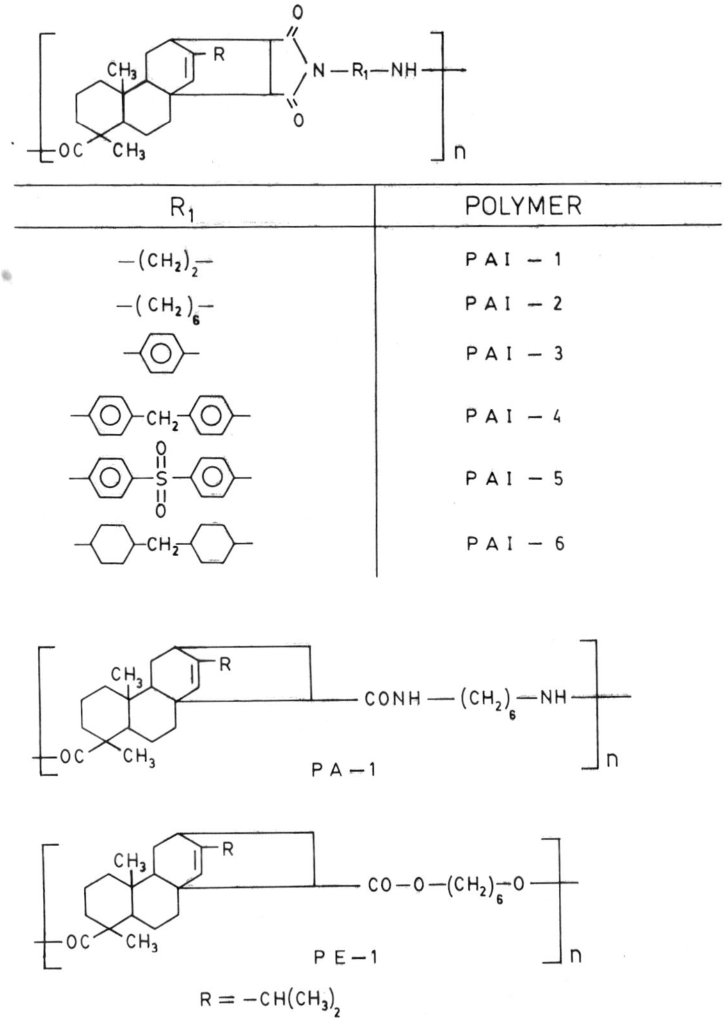

R$_1$	POLYMER
$-(CH_2)_2-$	P A I — 1
$-(CH_2)_6-$	P A I — 2
—⟨O⟩—	P A I — 3
—⟨O⟩—CH$_2$—⟨O⟩—	P A I — 4
—⟨O⟩—SO$_2$—⟨O⟩—	P A I — 5
—⟨O⟩—CH$_2$—⟨O⟩—	P A I — 6

P A —1

P E —1

$R = -CH(CH_3)_2$

Fig. 1. Structure of the polymers from rosin.

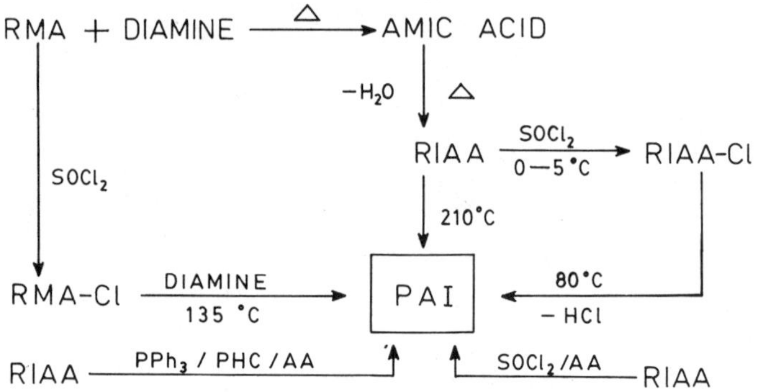

Fig. 2. Various routes for the synthesis of the polyamideimides from rosin.

insolubility is possibly due to a crosslinking reaction onset by
the unsaturation site of the rosin moiety at higher (210°C)
polymerization temperature.

In the solution polymerization of RIAA in NMP in the presence
of thionyl chloride at about 100°C the molecular weight of the
polymer is found to be higher (η_{inh} =0.26 dl/g)[13] than those
obtained earlier.

In order to improve both the yield and the molecular weight
of the rosin polymers, the technique of activated polymerization using triphenyl phosphine, polyhalocompound (PHC) and acid
acceptor (AA) Fig. 2) was adopted. The molecular weight of the
polymers obtained is higher than that of the polymers prepared
using thionyl chloride and acid acceptor; but the yield is
lower.[14,15]

The effect of various solvents and solvent-systems on the
yield and molecular weight of the polymers has also been investigated. It was found that a mixed solvent system of DMF:NMP
in the mole ratio of 3:1 containing 4 wt% LiCl offers the optimum
reaction medium. The improvement in the molecular weight of the
polymer is likely to be due to the increased solubility and high
polarity of the solvent system.[11,15]

Crosslinking Reaction

The polymers, on heating alone, do not crosslink even up to
150°C for extended periods of heating (~10 h). In the presence
of free radical initiators like azobisisobutyronitrile (AIBN)
and benzoyl peroxide (BPO) no crosslinking takes place up to
100°C for 10 h. At 150°C and above dicumyl peroxide (DCP) is
able to initiate crosslinking reaction. However, crosslinking
reaction occurs when the polymers are heated at 200°C even
without an added initiator (Table 2). The polymer can be
fully cured by heating at 250°C for 2 h in the presence of
DCP.

Characterization of Crosslinked Polymers

Solubility Characteristics

The virgin unsaturated polymers are found to be soluble
in highly polar solvents such as DMF, NMP, dimethyl sulfoxide
(DMSO), dimethyl acetamide (DMAC), m-cresol, conc. H_2SO_4
etc.[11-14] But the crosslinked polymers are insoluble in all
the above solvents and only swell in some highly polar solvents
and (10% w/v) sodium hydroxide solution. It is due to the fact
that irreversible covalent crosslinks in polymers are a much
stronger force than polarity or other secondary valence forces
which make the crosslinked polymer insoluble even in highly
polar solvents.

IR Spectra of Crosslinked Polymers

IR spectra of the crosslinked and uncrosslinked polymers
viz. PAI-2, CPAI-2, PAI-3, CPAI-3, PAI-4, CPAI-4, PAI-6, CPAI-6
are shown in Fig. 3. The absence of characteristic IR absorption
bands for olefinic double bond found in the case of unsaturated
polymers (Table 1) and the presence of other characteristic
bands (such as amide, imide, ester as the case may be) confirm
the structures of the crosslinked polymers as well as the
occurrence of crosslinking reaction. IR absorption bands near
725, 1725 and 1775 cm^{-1} are characteristic peaks for imide
groups for polymers (PAI-1 to PAI-6, near 1550 and 1640 cm^{-1}
for amide groups for polymers PAI-1 to PAI-6 and the polyamide
PA-1, near 1750 cm^{-1} for ester group for the polyester, PE-1.

Morphology of the Crosslinked Polymers

The relation between regularity of molecular structure
and crystallinity for polymers has been known earlier. The
extent to which a polymer will crystallize is basically determined by two important factors viz. i) the thermodynamic forces
favoring maximum potential crystallinity at equillibrium, and
ii) by the kinetic forces determining the rate and extent to
which the polymer ma- actually approach such maximum degree of
crystallinity. The factors responsible for the irregularity
in structure in amorphous polymers are loss of symmetry, flexibility of the chains, etc. The growth of crystal structures
from polymer solutions can result in more complex structures
sometimes reminiscent of features found in polymers crystallized
from the melt. However, in the crosslinked polymers further
complexity arise because a number of different structures are
possible and in some cases, crosslinking has a conflicting
effect in crystalline polymers. The introduction of crosslinks in crystalline polymers, generally, reduces the regularity
and crystallinity already present and thus produces a softening
of properties. In such cases, increasing crosslinking and
decreasing crystallinity may cancel each other out to some
extent in their effects upon properties. But for amorphous
polymers crosslinking generally induces some molecular order
thus increases the degree of crystallinity. Thus for amorphous
crosslinked polymers the degree of crystallinity is higher than
that of virgin uncrosslinked polymers. Since the rosin polymers are either almost completely amorphous or at best poorly
crystalline the increase of crystallinity, on heating preferably
in the presence of DCP, may be due to crosslinking. This offers
and indirect support to the onset of crosslinking theory for rosin
polymers.

X-ray diffraction diagrams are shown in Fig. 4 Since
there was no appreciable difference in the crystallinity of
PAI-5 and PAI-6 polymer samples on crosslinking, these were

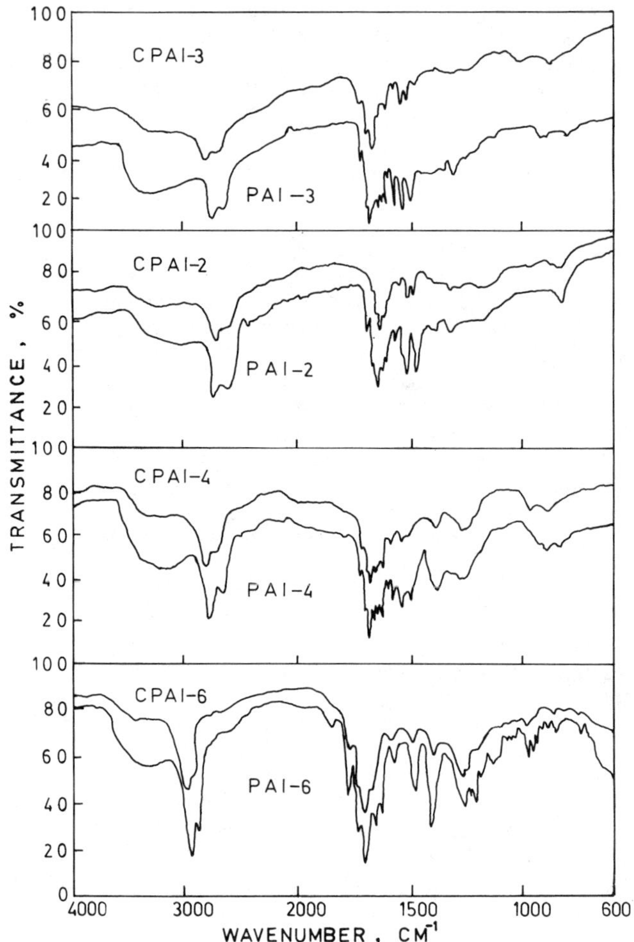

Fig. 3. IR spectra of the polymers from rosin. The crosslinked polymers have been represented by the prefix
'C' to the respective polymer.

excluded from Fig. 4. These polymers are found to be
amorphous even after crosslinking. The presence of bulky
sulfone group in PAI-5 and the possibility of different

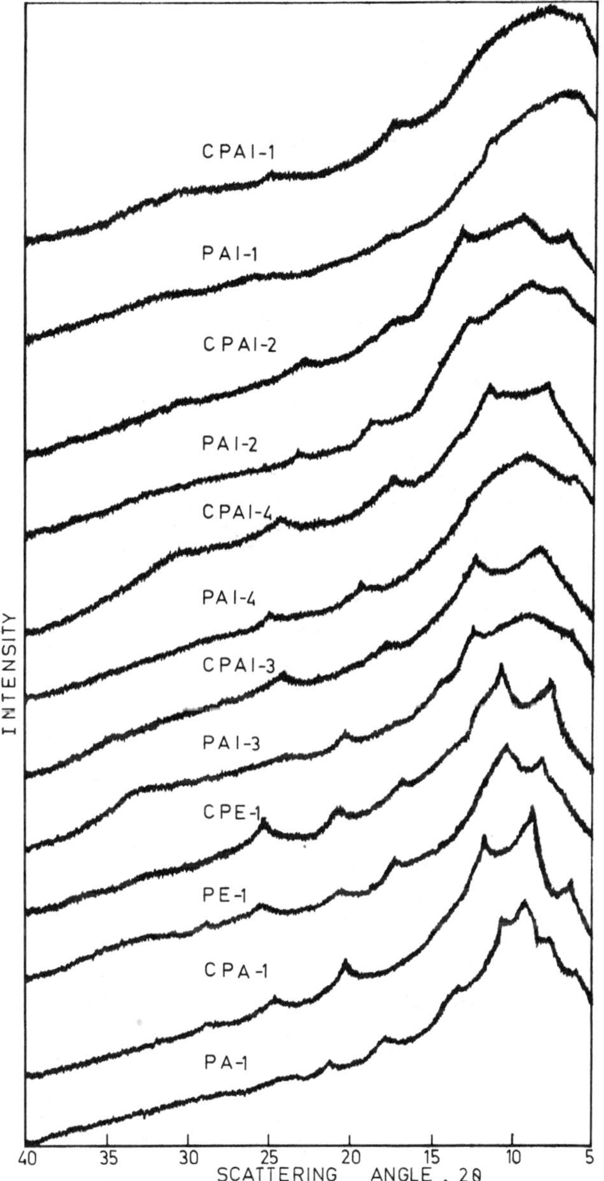

Fig. 4. X-ray diagram of the polymers from rosin.

conformation of cyclohexyl ring present in PAI-6 may prevent
the macromolecules from close packing. On the other hand,
PAI-1 is amorphous when uncrosslinked, but after crosslinking
improvement in crystallinity is noticed. Similar noticeable
increase in the crystallinity is observed after crosslinking
in the cases of PAI-2, PAI-3 and PAI-4 (Fig. 2: the prefix
'c' indicates crosslinked polymer).

Crosslinked polymers exhibit a variety of structures.
The interchain interaction is not only due to the crosslinking
but also due to the secondary valence forces such as
Van der Waal's force and particularly hydrogen bonds. Crosslink density has been determined by equilibrium swelling
method.[13] The effective crosslink density in the polymers,
υ/V, has been calculated from Flory-Rehner equation.[17]

$$\frac{\upsilon}{V} = -\frac{[\ln(1-Vr) + Vr + \chi Vr^2]}{V_0(Vr^{1/3} - Vr/2)} \qquad (1)$$

when υ refers to the number of network chains and V is the
volume of the unswollen network, V_r is the volume fraction of
the polymer in the swollen sample, V_O the molar volume of the
solvent and χ, polymer-solvent interaction parameter. The
volume fraction of the polymer (V_r) in the swollen state is
determined from the knowledge of density of the polymer (ρ_p),
density of the solvent (ρ_s), weight of the swollen specimen
(W_{SW}) and weight of the sample after drying (W_{AD}) using the
equation

$$Vr = \frac{1}{1 + \frac{\rho_P}{\rho_s}\left(\frac{W_{SW}}{W_{AD}} - 1\right)} \qquad (2)$$

χ is found out by the method of Rutkowaska et al.[16] by
measuring the variation of swelling degrees from swelling
measurements with temperature using Flory-Rehner equation[17]
for polymer samples.

$$\frac{-dVr}{dT} = \frac{\chi Vr/T}{\frac{5}{3}\chi Vr + \frac{2}{3} - \frac{1}{1-Vr} - \left[\frac{\ln(1-Vr)}{3Vr}\right]} \qquad (3)$$

The average molecular weight of the network chain between two
crosslinks, M_c, is calculated by using the relation

$$\overline{M}_C = \frac{\rho_P}{\nu/V} \tag{4}$$

The crosslink density, the polymer solvent interaction parameter for NMP solvent and the average molecular weight between
two consecutive crosslinks of the polymer are shown in Table 3.
$\overline{M}_C$ has also been calculated following the relationship[18]

$$T_g - T_g^\circ = \frac{3 \cdot 9 \times 10^4}{\overline{M}_C} \tag{5}$$

where T_g and T_g° are the glass transition temperatures of the
crosslinked and uncrosslinked polymer, respectively. The
results obtained by swelling equilibrium method and thermal
analysis method are summarized in Table 3. The values of $\overline{M}_C$
obtained by both these methods agree fairly well.

Table 3. Properties of Crosslinked Polymers Determined by
Swelling Equilibrium Method

Polymer	χ[a]	$\frac{\nu}{V} \times 10^4$ at 30°C mol/cm^3	$\overline{M}_C \times 10^{-3}$ g/mol Swelling method	Thermal analysis method
PAI-1	0.28	4.5	2.933	3.900
PAI-2	0.298	5.2	2.576	2.600
PAI-3	0.305	5.4 (6.4)[b]	2.500 (1.870)	–
PAI-4	0.33	5.2	2.615	2.600
PAI-5	0.335	5.0	2.640	2.600
PAI-6	0.365	4.8	2.708	2.600
PA-1	0.29	6.1	2.229	1.950
PE-1	0.325	5.9	2.228	1.950

[a]For NMP solvent at 30°C.
[b]Data in the parenthesis indicate the result obtained with
the polymer sample having the inherent viscosity value of
0.14 dl/g.

It is observed from the $\overline{M}_C$ values (Table 3) that on the
average one crosslink is formed per five repeat units of the

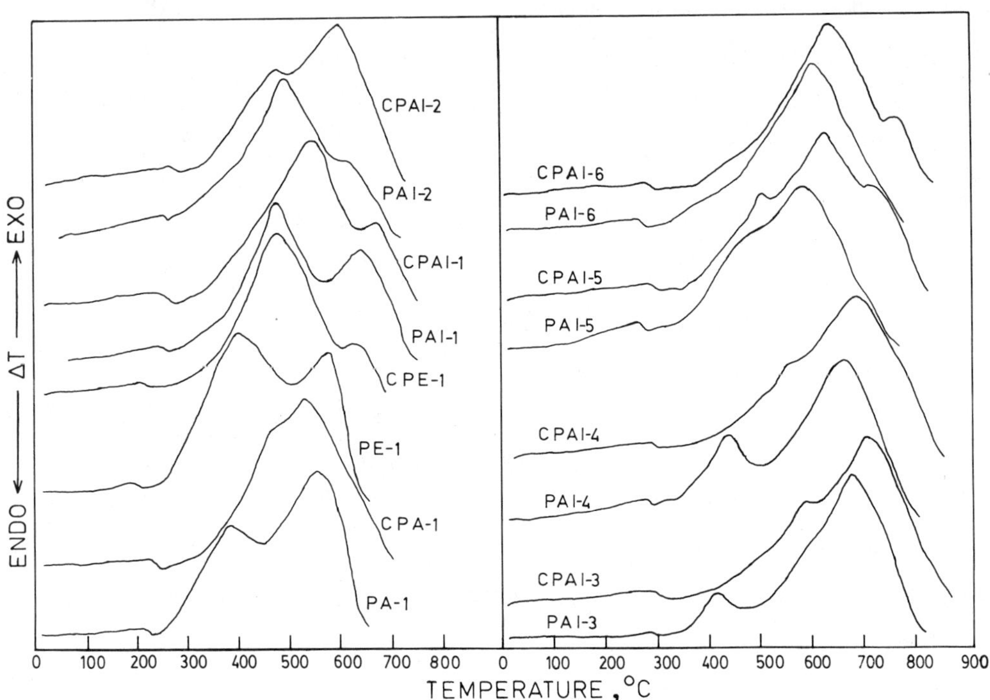

Fig. 5. DTA curves of the polymers from rosin.

polymers PAI-1 to PAI-6. If the $\overline{M}_c$ values can be reduced; i.e., the crosslink density be increased the thermal stability of the polymers is improved. We have studied this aspect further by heating two samples of crosslinked PAI-3 at 180°C for 350 h continuously. One of the samples was crosslinked by heating only at 180°C in the hot air oven while the other was cross-linked in the presence of DCP in the same oven. The weight losses were 6.5% and 4.5%, respectively. This difference in the weight loss values may be due to the higher crosslink density; i.e., lower $\overline{M}_c$ value of the sample crosslinked in the presence of DCP. However, this aspect needs further study before any definite conclusion is drawn.

Thermal Behavior of Crosslinked Polymers

Crosslinks between the polymer chains reduce the segmental mobility and thereby affect the thermal properties. It was found that the T_g values of the virgin uncrosslinked polymers are low compared to the crosslinked polymers. The increase in T_g in the crosslinked polymers may be due to the decrease of the segmental motion of the polymer chains subjected to cross-linking. It is interesting to note that the thermal stability of the crosslinked polymers is higher than the virgin unsaturated polymers.

Thermal behavior studies by DTA, TGA and isothermal aging shows that there is a significant increase in thermal stability of the polymers after crosslinking. DTA curves of the cross-linked polymers are shown in Fig. 5. Changes in the thermal behavior of the polymers before and after crosslinking are shown in Table 4. It is found that the T_g of the polymers is shifted by 15 - 20°C towards higher temperature due to cross-

Table 4. Changes in the T_g and Thermal Decomposition Temperature after Crosslinking of the Polymers from Rosin

Polymer	T_g of the polymers (°C)		Exothermic thermal decomposition started at °C	
	before cross-linking	after cross-linking	before cross-linking	after cross-linking
PAI-1	265	275	300	340
PAI-2	260	275	290	335
PAI-3	285	305	350	400
PAI-4	280	295	345	395
PAI-5	275	290	345	370
PAI-6	280	295	340	390
PA-1	225	245	260	300
PE-1	200	220	250	290

linking. Similarly the exothermic thermal decomposition is shifted by about 40°C towards higher temperature due to crosslinking.

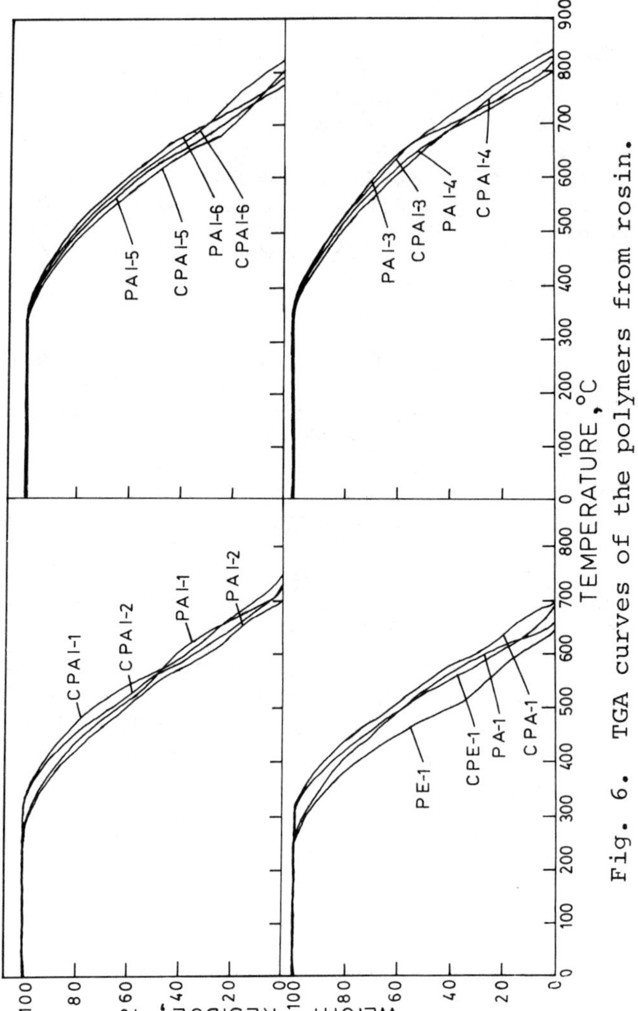

Fig. 6. TGA curves of the polymers from rosin.

    TGA curves of the crosslinked polymers are shown in Fig. 6.
The maximum temperatures at which no weight loss, 50% weight
loss and total weight loss occurred for uncrosslinked and crosslinked polymers are shown in Table 5. From this Table it is
found that thermal stability of the polymers has been increased
after crosslinking. The thermal stability of PAI-3 is found to
be the highest and that of PAI-2 the lowest among all the polyamideimides. This is consistent with the higher thermal stability of the benzine ring, the greater chain symmetry of para-
linked polymer and the lower thermal stability of aliphatic
chains, particularly of hexamethylene moiety.

Table 5. Thermal Stability of the Polymers from Rosin.

Polymer	No weight loss at °C		50% weight loss at °C		Total weight loss at °C	
	before cross-linking	after cross-linking	before cross-linking	after cross-linking	before cross-linking	after cross-linking
PAI-1	260	315	560	565	735	750
PAI-2	250	310	550	560	710	725
PAI-3	335	365	665	690	805	850
PAI-4	325	350	635	640	790	835
PAI-5	315	340	610	615	765	805
PAI-6	325	350	635	640	780	825
PA-1	240	300	535	545	665	695
PE-1	235	295	480	520	650	685

    Isothermal aging of the crosslinked polymers was carried
out to evaluate the thermal stability of the polymers. Isothermal aging of the crosslinked polymers at 250°C showed
1.95 - 2.95% weight loss after 2 h and 3.95 - 5.05% weight loss
after 36 h in air for PAI-1, PAI-2, PA-1 and PE-1. The crosslinked product of all the polymers showed no weight loss at
that temperature even after 36 h.

    Table 6 summarizes the results of the isothermal aging
of the polymers in air at various temperatures (>250°C). The
data in the Table show that crosslinking increases the thermal
stability considerably. The polymers synthesized from aromatic
or alicyclic diamines are thermally more stable than those
prepared from aliphatic diamines. This is consistent with
earlier observation.[19]

Table 6. Isothermal Ageing of the Polymers from Rosin in Air

Polymer[a]	Isothermal ageing Temp. °C	Cumulative weight loss (%) after					
		2 h	4 h	8 h	12 h	24 h	36 h
PAI-1	275	2.55	3.05	3.50	3.90	4.85	6.25
		(1.03)[b]	(1.03)	(1.03)	(1.56)	(2.05)	(2.34)
	300	5.15	6.24	6.98	7.56	9.11	11.24
		(2.25)	(2.68)	(3.01)	(3.26)	(3.85)	(4.50)
	350	(4.50)	(5.15)	(5.95)	(6.75)	(8.95)	(11.10)
PAI-2	275	2.80	3.30	3.90	4.75	5.95	7.45
		(1.08)	(1.11)	(1.11)	(1.65)	(2.12)	(2.40)
	300	5.70	7.10	8.04	8.96	10.90	13.25
		(2.42)	(2.85)	(3.25)	(3.54)	(4.05)	(4.95)
	350	(4.95)	(5.40)	(6.25)	(6.95)	(9.20)	(12.90)
PAI-3	300	2.70	3.05	3.45	3.70	4.55	5.05
	350	(1.75)	(2.05)	(2.45)	(2.85)	(3.75)	(4.95)
PAI-4	300	3.15	3.50	3.82	4.12	5.04	6.06
	350	(2.05)	(2.20)	(2.65)	(3.05)	(4.10)	(5.35)
PAI-5	300	3.50	3.87	4.42	4.80	5.75	7.05
	350	(1.95)	(2.15)	(2.40)	(2.90)	(3.80)	(5.05)
PAI-6	300	3.08	3.44	3.87	4.15	4.98	6.05
	350	(2.55)	(3.20)	(3.95)	(4.90)	(5.65)	(7.25)
PA-1	275	3.80	4.50	5.15	6.25	7.75	9.80
		(1.34)	(1.34)	(1.42)	(1.79)	(2.45)	(2.84)
	300	6.95	8.75	10.10	11.95	15.20	19.40
		(3.05)	(3.43)	(3.98)	(4.65)	(5.25)	(6.05)
	350	(6.55)	(7.20)	(8.45)	(10.10)	(13.75)	(18.55)
PE-1	275	4.85	5.65	6.20	7.50	9.45	12.40
		(1.76)	(1.84)	(1.90)	(2.02)	(2.90)	(3.34)
	300	8.10	11.05	13.10	15.20	21.40	27.10
		(4.02)	(4.64)	(4.98)	(5.65)	(6.50)	(7.15)
	350	(7.50)	(8.95)	(10.65)	(13.70)	(17.85)	(25.05)

[a]No weight loss occurred for other polymers before or after crosslinking at 275°C for PAI-3 to PAI-6.

[b]Values in the parenthesis indicate the result obtained for corresponding crosslinked polymers.

REFERENCES

1. C. E. Carraher, Jr., and L. H. Sperling (Eds.), Polymer Applications of Renewable-Resource Materials, Plenum Press, New York, 1983.
2. S. Maiti, S. Das, M. Maiti and A. Roy, in C. E. Carraher and L. H. Sperling, (Eds.), Polymer Application of Renewable-Resource Materials, Plenum Press, New York, 1983, p. 129.
3. S. Maiti, A. Ray, M. Maiti and S. Das, Org. Coat. Plast. Chem. Prepr. $\underline{45}$, 449 (1981).
4. S. Das, S. Maiti and M. Maiti, J. Macromol. Sci.-Chem., $\underline{A17}$, 1177 (1982).
5. M. Maiti and S. Maiti, J. Macromol. Sci.-Chem., $\underline{20}$,109 (1983).
6. W. H. Schuller, R. V. Lawrence and B. M. Culbertson, J. Polym. Sci. A-1 $\underline{5}$, 2204 (1967).
7. U.S. Patent 3, 5$\underline{2}$2, 211 (1970).
8. U.S. Patent 3, 503, 998 (1968).
9. U.S. Patent 3, 554, 982 (1967).
10. Pol. Patent 70, 466 (1974).
11. S. Sinha Ray, A. K. Kundu and S. Maiti, J. Appl. Polym. Sci. (communicated).
12. S. Sinha Ray, A. K. Kundu and S. Maiti, J. Polym. Mater. (communicated).
13. S. Sinha Ray, A. K. Kundu, M. Maiti, M. Ghosh and S. Maiti, Angew. Makromol. Chem., $\underline{122}$, 153 (1984).
14. S. Sinha Ray, A. K. Kundu, M. Ghosh and S. Maiti, European Polym. J. $\underline{21}$ 131 (1985).
15. S. Sinha Ray, A. K. Kundu and S. Maiti, J. Polym. Sci. Polym. Chem. (communicated).
16. M. Rutkowaska and A. Kwiatkowski, J. Polym. Sci. Polym. Symp. $\underline{53}$, 141 (1975).
17. P. J. Flory and J. Rehner, J. Chem. Phys., $\underline{11}$, 521 (1943).
18. L. E. Nielsen, Revs. Macromol. Chem. $\underline{4}$, 77 (1970).
19. M. Maiti, Ph.D. Thesis, Calcutta University, 1985.

SECTION V – PROTEINS AND LEATHER

THE OPTIMIZATION OF THE MECHANICAL PROPERTIES

OF REINFORCED COLLAGEN FILMS

Richard O. Mohring and Ferdinand Rodriguez

School of Chemical Engineering, Olin Hall
Cornell University
Ithaca, N.Y. 14853

INTRODUCTION

The general public has become well-exposed to the word collagen in
recent years. Unfortunately, the scientific significance of this protein
has not been the focus of attention. It is, rather, the world of adver-
tising, particularly of cosmetics, that has made collagen a household
word. A variety of skin creams and lotions containing collagen or collagen amino acids are aimed at those who would postpone the visual effects
of aging (cracks and wrinkles). The efficiency of the treatments is open
to question. However, collagen as an ingredient of biocompatible materials in sutures and skin grafts has a sound basis.[1] Collagen, a fibrous
protein, is a major constituent of skin and animal hides. It can also be
found in tendon, cartilage, and other connective tissues, and comprises
about 30% of all the proteins in mammals.[2-4]

The collagen molecule itself is made up of two major parts: tropocollagen and telopeptides. The tropocollagen section consists of three
individual polypeptide chains wrapped in a right-handed triple helical
structure with a molecular weight of about 300,000. It has an approximate
length of 300 nm and a diameter of 15 nm. The telopeptides are small non-
helical endgroups connected to the tropocollagen section, which comprise a
negligible amount of the total molecular weight of the collagen molecule.
It has been determined, however, that structural differences in these end
groups constitute the major differences between various collagens.[5,6]

The major amino acid constituents in collagen are glysine, proline,
hydroxyproline, and glutamic acid, together making up about 50% of the
total. The amino acids in the tropocollagen region follow a tripeptide
sequence Gly-x-y, where x very often is proline. Tyrosine is the main
component of the telopeptide group which is poor in hydroxyproline.[7]

Most collagen is water-insoluble due to covalent crosslinking between
individual molecules. This crosslinking effect tends to increase with the
age of the collagen.[8] By use of enzyme treatment (e.g. pepsin, proctase),
however, the cross-links can be broken and the collagen solubilized. The
telopeptides, which tend to label the particular collagen, are also removed in this treatment, forming what is called telopeptide-poor tropocollagen. These enzymes, unlike collagenase, specifically attack the

crosslinks and telopeptides, while leaving the main helical structure intact.[9,10] Solubilization studies of various hides have shown that young animal hide solubilize more readily than old and that steer hide is more soluble then cow.[8] Figure 1 shows a schematic diagram of crosslinked collagen and a collagen molecule before and after enzyme treatment.

In the past a great deal of work on collagen has been done indirectly in the leather industry through the study of the tanning of hides, of which collagen is the major constituent. Extracts from animal hides have also been used in the preparation of glues and gelatin.[3] In recent years, however, much work has been done in the field of collagen film-making for use in a wide variety of goods, including meat packaging materials, substrates for enzymes in catalytic reactions, moisture barriers, surgical dressings, and artificial organs and skin.

Medical applications of collagen are being studied in great depth, since collagen is a natural material which interacts well with living tissue. Collagen has been known to promote healing by providing a good covering for burns and wounds, and its absorption rate into the body can be controlled by its crosslink density. Cells have been observed to grow quite readily on collagen and it can also be used as a site for drug fixation.[8] Telopeptide-poor collagen has been shown to exhibit little antigenicity.[5,6] This occurs because antibodies, produced to attack implanted collagen, have sites which are specifically structured to attack the telopeptide region, which has been removed.[5] It is for these reasons that collagen is ideally suited for applications in the medical field.

The use of collagen in the preparation of membranes for hemodialysis has been studied by Kon, et al..[11] Collagen gels and membranes have also been utilized in both eye and aural surgery, and films have been used to coat synthetic plastics premolded in the shape of artificial organs.[8,12,13] The use of collagen fibers for suture material has also received a great deal of study.[14,15]

Treatment of burns and wounds through the use of synthetic materials, such as nylon velour laminated with polyurethane, has had a history of problems due to grafting difficulties and immune response to the graft.[16-18] These difficulties could be alleviated by use of enzyme-treated collagen. The collagen in this state can be stabilized and strengthened through various forms of cross-linking. Crosslinking with UV light and formaldehyde has received attention,[18-21] the latter having been used in arterial and valve grafts.[8] Studies by Cater, however, have shown that glutaraldehyde and acrolein are much more efficient crosslinking agents, and that the resulting collagen network was more stable than those tanned with formaldehyde.[22]

The chemistry of the collagen-glutaraldehyde reaction is not known. A mechanism for the reaction has been proposed by Richards and Knowles, however, based on evidence from NMR spectroscopy.[23] Glutaraldehyde reacts with itself by aldol condensation to form α,β unsaturated aldehydes. It is these unsaturated aldehydes which react with the amino groups in collagen, giving secondary amines (Figure 2). This Michael-type addition should give crosslinks which are stable to hydrolysis.

Since collagen crosslinked with glutaldehyde is stable and more efficient than with formaldehyde, glutaraldehyde would, therefore, be an ideal choice for use in film preparation for burn treatment. Studies by Chapman show promising tensile strengths for these films.[24]

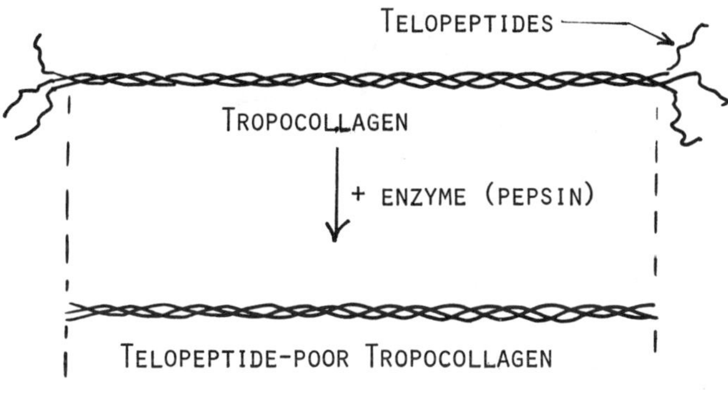

Figure 1.   Collagen structures. (a), Fibrils are assembled by the alignment of individual triple helices (lines with arrows at ends). Most crosslinks are among the endgroups. The endgroups are removed by enzyme digestion (b), which also removes most of the cross-links in the case of calfskin collagen.

(A) Aldol Condensation of Glutaraldehyde (G)

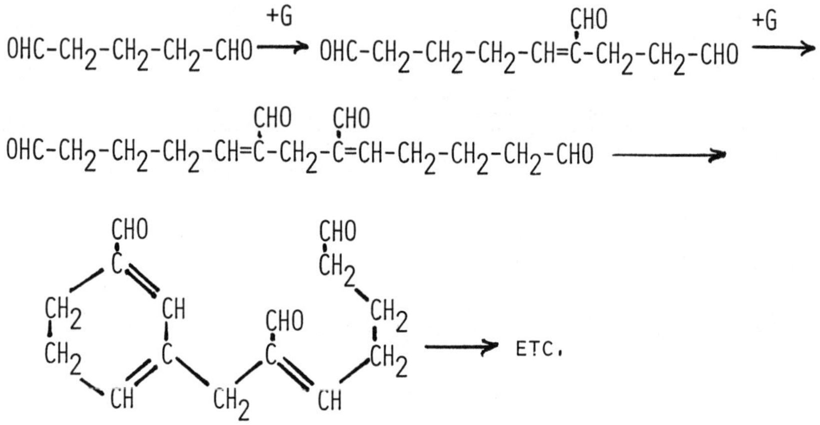

Compound I

(B) Crosslinking Reactions

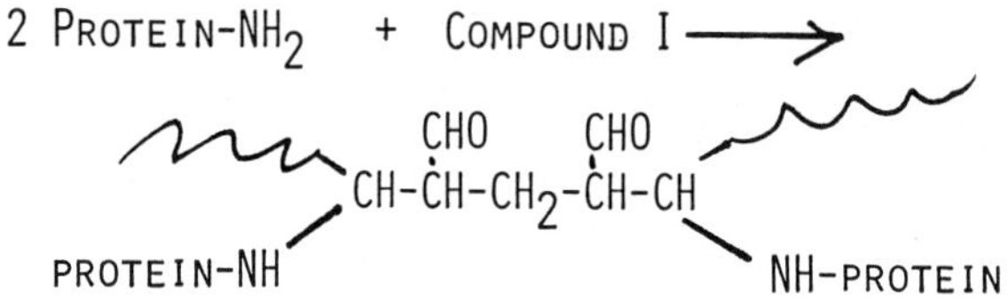

Figure 2. (a) The aldol condensation of glutaraldehyde proceeds in the normal fashion while the (b) crosslinking reaction proceeds with amino groups on the protein.

It has been the purpose of this research to enhance further the mechanical properties of collagen films to obtain strengths applicable to human skin replacement. It is hoped that by investigation of the variables involved in the crosslinking step and by addition of various types of fillers to the collagen prior to crosslinking that these goals can be reached.

EXPERIMENTAL PROCEDURES

Collagen Preparation

A 500 g sample of chopped calf hide containing about 200 g of solids was placed in three liters of water and run through a colloid mill to obtain a homogeneous mixture. The pH of the slurry was adjusted to 3 and kept there with 1 N HCl during treatment with 1 g of pepsin for a period of four days at 4°C for telopeptide removal. The mixture then was pumped through a die punched with holes into a concentrated NH_4OH coagulating bath. The pump and die procedure was used in an attempt to maximize the collagen surface area during the coagulation step. The mixture was allowed to remain at the high pH for at least 12 hours to stop the enzyme reaction. The collagen was then washed with distilled water several times, until a pH of 8 was obtained, and redissolved with 1N HCl to give a pH of 3. The solution was vacuum-filtered, once with eight layers of cheesecloth and a second time with a tightly woven polyester cloth, reprecipitated with 1N caustic and washed with acetone to remove fats. The collagen was then filtered, dried, and placed back into acid solution at a concentration of 5% and a pH of about 3. It was finally placed under vacuum to remove air bubbles, and bottled for storage at 2°C.

The procedure was performed four times, obtaining about a 50% yield of soluble collagen each time. The procedure was adapted from a collagen treatment method described by Miyata.[8]

Emulsion Polymerization

Ethyl acrylate and methyl methacrylate monomers were washed once with 5% caustic saturated with NaCl and three times with saturated NaCl solution for inhibitor removal. Eighty ml of monomer (ethyl acrylate, methyl methacrylate, and triethylene dimethacrylate in varying proportions depending on the latex desired) were charged into a stirred reaction flask already containing 300 ml of distilled water and a varying amount of 10 weight percent sodium lauryl sulfate solution (10 ml unless otherwise specified). Three initiator solutions (five ml of five weight percent potassium persulfate, five ml of five weight percent sodium metabisulfite, and one ml of one weight percent ferrous sulfate) were then charged into the vessel at 30 second intervals.

The progress of the reaction was followed on a temperature recorder until a peak in temperature is reached. For a more detailed description of the equipment and procedure used in emulsion polymerization, see Rodriguez.[25] Table 1 gives a complete listing of all the latexes prepared and used in this research.

It is known that small amounts of acrylic latexes added to collagen films can improve ultimate tensile strength.[24] Poly(ethyl acrylate) latex was prepared in various particle sizes and added to the films. Particle size can be assumed to decrease with the amount of surfactant (sodium lauryl sulfate) charged to the emulsion polymerization.[26]

                              Table 1
                           Latexes Prepared

Basic recipe:

     300 ml water
      10 ml 10% solution of sodium lauryl sulfate
      80 ml ethyl acrylate
       5 ml 5% solution of potassium persulfate
       5 ml 5% solution of sodium metabisulfite
       1 ml 1% solution of ferrous sulfate heptahydrate

Initial temperature 22 to 23°C

Series A: Variations in surfactant concentration

     1. 1/9 basic amount of surfactant
     2. 1/3 basic amount
     3. Twice the basic amount
     4. Seven times the basic amount

Series B:

     1. 10 wt% of ethyl acrylate substituted by triethyleneglycol di-
         methacrylate (TEDMA)
     2. 30 wt% substituted

Series C: Stiffening agent

     1. 10 wt% of ethyl acrylate substituted by methyl methacrylate
     2. 30 wt% substituted
     3. 60 wt% substituted
     4. 100 wt% substituted
     5. Combination of 60 wt% substituted by methyl methacrylate plus 10
         wt% substituted by TEDMA

Film Preparation

     The five percent collagen solutions and the solutions mixed with
various additives were cast directly into the rubber molds and coagulated
in 1N NH_4OH for one hour. All the films were then crosslinked with glutaraldehyde for 24 hours, dried, and tested for mechanical properties.

     Tensile strength testing of the films was performed on the Instron
machine at an initial jaw separation of 4.4 cm, a jaw separation rate of
2.5 cm per minute, and a chart speed of 25 cm per minute. The films were
sprayed on both sides with a water atomizer ten seconds prior to testing
to simulate the natural state of skin. All tests were performed at approximately 24°C and 55% relative humidity.

RESULTS AND DISCUSSION

Poly(ethylacrylate) Particle Size

     As mentioned previously, Chapman has established that acrylic latexes
can be used to enhance the tensile strength of collagen films. It should
be noted also that the general topic of polymer-leather composites has
been reviewed recently.[27] The particular interaction of methyl methacrylate with leather also has been reported.[28]

In the present study, an optimum particle size has been obtained with a
surfactant charge of 8.77 x 10^3 moles surfactant/liter product giving a
tensile strength of 34.1 MP, a 15% improvement over the 29.6 MP, a value
obtained for pure collagen (Fig. 3). A very large decrease in modulus
with increasing particle size was also observed.

212

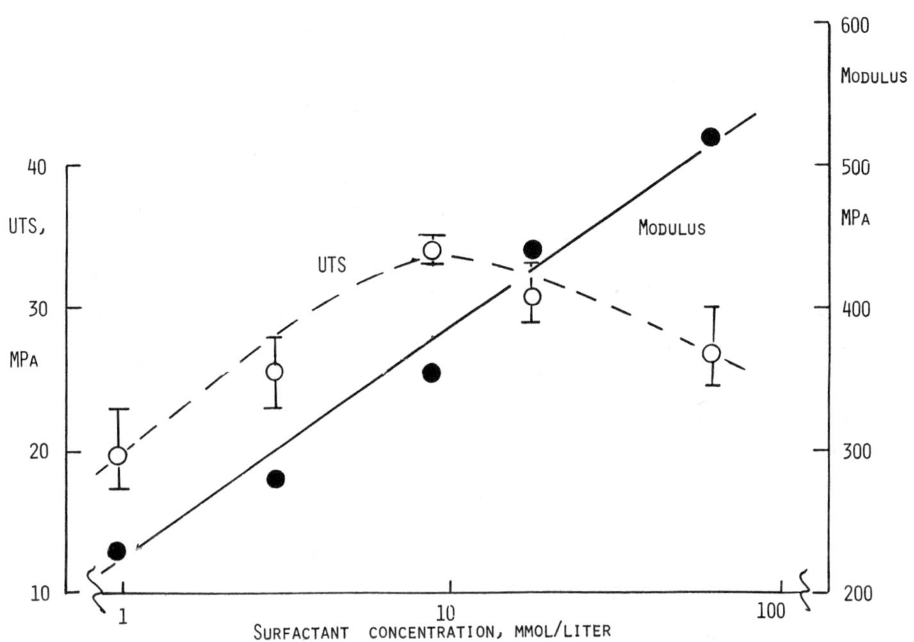

Figure 3. The modulus of collagen films containing 25 wt.% latex (75 wt% collagen) increases with surfactant concentration while ultimate tensile strength, UTS, goes through a maximum. It is assumed that particle size of the poly(ethyl acrylate) latex decreases with increasing surfactant concentration during polymerization.

These results can be most readily explained by assuming that ethyl acrylate is acting as a filler in this system.

There are three ways collagen and ethyl acrylate latex are most likely to interact:

1) simple physical inclusion of the filler particles (and agglomerates) in a matrix of polymer
2) physical inclusion of the filler along with wetting of the filler surface by the polymer
3) Definite physical adhesion of the polymer to the surface of the filler particles.[29]

The first case, the filler only acts as a diluent and can interrupt the polymer structure and would, therefore, lower tensile strength. In cases 2 and 3, a rise in tensile strength would be expected by relieving the individual stress placed on a collagen molecule by spreading it over an aggregate. In this case the ethyl acrylate also acts as an additional cross-link. This phenomenon is called "stress-relief" or "stress-sharing."[30,31] In the case of ethyl acrylate-collagen, this effect would explain the rise in tensile strength. The optimum can be explained by the fact that polymer-filler interaction is highly dependent on particle geometry, or in this case, particle size. The optimum particle size will be where the particle is about the same size as the growing edge of a crack, say about 100 nm. Particles which are smaller will not spread the stress over as many molecules, and larger particles will tend to disrupt the collagen structure itself. Both, therefore, would decrease the tensile strength of the film. It also seems intuitively correct that the larger particles which disrupt structure would decrease tensile strength much more drastically than the smaller particles which only have lower "stress-relief" capabilities.

In studies done by Alter, decreasing particle size was accompanied by increases in modulus for filler particle diameters of less than 0.2 micrometers.[32] This behavior is consistent with the collagen-latex results and seems to indicate a polymer-filler type interaction.

Crosslinked Latexes

If the rigidity of a filler particle is increased without decreasing filler-polymer adhesion, the tensile strength of the mixture should improve. The rigid structure of ethyl acrylate was increased by crosslinking with varying amounts of triethylene dimethacrylate (TEDMA). These latexes were then added to the collagen to determine the effect of crosslinked latexes on tensile strength and modulus (Fig. 4). Addition of 10% TEDMA to the latex was found to increase tensile strength by 10% and modulus by 60%. Since the ethyl acrylate particle acts as a major crosslink between several collagen molecules, an increase in modulus would be expected along with an increase in latex particle rigidity.

The sudden decline in both tensile strength and modulus at higher latex crosslink densities is most likely due to a breakdown of the collagen-ethyl acrylate "bond". The collagen is no longer able to adhere to the highly rigid surface of the ethyl acrylate, and thus the "stress-sharing" ability of the latex breaks down. With the collagen no longer able to "wet" the latex, the ethyl acrylate particles can act only as a diluent. With a highly rigid structure, the latex particles can no longer

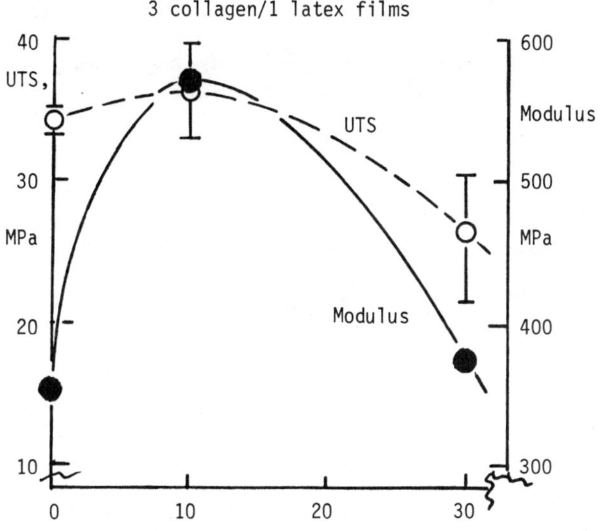

Figure 4. Crosslinking the latex by copolymerization
with triethyleneglycol dimethacrylate,
TEDMA, can increase UTS somewhat. The effect
on modulus is more pronounced.

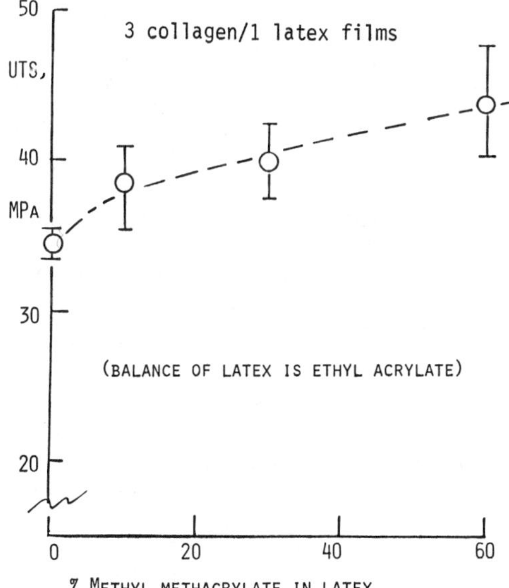

Figure 5. Increasing the T_g of the latex by
copolymerization with methyl methacrylate
up to 60 wt.% is seen to increase the
strength of the collagen-latex film.

easily deform into the collagen void spaces and could thereby cause disruptions in the collagen structural network. A trade-off between increased rigidity and polymer-filler adhesion occurs, resulting in a rapid decline in both tensile strength and modulus. Because the polymer particles are crosslinked before being dispersed in the collagen network, the system resembles a filler-rubber model more than it does an interpenetrating network.

Copolymerization with Methyl Methacrylate

After the favorable results obtained with the addition of poly(ethyl acrylate) (EA), it was decided to optimize the latex further by copolymerization with methyl methacrylate (MMA). Poly(methyl methacrylate) films in general have higher glass transition temperatures than ones prepared from ethyl acylate, so it was thought that copolymers would give stronger filler particles than just pure poly(ethylacrylate) filler.

Copolymers of 10, 30, and 60% MMA were prepared an added to the collagen. It is observed that tensile strength varies linearly with percent MMA in the latex (Fig. 5). This result is in accordance with the theory that MMA increases the strength and rigidity of the filler particles without altering their adherence to the collagen, and thereby increases tensile strength.

A latex consisting of 100% MMA was also prepared and tested in the collagen. The films were rough and brittle and therefore unusable. This can be explained by the fact that 100% MMA has a glass transition temperature of 110°C, causing the films to be brittle and un-usable at room temperature.

Water Absorption Tests

Increased tensile strength obtained from the latex-filled films can be caused by insufficient wetting prior to testing. It can be postulated that the addition of the latex lowers the water absorption, giving drier films which fail at higher stresses. This effect would also fit in with the increased modulus obtained with the latex-filled films. To test this possibility, pure collagen and 60% MMA-40% EA latex filled collagen films were weighed, placed in distilled water for varying amounts of time, blotted and weighed again. The additional weight was assumed to be total water absorption and a percent water absorption was calculated based on the dry weight of the film. The results (Fig. 6) show that water absorption ability, within the limits of the experiment, does not change when latex filler is added to the films.

Methyl Cellulose

Addition of up to 20% methyl cellulose was found to enhance tensile strength of the films by as much as 60% in some cases (Fig. 7). Additions of greater than 20%, however, drastically reduced tensile strength.

Consistency of the tensile test results also seemed to be improved and outward appearance of the films also tended to be smoother and more consistent than the pure collagen films. An optimum in tensile strength was found at a 5% methyl cellulose concentration.

The fact that small amounts of methyl cellulose enhance tensile strength is probably the result of two different effects. One is that methyl cellulose added to the collagen makes the solution much easier to

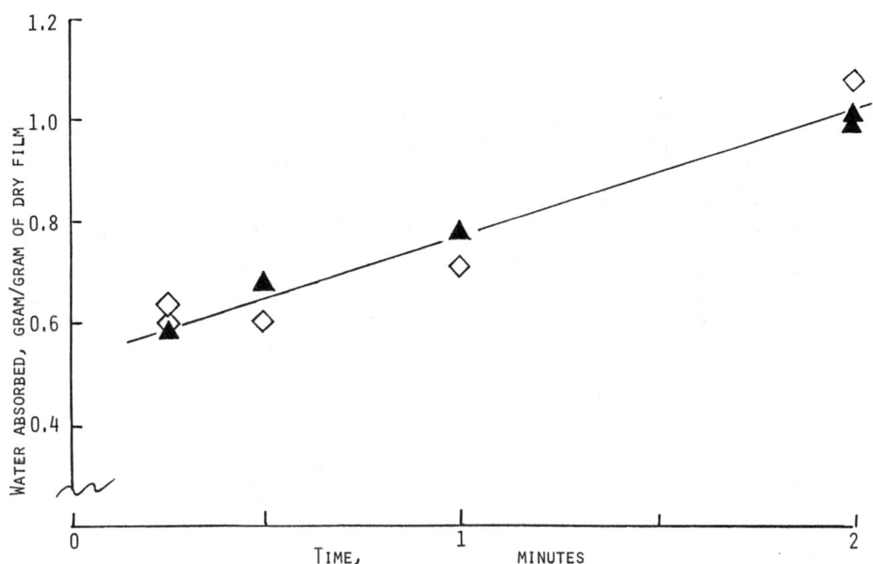

Figure 6. The water absorption for a film of corsslinked collagen alone (solid triangles) is about the same as for a film composed of 3 parts collagen and 1 part of a 60 wt.% methyl methacrylate, 40 wt.% ethyl acrylate latex (hollow diamonds).

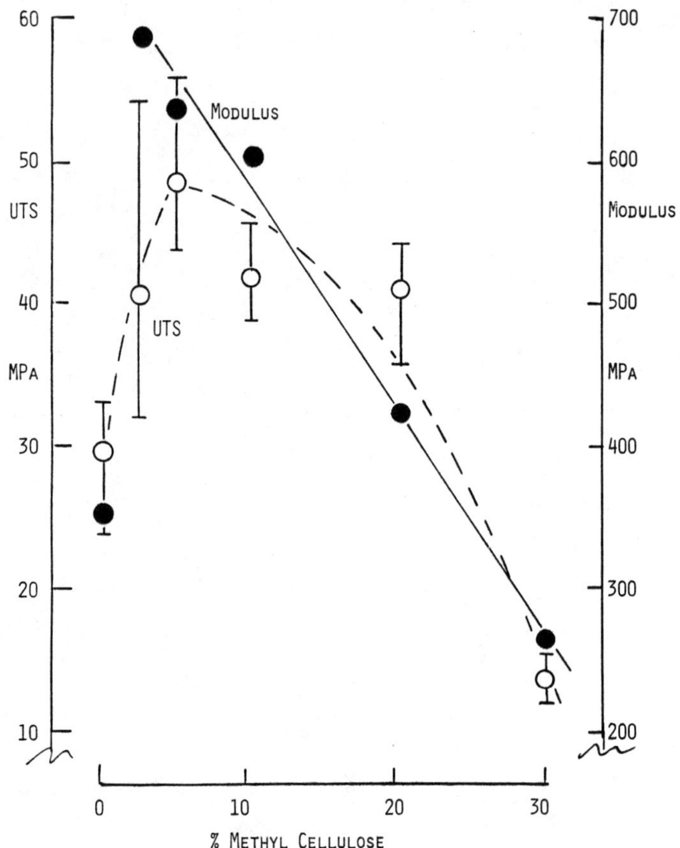

Figure 7. The addition of a small amount of methyl cellulose to a collagen-latex (3:1) film can enhance the strength markedly. The effect appears to peak at about 5% methyl cellulose (based on the weight of collagen + latex). The latex used was made from the basic recipe of Table 1.

handle and spread, giving films with fewer macroscopic imperfections and
thus higher strengths. The other effect is probably a filler interaction
between the methyl cellulose and the collagen. Although methyl cellulose
does not act as a particulate filler in the manner that latexes seem to,
it is still capable of connecting collagen molecules by physical adhesion
and of filling void spaces which could have become initiation sites for
failure of the films.

The drop-off in tensile strength beyond the 20% level is most likely
due to the start of a dilution effect whereby the higher amount of methyl
cellulose causes a lower crosslink density per unit volume of film. The
larger distance between neighboring collagen molecules lowers the proba-
bility of the crosslinking reaction. This dilution effect also counter-
balances improved macroscopic consistency obtained with increasing amounts
of methyl cellulose, causing the wide band of concentrations at which
methyl cellulose enhances tensile strength.

Brennan[15] studied the possible crosslinking of methyl cellulose
placed in 0.2% glutaraldehyde for 24 hours, and found that although some
cross-linking does occur, the resulting structure is extremely weak and
not comparable in strengths to crosslinked collagen. This effect would
tend to enhance tensile properties of films with low concentrations of
methyl cellulose by strengthening the structure filling the void spaces.
At higher methyl cellulose concentrations, however, these weak crosslinks
would not be able to make up for the collagen-collagen crosslinks lost due
to the dilution effect.

Modulus was found to decrease with increasing methyl cellulose concentration, which is consistent with the dilution effect and decreased
crosslink density. Although it is not quite clear in Figure 7 the modulus
of the films probably reaches a maximum somewhere in the range of 2.5 to
5.0% methyl cellulose.

<u>Combined Effects</u>

Several attempts at optimizing tensile strength by combination of
previous experiments and by looking at the possibility of annealing were
studied. The results for these studies are found in Table 2. It is seen
that, in many cases, improvements in tensile strength from different effects are not direcly additive, and that at times may even detract from
each other.

Addition of methyl cellulose to collagen containing crosslinked
poly(ethylacrylate) seems to do nothing for tensile strength. Methyl cellulose added to collagen containing 60-40 MMA-EA latex, on the other hand,
gives films with high strengths, although the effects are not additive.
Methyl cellulose-poly(ethylacrylate) collagen films also exhibit a small
improvement in strength compared to films without methyl cellulose. Again
the effects are observed to be far from additive.

In general, interactions between additives are complex and most
likely new optimum concentrations for the combinations would have to be
obtained. It is probable that methyl cellulose added to a collagen

solution which is already 25% latex, begins to exhibit a dilution effect
which competes with the favorable increase in smoothness and consistency.
The relative importance of these two competing effects seems to vary with
the type of latex filler added.

Crosslinking the 60-40 MMA-EA latex seems to do nothing to increase
film tensile strength. This is probably caused by a trade-off between
increased rigidity of the already rigid latex particles and lowered adhesion between latex and collagen.

Annealing is used especially in metallurgy to make materials more
ductile and relieve strain hardening by allowing the material to flow and
recrystallize.[33] Annealing is also used in the polymer industry to remove
points of stress in extruded films to increase strength and optical clarity.[34] Before the collagen films could be annealed, the shrinkage temperature of the films had to be found, so that a reasonable annealing temperature could be found. The shrinkage temperature was 77.5°C for collagen
and 78.0°C for collagen containing 60-40 MMA-EA latex. These results are
consistent with values found by Cater for cross-linked kangaroo tail tendon collagen.[22]

The films were annealed in the dry state, in an oven at 61 to 62°C
for thirty minutes. Tensile strength results for both annealed and control films are found in Table 2. The results look quite promising for
both plain collagen and for films containing 60-40 MMA-EA latex, with
tensile strength increases of 47 and 36 percent respectively. It is most
probable that these increases are due to the disappearance of a large
number of stress points and imperfections through the annealing step. It
is also likely that some fusion of collagen to the latex is occurring also
causing increases in the strength of the films.

Table 2

Selected Optimization Experiments

All strengths reported as ratio to UTS for collagen film (4,290 psi, 29.6
MPa). All except the first have 25 wt% latex.

Basic film for comparison	UTS ratio	Attempted change	UTS Ratio
Collagen alone	1.00	Annealed	1.46
+ basic latex	1.15	+ 5% methyl cell.	1.26
+ 10% TEDMA latex*(Ser.B-1)	1.23	+ 5% methyl cell.	1.27
+ 60% MMA latex*(Ser. C-3)	1.49	+ 5% methyl cell.	1.70
"		+ 10% TEDMA	1.48
"		Annealed	2.02

*See Table 1.

CONCLUSIONS

Tensile strengths of enzyme-treated collagen films can be greatly increased by some form of crosslinking. Crosslinking with glutaraldehyde is
attractive since it has been found to make collagen more resistant to
attack from collagenase and in vivo absorption after implantation.[8,35]
The results of the present research, and that of Chapman[24], however, show
that there is an optimum crosslink density.

Tensile strength of the films depends on crosslinking conditions and
on pH during the crosslinking reaction. A crosslinking bath which is not
a solvent for collagen is needed to inhibit film swelling and allow the
collagen molecules to be close enough to react.

The addition of acrylic latex fillers enhances tensile strength of
the films. Optimum tensile strength depends on the particle size of the
latex. Collagen-latex interaction and good adhesion is also important for
ultimate tensile strength of the films. Copolymerization of ethyl acrylate latex with methyl methacrylate increases particle modulus and thereby
the strength of the films. Crosslinking the latexes with triethylene
dimethacrylate gives films with both higher modulus and tensile strength,
until a tradeoff with decreased collagen-latex interaction is reached.

The addition of 5% methyl cellulose gives films containing fewer visible imperfections and higher tensile strengths. Tensile strength of all
films are higher when 5% methyl cellulose is added. Annealing of the
films also gives promising results in relieving imperfections and thereby
increasing strength.

In general, it can be concluded that glutaraldehye crosslinked collagen films, with the addition of small amounts of fillers, can now be
prepared in the strength ranges of natural skin and offer a promising skin
substitute for medical applications.

ACKNOWLEDGMENT

The authors are grateful to Dr. T. Miyata (Japan Leather Works) and
Dr. Kurt H. Stenzel (Cornell Medical College) for advice and
encouragement.

REFERENCES

1. Rodriguez, F., _Polymer News_ 9, 262 (1984).

2. Ramachandran, G.N., "Treaties on Collagen," Vol. 1, Academic Press,
    New York, 1967.

3. Gustavson, K.H., "The Chemistry and Reactivity of Collagen," Academic
    Press, New York, 1956.

4. Rigby, B.J., _Advan. Chem. Phys._, 21, 537 (1971).

5. Schmitt, F.O., L. Levine, M.P. Drake, A.L. Rubin, D. Pfahl, and P.F.
    Davison, Proc. Natl. Acad. Sci. U.S., 51, 493 (1964).

6. Davison, P.F., L. Levine, M.P. Drake, A.L. Rubin, and S. Bump, _J.
    Exptl. Med._, 12, 331 (1967).

7. Zvanut, Carl W., Ph.D. Thesis, Cornell University (1977).

8. Personal Communications, Dr. T. Miyata, (1976).

9. Nishihara, T. and T. Miyate, _Collagen Symposium_, 3, 66 (1962).

10. Rubin, A.L., I. Pfahl, P.T. Speakran, P.F. Davis, and F.O. Schmitt,
    Science, 139, 37 (1963).

11. Kon, T., G.L. Mrava, D.C. Weber, and Y. Nose, J. Biomed. Mater. Res., 4, (1970).

12. Hamed, G., Masters Thesis, Cornell University (1973).

13. Personal Communications, Dr. T. Miyata, (1976). Also described in Japanese patent No. 7,1,28,193.

14. Schimpf, W.C., Masters Thesis, Cornell University (1976).

15. Brennan, J.M., Masters Thesis, Cornell University (1977).

16. Lynch, J.B. and S.R. Lewis, ed., "Symposium on the Treatment of Burns," Chapter 28: Spira, M. and C.W. Hall, "Synthetic Materials in the Treatment of Burns," C.V. Mosley Co., St. Louis, 1973.

17. Spira, M., J. Fissette, S.B. Hardy, and F.J. Gerow, J. Biomed. Mater. Res. 3, 213 (1969).

18. Rubin, A.L., T. Miyata, and K.H. Stenzel, J. Macromol. Sci.-Chem., A3(1), 113 (1969).

19. Gustavson, K.H., "The Chemistry of the Tanning Process," Chapter 7, Academic Press, New York, 1956.

20. Highberger, J.H. and I.J. Salcedo, J. Am. Leather Chem. Assoc., 35, 11 (1940); 36, 271 (1941).

21. Highberger, J.H. and C.E. Retzch, J. Am. Leather Chem. Assoc., 34, 131 (1939).

22. Cater, C.W., J. Society Leather Trades' Chemists, 47, 259 (1963).

23. Richards, F.M. and J.R. Knowles, J. Mol. Biol., 37, 231 (1968).

24. Chapman, E.W. and F. Rodriguez, Polymer Engineering and Science, 17, 5, 282 (1977).

25. Rodriguez, F., "Principles of Polymer Systems," 2nd Ed., McGraw-Hill, New York, 1982, p. 517.

26. Ibid., p. 118.

27. Jordan, E.F., Jr., B. Artymyshyn, M.V. Hannigan, R.J. Carroll, and S.H. Feairheller, in Polymer Applications of Renewable-Resource Materials (C.E. Carraher, Jr. and L.H. Sperling, eds.), Plenum Press, New York, 1983.

28. Harris, E.H., H.A. Gruber, P.R. Buechler, and S.H. Feairheller, in Polymer Applications of Renewable-Resource Materials (see previous reference).

29. "Encycl. of Poly. Sci. and Tech.," Vol. 6, Wiley, New York, 1967.

30. "Encycl. of Poly. Sci. and Tech.," Vol. 12, Wiley, New York, 1967.

31. Bruins, P.F., Polyblends and Composites, Interscience, 1970.

32. Alter, H., _J. Applied Pol. Sci._, 9, 1525 (1965).

33. Van Vlack, L.H., "Materials Science for Engineers," Addison Wesley, New York, 1970.

34. Sweeting, O.J., "The Science and Technology of Polymer Films," Vol. 1, Wiley, New York, 1968.

35. Harris, E.D., Jr. and M.E. Farrell, _Biochem. Biophys. Acta_, 278, 133 (1972).

TITANIUM-CONTAINING POLY-ALPHA-AMINO ACIDS FROM DIPEPTIDES

Charles E. Carraher, Jr.[a,b], Louis G. Tisinger[b], and
William H. Tisinger[b]

Department of Chemistry[a], Wright State University, Dayton, Ohio
45435 and Department of Chemistry[b], Florida Atlantic University
Boca Raton, Florida 33341

ABSTRACT

Titanium-containing poly-alpha-amino acids were synthesized employing
both the classical (aqueous) interfacial and aqueous solution polyconden-
sation reaction systems. Product yields are moderate, but are greatly
affected by the amount of sodium hydroxide present. Further, yield
is increased by addition of catalytic amounts of a silicon antifoaming
agent.

$$Cp_2TiCl_2 + H_2N-R-\overset{\overset{O}{\|}}{C}-O^- \longrightarrow \left(\overset{\overset{Cp}{|}}{\underset{\underset{Cp}{|}}{Ti}}-NH-R-\overset{\overset{O}{\|}}{C}-O\right)$$

where Cp is cyclopentadiene.

Product structure was confirmed by control reactions, mass spectrophotometry, FT and classical infrared spectroscopy, elemental analysis
and light scattering photometry.

INTRODUCTION

The term amino acid in its broadest sense includes all compounds
containing the amine and acid functional groups. Here coverage is
restricted to include only naturally occurring alpha-amino acids.

Cystine was isolated by Wollaston in 1810 from a urinary calculus.
Since then many other amino acids have been isolated and identified.

Free amino acids occur in nature but only to a limited extent.
Natural amino acids occur largely in combined forms as amides (glutamine
and asparagine) and as polypeptides or proteins. As enzymes and hormones,
proteins catalyze and regulate most body functions; as tendons and muscles
they provide the means for movement, containment and construction of
our inner body; as skin, hair and nails they provide the body with an
elastic, semipermeable covering; as major components of the hemoglobins
they act as a key ingredient in oxygen exchange; as antibodies they
provide the body with an intricate self-defense system, etc.

Proteins come in a wide variety of shapes, sizes, functions and
chemical composition, yet in spite of this variety, they all contain
mainly simple amino acids. Natural proteins yield on hydrolysis up
to 22 different amino acids, all of the alpha variety and all (except
glycine) of the L configuration. (Some D-amino acids have been obtained
from the cell walls of certain bacterial and from certain antibiotics.)

While amino aids have been synthesized employing a possible primitive
atmosphere containing ammonia, hydrogen, methane, water and electrical
discharges, as well as a variety of more closely controlled synthetic
techniques (such as the Strecher synthesis, from aldehyde condensations
with amides containing active methylene groups, ammonolysis of alpha-
halogen acids and from N-acetyl aminomalonic esters), of interest here
is the generation of a wide variety of mono, di, tri, ...peptides from
natural sources, mainly proteins.

Direct thermal polymerization of alpha-amino acids is complicated
by their tendency to undergo cyclic dimerization forming diketopiperazines.

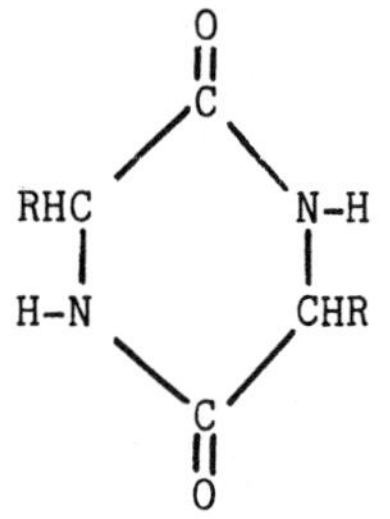

$$
\begin{array}{ccc}
 & \overset{\displaystyle O}{\underset{\displaystyle \|}{C}} & \\
RHC & & N\text{--}H \\
H\text{--}N & & CHR \\
 & \underset{\displaystyle O}{\overset{\displaystyle \|}{C}} & \\
\end{array}
$$

Even so, copolymers have been formed through heating at 160-210°C
dry mixtures of glutamic acid, aspartic acid or lysine with other alpha-
amino acids (1-3). Over three decades ago Woodward and Schramm produced
a high polymer from the N-carboxyanhydrides of L-leucine and DL-phenylala-
nine (4-6). Films cast from the products were tough and clear. Even
so non-alpha polypeptides, nylons have thus far dominated the marketplace
due in large measure to the ready availability of the necessary feedstock.

A large number of techniques are available for the separation of
alpha-amino acids including a wide variety of chromatographic methods.
These chromatographic methods include HPLC, ion-exchange chromatography
and GPLC.

Structure-property control is advantageous and can be obtained
from employing purified single di, tri, etc., peptides. Even so, for
many less structurally demanding applications, use of feedstocks containing
mixtures of peptides may be acceptable. It is important to note that
amino acids containing more than two Lewis bases (including tryptophane,
tyrosine, serine, lysine, histidine, glutamic acid, cystine, cystein,
aspartic acid and arginine) will probably form crosslinked materials
if reacted with difunctional metal-containing reactants such as Cp_2TiCl_2
and R_2SnCL_2. Additional aspects of alpha-amino acids are covered in
recent reviews and books (for instance 7-9).

Amino acids can exist in a cationic, dipolar (zwitterion) and anionic
form depending on the pH. Under the presently employed reaction conditions,

where base is added to neutralize the acid functional group and to act
as a scavenger of HCl generated from the reaction of the metal-chloride
and amino group, the amino acid exists in the anionic form.

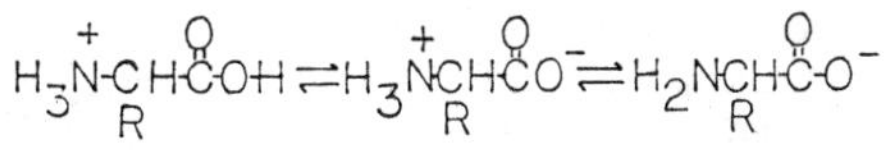

$$H_3\overset{+}{N}\text{-CHCOH} \rightleftharpoons H_3\overset{+}{N}\text{NCHCO}^- \rightleftharpoons H_2\text{NCHCO}^-$$

High molecular weight titanium-containing polyamines $\underline{1}$ (10-15)
and polyesters $\underline{2}$ (11-18) have been previously synthesized. Here the
initial synthesis of the corresponding titanium-containing polyamine
acids ($\underline{3}$) is reported.

Use of a monopeptide alpha-amino might lead to the formation of
unwanted five-membered cyclic products ($\underline{4}$). Thus dipeptides were chosen
for initial study since they are readily available in modest quantity
and they offer enough separation between the amino and carboxylic reaction
sites to preclude much cyclization.

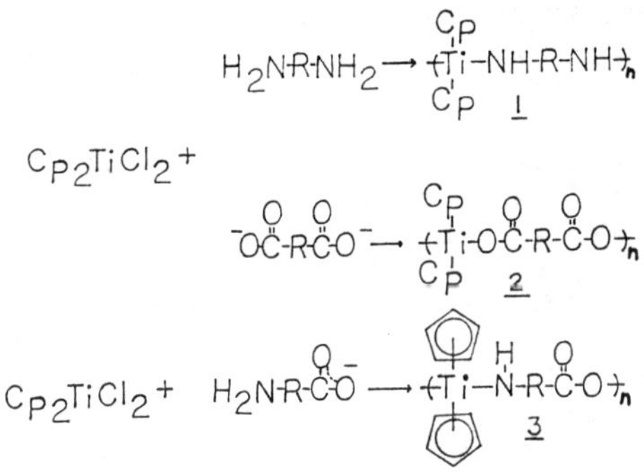

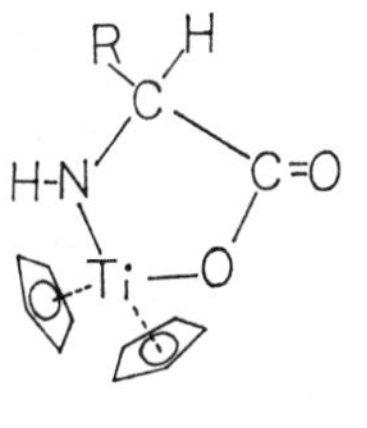

$$H_2N\text{-R-NH}_2 \longrightarrow \left(\underset{\underset{\text{Cp}}{|}}{\overset{\overset{\text{Cp}}{|}}{\text{Ti}}}\text{-NH-R-NH}\right)_n \quad \underline{1}$$

$$Cp_2TiCl_2^+$$

$$^-\text{OC-R-CO}^- \longrightarrow \left(\underset{\underset{\text{Cp}}{|}}{\overset{\overset{\text{Cp}}{|}}{\text{Ti}}}\text{-O-C-R-C-O}\right)_n \quad \underline{2}$$

$$Cp_2TiCl_2^+ \quad H_2N\text{-R-CO}^- \longrightarrow \left(\text{Ti-N-R-C-O}\right)_n \quad \underline{3}$$

$$\underline{4}$$

A one quart Kimex emulsifying jar was employed as the reaction vessel. It is placed on a Waring Blendor (Model 1120) whose no-load speed is about 18,000 rpm (120 volts). The reactants enter the reaction vessel through a hole in the emulsifying jar lid. The liquid containing the amino acid is added first. Stirring is begun and the liquid containing the Cp_2TiCl_2 is rapidly added. The product is a precipitate which is recovered employing suction filtration, repeatedly washed, transferred to a glass petri dish and allowed to dry.

Mass spectrophotometry was performed utilizing a direct insertion probe in a Kratos MS-50 Mass Spectrometer operating in the EI mode, 8 KV acceleration and a 10 second/decade scan rate with variable probe temperature.

Infrared spectra were obtained using potassium bromide pellets employing a Perkin-Elmer Model 457 and a Nicolet 5DX Fourier Transform infrared spectrophotometer.

Molecular weight determinations were performed using a Brice-Phoenix Model BP-3000 Universal Light Scattering Photometer. Refractive index increments, dn/dc, were determined using a Bausch and Lomb Abbe Refractometer Model #3-L.

Elemental analysis was carried out using a Perkin-Elmer 240 Elemental Analyzer for C, H and N and thermal analysis for titanium.

RESULTS AND DISCUSSION

Product yield (Table 1) depends on the amount of added base with moderate yields obtained employing two equivalences of base per mole of amino acid. One equivalence of base acts to neutralize the acid functional group. For some reactions excessive foaming occurs. Thus silicone oil (Dow Corning; 200 centistrokes) was added. Yields improved. It is not known whether the principle reason for improved yields is due to the silicon oil's antifoaming ability or to some other factor. The titanium-containing polypeptides were also synthesized employing a simple solution system where both the Cp_2TiCl_2 and dipeptide are added to give separate aqueous phases and these are subsequently mixed together.

Synthesis of the titanium-containing polyamines (1) and polyesters (2) occurred employing both the classical interfacial and aqueous solution condensation systems (10-18). Because of the aqueous solubility of Cp_2TiCl_2 reaction employing the interfacial reaction system can occur in either the aqueous or organic phases. It is believed that reaction with the diamines occurs in the organic phase while reaction with the dicarboxylic salts occurs in the aqueous phase (10). Thus, there was no guarantee that reaction with a Lewis base containing both an amino and a carboxylic functional group would occur appreciably since it is possible that the reactants may have to continually (often) pass from one phase to another, slowing the reaction to greater than that of the reaction schedule interval. Currently the major site(s) of reaction is unknown.

The products exhibit the typical poor solubilities characteristic of many organometallic polymers (19). Dilute solutions of polymers are obtained in DMSO so as to permit molecular weight determinations. Sample results are cited in Table 1. These results are preliminary but do show that at least some of the products are high polymers. Effort is now focused on synthesis of products with greater solubilities.

Elemental analysis results are in partial agreement for the proposed structure. Poor combustion of many organometallic polymers is believed responsible for the generally poor elemental analysis results obtained (20). For the product from Gly-Gly the calculated percentage of Ti is 16.3, found is 15.2.

Table 1. Sample results for the synthesis of titanium-containing peptides.[d]

Yield (%)

Dipeptide	Concentration of Base (mmole)				dn/dc	Mol. Wt.
	2.0	2.0	2.0[a]	20.0	(ml/g)	(Daltons)
Gly-Gly	2	23[b]	30	Trace	1.6	1.2×10^5[b]
Gly-L-Leu	2	20	34[c]	zero	0.33	5.0×10^2[c]
Gly-L-Val	4	-	20	zero	-	-

a. Three drops of silicone oil (200 centistrokes) added.
b. Weight-average molecular weight in DMSO for a 0.12% solution.
c. Weight-average molecular weight in DMSO for a 0.27% solution.
d. Reaction conditions: Cp_2TiCl_2 (1.00 mmole) in 50 ml; chloroform added to stirred (18,00 rpm no-load) aqueous (50 ml); solutions containing the dipeptide (1.00 mmole) for 180 seconds stirring time at about 25°C.

The results of the infrared spectroscopy are consistent with the proposed structure. For instance, for the product from glycyl-glycine bands due to the presence of the Cp moiety are found at about 1010, 1025 cm^{-1} are not present in the dipeptide itself. Glycyl-glycine has two bands within the 1375-1450 cm^{-1} region whereas the products have three bands, the third band indicating the presence of the Cp moiety. Other bands characteristic of the presence of the dipeptide moiety are present. For instance, there is a broad, strong band between 3350-3500 cm^{-1} characteristic of the N-H (free, nonhydrogen bonded) stretching in secondary amides. Bands present within the 1710 to 1500 cm^{-1} region are characteristic of ester and amide-carbonyl stretching. The absence of a strong band within the 3100-2600 cm^{-1} region is consistent with the absence of the NH_4 zwitterion-associated grouping (21).

Mass spectra (Tables 2-4) were obtained for a number of samples and were in agreement with the product containing both the dipeptide and Cp_2Ti moieties (Tables 2-4). For the product from glycyl-leucine ion fragments characteristic of the Cp moiety are present at 67, 66, 65, 64 and 40 m/e (Table 2). Ion fragments characteristic of the dipeptide are present and numerous. For this compound, ion fragments are found at m/e 36 and 38 believed derived from Cl endgroups and/or trapped chloride ion. The assignment of these fragments as being derived from the chloride ion is supported by the ratio of 36:38 being 3:1, the ratio predicted from the relative isotopic abundance of Cl-35 to Cl-37. The presence of large amounts of Ti-Cl moieties is ruled out due to the color of the products. The products are a brown to brown-yellow and the presumably high polymers should be long enough to minimize color contributions by endgroups, thus the chlorine may well be present as endgroups.

While the overall structure can be depicted as 3, the actual structure may be alternating as depicted in 3 or may be an alternating copolymer as depicted in 5 where R is the hydrocarbon portion of the alpha amino acid. Eventually FT-NMR can be employed to assist in determining

the proportions of alternation to block formation. Further, it is reasonable that the products formed from the aqueous solution systems may
have different proportions of alternation/block formation resulting
in the actual physical properties varying according to the synthetic
approach employed.

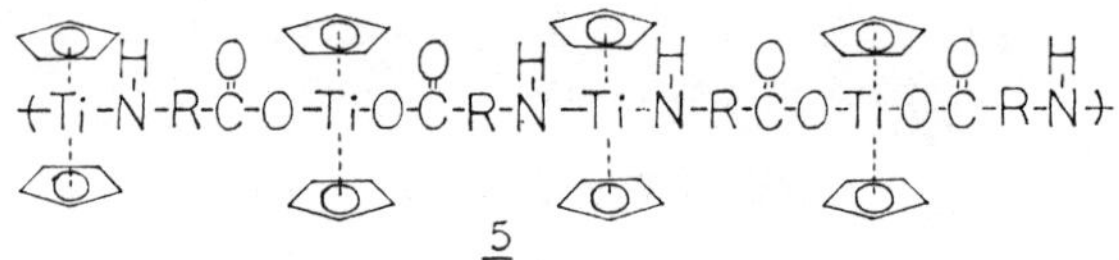

$$\begin{array}{c}\text{(Cp)} \quad \text{(Cp)} \quad \text{(Cp)} \quad \text{(Cp)} \\[4pt] \overset{\text{(Cp)}}{\underset{\text{(Cp)}}{\text{Ti}}}-\text{N-R-}\overset{\text{H}\;\;\text{O}}{\text{C}}\text{-O-}\overset{\text{(Cp)}}{\underset{\text{(Cp)}}{\text{Ti}}}\text{-O-}\overset{\text{O}\;\;\text{H}}{\text{C}}\text{-R-N-}\overset{\text{(Cp)}}{\underset{\text{(Cp)}}{\text{Ti}}}\text{-N-R-}\overset{\text{H}\;\;\text{O}}{\text{C}}\text{-O-}\overset{\text{(Cp)}}{\underset{\text{(Cp)}}{\text{Ti}}}\text{-O-}\overset{\text{O}\;\;\text{H}}{\text{C}}\text{-R-N}\end{array}$$

<u>5</u>

Table 2. Ion fragmentation correlation of cyclopentadiene.

m/e	66	65	39	40	48	64	67	63	31
Source									
John Wiley[a]	100	47	32	27	8.2	8.2	6.3	5.8	4.9
Gly-L-Val	100	52	63	31	–	–	7.7	2.3	2.9
Gly-L-Leu	100	45	–	42	–	8.9	5.3	–	–

a. E. Stenhagen, S. Abrahamsson and F. McLafferty (Eds.),
    "Atlas of Mass Spectral Data, Wiley, N.Y., 1969.

    There are several reasons for desiring the synthesis of titanium-
containing polyamino acids. These include enlarging the cadrei of non-
petroleum-based feedstocks for polymer applications, synthesis of a
new family of polymers, synthesis of organometallic polymers which exhibit
better solubility properties (derived from dissymmetry of the peptides)
and synthesis of bioactive materials. Concerning the latter factor,
there are a number of low molecular weight polypeptides that perform
specific biological tasks including acting as homing agents and catalysts.
Further, the Cp_2Ti moiety is known to be an active antitumoral agent
and a coupling of this moiety with oligomeric peptides may provide some
useful biological properties.

Table 3. Ion fragments from the mass spectrometry of the product of Cp_2TiCl_2 and Gly-Leu (relative intensities > 3%; to $m/e = 200$)

m/e	Relative %	Fragment (possible)
36	12	HCl(35)
38	4	HCl(37)
39	52	Cp
41	58	$CONH(43)$; $HC(CH_3)_2$ (43)
42	13	
42	5	
43	9	
43	40	
43	40	
44	20	CO_2, $NHCHCH_2$
53	8	
55	16	$NHCHCH_2CH(55)$, $HC(CH_3)_2CH_2(57)$
55	17	$CH_2CONH(57)$, $OCNHCH(57)$, $OCNHCH(56)$, $HCCO_2(57)$
56	5	
57	17	
57	15	
65	8	Cp
66	17	Cp
67	8	Cp
69	15	$NHCH_2CONH(69)$, $H_3C\ C(CH_3)_3(70)$
70	6	
71	4	
81	7	$NHCH_2CONHCH(82)$
84	16	$HC(CH_3)_2CH_2CHNH(85)$; $NHCH_2CONHCH(85)$; $CH_2CONHCHCH_2(84)$
85	30	
86	5	
99	5	$HC(CH_3)CH_2CHCO_2(99)$
113	5	
114	100	$HC(CH_3)_2CH_2CHCO_2(114)$; $HC(CH_3)_2CH_2CH\ NHCO(113)$
115	6	
127	5	$HC(CH_3)_2CH_2CHCO_2(129)$; $NHCH_2CNHCHCO_2(129)$
137	4	

Brackets indicate that the ion fragments may arise from common fragments differing only in the number of hydrogens.

Table 4. Ion fragments from the mass spectrometry of the product
of Cp_2TiCl_2 and Gly-Val (m/e 37 to 200; relative
intensities > 3%)

m/e	Relative %	Assignment
37	6	Cl(37)
38	16	HCl(37)
39	62	C_3H_3
40	31	Cp
41	44	CH_2CO
42	8	$NH_2C_2H_2$
43	31	C_3H_7
44	72	CO_2, $CONH_2$
54	4	C_4H_6
55	31	G, $CHCO_2$
56	9	C_4H_8
57	26	
62	5	
65	52	Cp
66	100	Cp
67	8	Cp
69	17	
69	5	$CONH_2CHCH$(69)
69	4	CH_2CONH_2CH(70)
70	6	CO_2CHCH(70)
71	5	$CHCHCO_2$(72)
71	4	
83	12	$CH_2CONHCHCH$
119	6	
125	3	GV$-C_3H_7$
130	7	
136	7	GV$-C_2H_6$
137	8	
141	3	GV$-C_2H_3$
150	3	GV$-O$
156	4	GV$-C$
168	4	GV
169	4	

where GV = Gly-Val minus 2 H

Brackets indicate that the ion fragments may arise from common
fragments differing only in the number of hydrogens.

<u>Acknowledgements</u>

The authors are pleased to acknowledge support from NSF Grant CHE 8310555 that allowed purchase of the FT-IR, ACS PRF Grants 15508-B7 and 13084-B3-C that allowed student support and NSF Grant CHE 78-18572 for support of the Regional Instrumentation Facility at the University of Nebraska.

REFERENCES

1. S. Fox, K. Harada and D. Rohlfing, "Polyamino Acids, Polypeptides and Proteins," (M. Stahmann, Ed.), U. Wisconsin Press, Madison, 1962.
2. S. Fox and K. Harada, U.S. Pat. 3,076,790 (1963).
3. S. Fox and K. Harada, J. Amer. Chem. Soc., $\underline{82}$, 3745(1960).
4. R. Woodward and C. Schramm, J. Amer. Chem. Soc., $\underline{69}$, 1551 (1947).
5. H. Leuchs and W. Geiger, Chem. Ber., $\underline{41}$, 1721(1908).
6. R. Woodward, U.S. Pat. 2,657,972(1953).
7. M. Bodanszky, Y. Klausner and M. Ondetti, "Peptide Synthesis," Second Ed., Wiley, N.Y., 1976.
8. M. Goodman and J. Heienhofer (Eds.), "Peptides," Wiley, N.Y., continuing.
9. C. Anfinsen, J. Edsall and F. Richards (Eds.), "Advanced in Protein Chemistry," Academic Press, N.Y. continuing.
10. C. Carraher, J.L. Lee, Organic Coatings and Plastics, $\underline{34}$(2), 478(1974).
11. C. Carraher, Makromolekulare Chemie, $\underline{166}$, 31(1973).
12. C. Carraher, J. Polymer Sci., $\underline{A-1}$, $\underline{9}$, 3661(1971).
13. C. Carraher and J.L. Lee, J. Macromol. Chem., $\underline{A9}$(2), 191(1973).
14. C. Carraher and J.E. Sheats, Makromolekulare Chemie, $\underline{166}$, 23(1973).
15. C. Carraher, Chemtech, 741(1972).
16. C. Carraher and S. Jorgensen, J. Polymer Sci., $\underline{16}$, 1965 (1978).
17. C. Carraher, R. Pfeiffer and P. Fullenkamp, J. Macromol. Sci.-Chem., $\underline{A10}$(7), 1221(1976).
18. C. Carraher, Eup. Polymer J., $\underline{8}$, 1339(1972).
19. C. Carraher, J. Chem. Ed., $\underline{58(11)}$, 92(1981).
20. C. Carraher, J. Macromol. Sci.-Chem., $\underline{A17-(8)}$, 1293D(1982).
21. R. Silverstein, G. Bassler and T. Morrill, "Spectrometric Identification of Organic Compounds," 4th Ed., 1981, Wiley, N.Y., page 129.

VISCOELASTICITY OF CALF HIDE IMPREGNATED WITH RADIATION-POLYMERIZED
POLYHYDROXYETHYL METHACRYLATE

Paul L. Kronick, Bohdan Artymyshyn,
Peter R. Buechler, and William Wise

Eastern Regional Research Center

U.S. Department of Agriculture

Philadelphia, PA   19118

INTRODUCTION

In order to combine the desirable properties of synthetic polymers
(elasticity and water, chemical, and biological resistance) with those of
animal hide (strength, flexibility, and dyability), composites can be
formed between the two types of material. Actually, adding materials to
hide has long been an aspect of traditional leathermaking, usuaslly to
increase thermal stability and to make the hide more hydrophobic. Newer
uses for hide material require it to be stable but compatible with aqueous
environments. Appropriate compositions can be formed from leather and
hydrophilic polymers. Composites of collagen with poly(hydroxyethyl
methacrylate) (polyHEMA or pHEMA) (1), with starch (2), or with polyacrylamide (3) have been described for surgical implants with blood compatibility and lack of tissue inflammatory reactions. Other uses for these
compositions may be found as substrata for cells and enzymes in biotechno-
logy.

The previously described composites were prepared from collagen
fibers from comminuted skin, in which the strength of the original skin
has been lost. Calfskin dermis, about 98% collagen, has a fibrous texture with a density of 0.5 g/cm^3. Since collagen has a density of
1.4g/cm^3, there is a considerable amount of free space between and within
the fibers. These have irregular cross sections about 5 μm thick and are

collected into bundles of various thicknesses, around 100 μm. The fibers
are subdivided into fibrils with regular round cross sections of 100 nm
diameter. The free space among the fibrils and among the fibers account
for the flexibility of the material; it stretches with difficulty, however,
because the fiber bundles are only slightly extensible. Note that if the
voids are collapsed by air drying the skin or if they are filled with
polymer, shear between fibers is restricted; the flexibility is lost; and
a hard but very strong material results.

The 100nm fibrils are stable structures that persist in comminuted
preparations even after treatment with dilute acid and base. The lowest
level of fiber structure is that of the Type I or type III collagen
molecule, about 1.5 μm thick and 300 μm long, packed together in regular
array to make up these fibrils. When stained with phosphotungstate,
regular bands appear along the fibrils, which reflect the regularity of
the molecular packing, evidently a genetic adaptation of the collagen
protein chain.

Consideration of these structures indicates that impregnation of
animal skin should occur between fibers or fibrils, not between molecules,
if the structure is not to fall apart. Insertion of polymer into the
fibrils among the molecules would be expected to weaken the structure,
which is held together by cooperative electrostatic and hydrophobic
bonds. The desired structure is then a composite of the interpenetrating-
network type with cocontinuous phases.

Electron micrographs published earlier (4) showed the disposition of
polymethylmethacrylate (pMMA) introduced into leather by _in-situ_ emulsion
polymerization. The polymer entered the hide, coated the fibers and
bundles, and filled the space among them. Very little polymer entered
the spaces among the fibrils. The spaces within the fibers may be about
the same size as or smaller than the micelles of the polymerization,
inhibiting their diffusion. The result was a hard, strong composite,
with few of the properties of leather. Here we have attempted to prepare
structures with the polymer within the fibers, not between them, so that
the deformation characteristics of the open, woven structure would be
preserved, but with greater stability to chemical degradation and modified
surface properties. For compatibility with aqueous systems and maximum
capillary penetration of the collagen fibers, HEMA was used as monomer.

It has a surface tension of 40 dyne/cm, close to the critical surface tension of collagen (5). It can be shown theoretically that capillarity is a maximum when the penetrant has a critical surface tension slightly lower than that of the walls of the capillaries. We deposited HEMA from acetone solution into specimens of chrome tanned calf hide and polymerized it with γ-radiation. The paramagnetic Cr^{3+} ions, by relaxing all nearby protons from excited magnetic states, allowed us to analyze the spatial distribution of the polymer with respect to the Cr^{3+}-containing collagen fibrils. Our measurements show that pHEMA in the product penetrates into the 10-μm fissures. The dynamic mechanical properties of the pHEMA were imparted to the composite without loss of the flexibility of the open-fiber structure.

METHODS

HEMA WAS PURCHASED FROM Polysciences, Inc. (Warrington, PA)[*] and used without further purification. It was observed to contain some polymer, which precipitated from water but not from acetone.

Calf skin was chrome tanned according to the USDA standard process (6), which involves removing epidermis and hair with basic sodium hydrosulfide, swelling with limewater containing sodium sulfide, neutralizing to pH 8 and treating with trypsin, then treating with chromic sulfate and 6% sodium choride at pH 1.5, and neutralizing to form chromic carboxylate crosslinks in the collagen. The product contains 3.2 to 3.7% chromium as Cr_2O_3, based on dry weight.

Undried tanned skin samples, about 5 x 5 x 0.5cm^3, were extracted with water and then with acetone and then rolled in acetone solutions of HEMA for 8 hours. The acetone was evaporated, leaving the residual HEMA to penetrate the fibers of the leather. The samples, containing different amounts of HEMA, were sealed into a jar under N_2 and irradiated for 3 Mrad (5 hours) at room temperature, by means of the 0.67-MEV γ-radiation from ^{137}Cs. An 0.2-mm thick film of outgassed HEMA, contained between glass plates with spacers, was included in the jar.

---

[*]Reference to brand or firm name does not constitute endorsement by the U.S. Department of Agriculture over others of a similar nature not mentioned.

A measure of the interpenetration of synthetic polymers among the collagen fibrils of hide was sought through NMR studies of the polymer. The ^{1}H spin-lattice relaxation times of collagen crosslinked with chromic ions (tanned leather) are significantly shortened with respect to untanned collagen, 3 ms versus 600 ms. Consequently chrome tanned leather constitutes an ideal matrix for probing the spatial distribution of slower relaxing (T_1 _ca._ 100 ms) polymer protons through the induction of faster relaxation times in the polymer by contact with paramagnetic centers. A pulse sequence (7) previously devised to examine the phase homogeneity of pMMA in chrome tanned leather, (8) was used. This pulse sequence combines inversion-recovery of the protons (9) with cross-polarization (10) and interrupted ^{1}H decoupling (11) to measure the proton T_1 values of the polymer. The pulse sequence is illustrated in Fig. 1 along with the CP-MAS Fourrier-transformed spectra of the 1:4 polymer:leather composite obtained at selected values of the pulse interval. The NMR studies were performed on a JEOL FX60QS spectrometer modified to carry out CP-MAS experiments. Instrument operating conditions were 5000 scans acquired in 2K memory using a 8 KHz spectral window; the cross-polarization contact time, 0.8 ms; the period of interrupted decoupling, 40 µs; and the recycle delay, 0.5 s.

The real and imaginary parts of the complex modulus were determined at two frequencies (3.5 and 110 Hz) with a Rheovibron II (for the pHEMA film) or a Rheovibron III (for leather samples) dynamic viscoelastometer (Toyo Baldwin Co.)*, operated at an oscillating relative deformation of 3.8×10^{-4}. The samples were mounted in a chamber within an insulated copper block. The chamber was rapidly cooled to -100°C with boiling liquid nitrogen and then warmed at a rate of 2-3°C/min with electric heaters in contact with the copper block. The temperature was measured with a thermocouple inside the chamber very close to the sample. Temperature uniformity within the chamber was achieved by a flow of precooled dry nitrogen. Dimensions of a typical leather sample were $50 \times 5 \times 3$ mm^3; those of a pHEMA film sample, $50 \times 5 \times 0.1$ mm^3. This instrument (12) measures the amplitudes of the harmonic force and amplitude signals and determines their phase, δ, by means of a phase meter of high precision. The phase data, together with the force and deformation amplitude signals, give the real and imaginary parts, E' and E", of the complex modulus:

$$E^* = E' + iE'', \text{ with } \tan \delta = E''/E'.$$

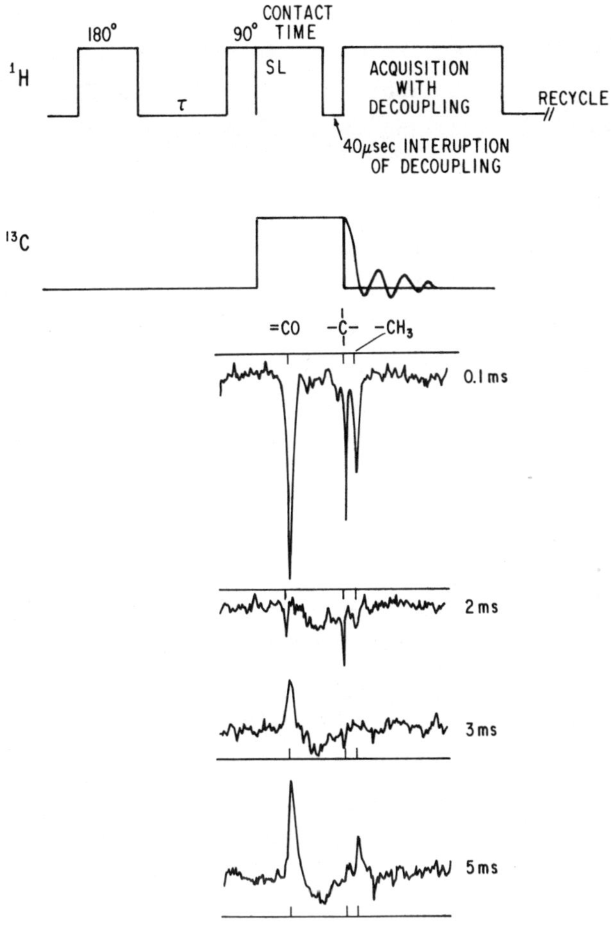

Fig. 1. Proton T_1 interrupted decoupling/inversion-recovery pulse sequence and spectra at selected time intervals for the hide-pHEMA composite containing 20% pHEMA.

Differential scanning calorimetry was carried out with a Perkin Elmer DSC instrument at a scanning rate of $10°$/min from $30°$ to $150°$ for hide-pHEMA or from $150°$ to $270°$ for neat radiation-polymerized pHEMA.

RESULTS

The pHEMA film prepared by irradiation was very highly crosslinked, swelling only to 25% in water. When pHEMA is prepared by chemical initiation, its swelling is not sensitive to the degree of crosslinking, usually swelling to 30-40% (13). The reduction of water uptake in these samples is therefore striking.

By means of a Du Nouy tensiometer, the surface tension of our HEMA reagent was determined to be 40 dyne/cm, matching the critical surface

tension reported for collagen (5). The ability of the HEMA to wet and
penetrate the collagen fibers of the calf skin was obvious from the
spreading behavior, with no measurable wetting angle, and from the dry
appearance of the samples containing up to 40% monomer.

Carbon-13 resonances are observed in three well separated spectral
regions: carbonyl, quaternary carbon, and methyl (Fig. 1). The carbonyl
and methyl arise from carbons in both the collagen and polymer, while the
remaining resonance has been demonstrated to arise only from quaternary
carbons in pHEMA. The more rapid spin-lattice relaxation of the chrome
treated collagen relative to the polymer is also evident in these spectra,
the carbonyl resonance being partially recovered at 3 ms while the polymer
peak remains inverted.

Spin-lattice relaxation times of pHEMA were obtained by measuring
the intensities of the quaternary-carbon signals. Attempts to fit the
data to a single relaxation time were unsuccessful; it was therefore
decided to assume that spin-lattice relaxation was occurring in the polymer at two different rates, a fast one for polymer penetrating among the
chrome treated fibrils, and a slow one for polymer segregated among the
fiber bundles. The relaxation of the quaternary carbon signal intensity
a(t) would be described by the following equation (1):

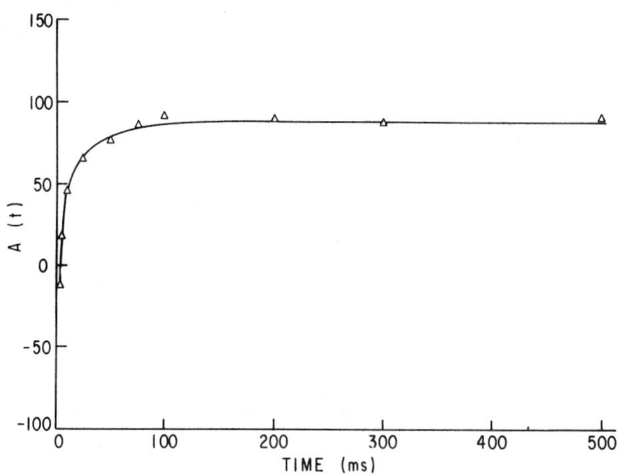

Fig. 2. Relaxation of quaternary protons in the hide-pHEMA-leather
composite containing 20% pHEMA.

TABLE I

Wgt. % pHEMA	P_f	T_f (ms)	P_s	T_s (ms)
20	0.72 (±.07)	3.55	0.28 (±.07)	30.55
33	0.66 (±.16)	3.55	0.37 (±.15)	30.55
37	0.58 (±.14)	3.55	0.48 (±.11)	30.55
52	0.32 (±.04)	3.55	0.68 (±.07)	30.55

$$a(t)=p_f[1-2\exp(-R_f t)] - p_s[1-2\exp(-R_s t)] \quad (1)$$

where p_f and p_s are the fast and slowly relaxing fractions of the polymer, and R_f and R_s are their respective rates (relaxation times are inverses of the rates). Intensity data were fitted to the above equation using a least-squares computer program. While the long relaxation time might be expected to equal that of neat polymer, 102 ms, a better fit to the data for all the samples was obtained when it was shorter; 30.6 ms (Fig. 2).

The shorter value might be due to the proximity of a portion of the interfibrous polymer to the surfaces of the Cr^{+3} bearing fibers. The data for each sample was therefore fitted to equation (1) with $R_s=1/30.6$ ms^{-1} and $R_f=1/3.6$ ms^{-1} to obtain fractions, p_f, for the polymer in the faster relaxing regions (Table I).

The table shows that, at low pHEMA content, the fractions of fast-relaxing polymer protons are high, most of the polymer being in the proximity of Cr^{+3} in the collagen fibrils. As the content increases above 37%, the proportion of slowly relaxing polymer protons increases rapidly, as all the added polymer enters regions of the composite distant from the Cr^{+3}-containing collagen. Therefore, the NMR data show that the capacity of the collagen for the pHEMA is about 37%, above which the value of p_f decreases rapidly. The end point does not appear to be sharp.

The dynamic mechanical properties of the pHEMA film are shown in Fig. 3. The main transition is a strong one, with a large maximum of tan δ. The temperature of the transition, 140°C, is much higher than that at 1 Hz reported for pHEMA prepared by chemical initiation, 109°C (14). This might be expected from the high degree of crosslinking which we infer from the low degree of swelling. Above 200°C, a rubbery region

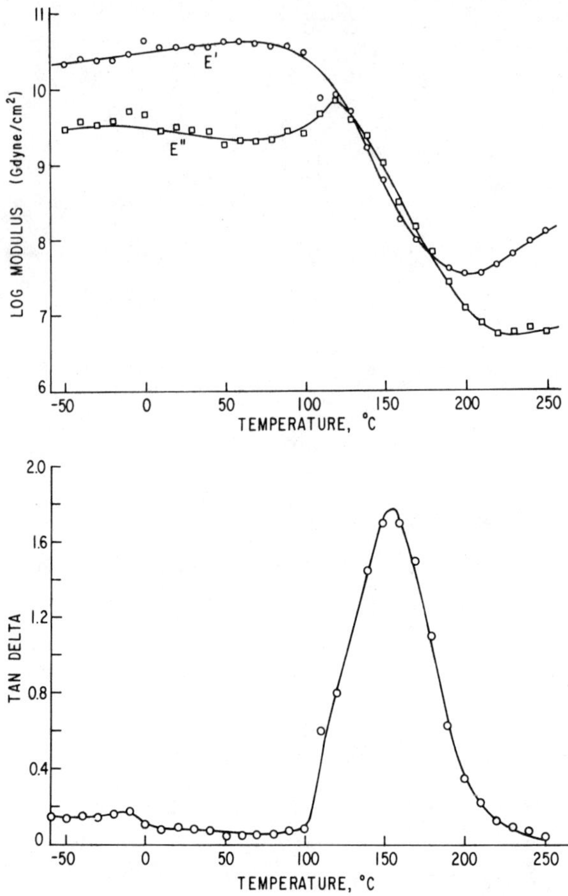

Fig. 3.   Dynamic mechanical properties of radiation-polymerized pHEMA film.

appears, E' increasing with temperature as expected from rubber elasticity theory. The molecular weight between crosslinks calculated from the slope of E' vs. temperature in this region by means of the classical equation $M_c = RT\rho/(slope)$, however, equals only 250 Daltons (ρ = density). This is below the range of validity of the equation, but we have already mentioned the small swelling in water, confirming a high crosslinking density. Chemical reaction during the measurement, progressively introducing substantial numbers of crosslinks (more than 1 per 1000 residues) at the high temperatures at which the positive slope of the modulus-temperature curve is observed, is unlikely, since the material gave no exotherm

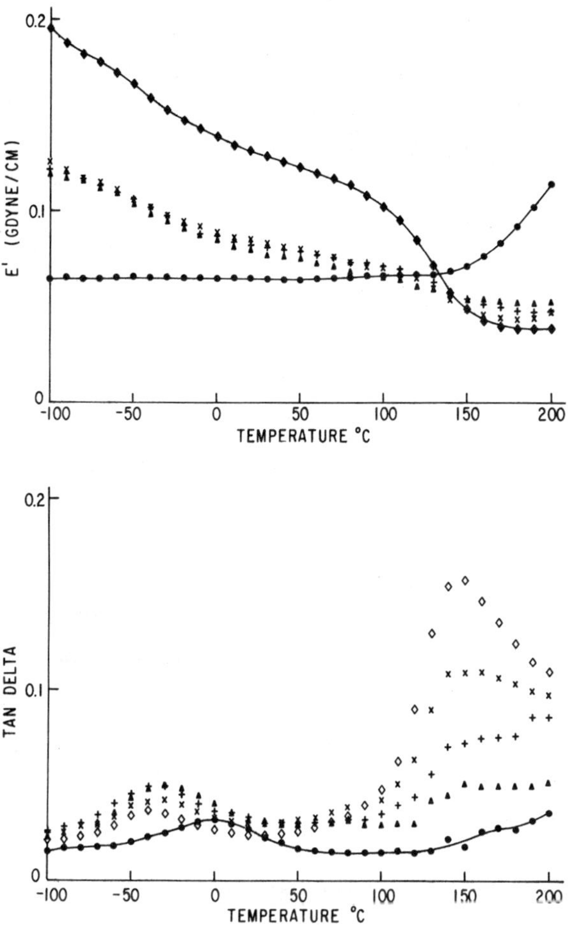

Fig. 4. Dynamic mechanical properties of pHEMA-hide composites. % pHEMA from bottom: 0, 8, 20, 33, and 37.

or endotherm (less than 0.05 kcal/mol) in differential scanning calorimetry. It should be observed that, although the degree of crosslinking is very high, the polymer, which imbibes water up to 25% of final volume, still hydrates to a hydrogel as defined by Wichterle and Lim (13).

The data for the polymer were obtained at two frequencies, 110 and 3.5 Hz, and can be used to calculate a heat of activation of 17 kcal/mole for deformation.

The mechanical behavior of the composites is shown in Fig. 4. The compositions with up to 37% polymer, which had nearly homogeneous polymer protons as indicated by the ^{13}C spectra, have nearly identical E' behavior.

The polymer stiffens the assembly below the glass transition and softens
it above in all compositions, the softening increasing with the amount of
polymer present. The increase in E' of the unimpregnated collagen above
120°C is unexplained, but may be due to dehydration, as observed before
(15). The mechanical relaxation of the pHEMA is clear, with a large maximum in tan δ at exactly the same temperature as observed in the neat
polymer (Fig. 3). As in the case of E', increasing the polymer content
alters the mechanical loss only above temperatures where the polymer is
glassy.

The height of the main loss peak increases with the polymer content
of the composite. That of the low-temperature peak, usually named the γ
peak, decreases with polymer content. Since this peak is not present in
the loss curves of the leather or of the dry polymer film, this behavior
of the composite cannot be understood by superposing the properties of
the components. It has been reported that the γ loss peak of pHEMA is
supressed by small amounts of low molecular-weight material, such as
water. The small decrease of the γ peak might be due to the water content
increasing slightly with polymer content. The water content must be
small, since the position of the main transition is the same for all the
compositions.

Another surprising feature of the data at low temperatures is the
small increase in modulus of these composites over that of leather, considering the relatively large amount of polymer (over 30%) with its modulus over 100 times as great as that of the leather. Above 37% polymer
the modulus does not increase regularly, and segregated aggregates of
polymer are seen in the impregnate. The behavior is not that of a composite of two continuous phases, but rather that of a continuous collagen
phase containing discontinuous bodies of polymer held among the fibrils.

At high polymer contents, above 37%, pHEMA was observed to segregate
into interstices among the fibers. We could see this clearly by fluorescent staining with Sandocryl Brilliant Red by the method of Lowell and
Buechler, but omitting their counterstain (16). Observed under a fluore-
scence microscope, samples containing 37% pHEMA or less showed dimly
fluorescent fibers demarcated by dark lines corresponding to the interstitial
regions; those containing more than this amount were brightly fluorescent
between the fibers. As the polymer becomes less evenly dispersed, the

244

transition temperature moves upward, so that a sample with 50% polymer
had a transition at 160°C. Samples with more than 52% polymer had deposits of polymer that were observable without a microscope and had poorly
reproducible dynamic mechanical properties.

DISCUSSION

We have previously described leather grain impregnated with a polymer prepared in situ from a urethane oligomer and vinyl monomers (15).
By staining and microscopy the polymer was located in the interstices of
the fibrous matrix; the dynamic mechanical properties could be described
by suitable averaging of the properties of the two components. The
particular average used was that of the Takayanagi model (17), which
combines the two viscoelastic components partly in parallel and partly in
series. There was no evidence for chemical or local interaction between
the two components, but there was mechanical interaction which prevented
description of the system either as an assembly of parallel rods of the
two components or as a series of sheets piled transverse to the deformation.

The intimacy of polymer-fiber interaction in the present system was
examined with an NMR technique used previously (8) to explore composites
of leather with hydrophobic polymers prepared from emulsions. Those systems also contained Cr^{+3} ion bound intimately to the protein of the fibrous matrix, which caused the protons of the protein and nearby polymer
to relax rapidly. They, however, could contain only about 2% polymer
closely enough associated with the collagen to be relaxed by the chromium.
The protons of polymer in excess of that amount relaxed much more slowly,
showing that the excess polymer was not located in intimate contact with
the chromium-protein complex. Since all the protein in the composite is
packed into fibers, most of the polymer must have been located between
rather than within these fibers.

In the present work, the polymer was chosen so that it would spread
spontaneously into the pores between collagen fibrils. The surface tension of the polymer was considerably higher than those of the polymers
used previously (4,8,15), about 40 dyne/cm, which matched that of collagen
(5). Unlike the nonpolar monomers, HEMA could penetrate the fibers to
the level of the fibrils. Since it does not disturb the mechanical

integrity of the hide or even soften it, as water does, it must be inferred
that there is no penetration to the molecular level within fibrils.
Again, since the interfibrillar polymer did not affect the elastic modulus
much, interfibrillar friction is not of great importance in determining
the resistance to deformation.

The NMR results show that the first monomer added to the hide penetrates the fibers and that this can rise to about 37% of the total mass.
This material when polymerized has less effect on the storage modulus E'
than has polymer in excess of that. It is concluded that the effects of
impregnant on the modulus of leather that are usually reported (4, 15)
are mostly on the interfiber level, since only this excess polymer raises
the storage modulus substantially. The loss modulus maximum increases
uniformly with increased pHEMA content. Unlike in the system that we
reported earlier (15), E" of the present type of composite does not run
parallel to E' as the temperature is varied, but instead has maxima at
the mechanical transition temperatures of the neat polymer, with rather
large maximal values of tan δ. This is to say that here E" is controlled
by the properties of the interfibrillar polymer, which dominates the
interfiber adhesion or friction. The pHEMA acts as a plasticizer at high
temperature and stiffens the structure at low temperature without increasing
the interfiber friction.

Unlike the hydrophobic ethylhexyl acrylate/vinyl pyrrolidone/ oligoure-
thane mixture impregnated into leather grain described in the previous
paper (15), radiation-polymerized pHEMA in a composite with leather
behaves like that polymerized as a film. Its effect on the leather is to
protect it from chemically induced stiffening above 150 $^\circ$C, while only
slightly stiffening it at lower temperatures. The polymer does penetrate
between the fibrils, to the penultimate level of the fibrous structure,
as shown by proton relaxation of the polymer by the collagen-bound chromium.
The temperature of the transition of pHEMA is sensitive to its water
content. The presented data describes samples which we tried to keep
dry, but they must have contained very small amounts of water that were
difficult to remove.

It is shown that it is possible, by choosing monomers with high
surface tension, to impregnate the fibers of hide to the level of the
fibrils. The result, even when a glassy polymer is used with a strong

transition and a very high modulus at low temperature, is a composite
with the mechanical properties little different from those of the starting
leather at all temperatures. Composites were also prepared similar to
those discussed here using untanned calfhide. The polymer does raise the
denaturation temperature of the collagen in these by 10°C as determined
by differential scanning calorimetry, probably by stabilizing the fibrils,
which in turn stabilize the cooperative forces among the molecules. It
should be mentioned that the radiation itself lowered the denaturation
temperature of the hide by that much judging from the control, so the
dose of 3 Mrad was apparently excessive, even for polymerization. Interfibrillar pHEMA should permit small molecular-weight materials to diffuse
to the collagen and vapors to diffuse among the fibers, while, we believe,
still preventing microbial degradation of the protein.

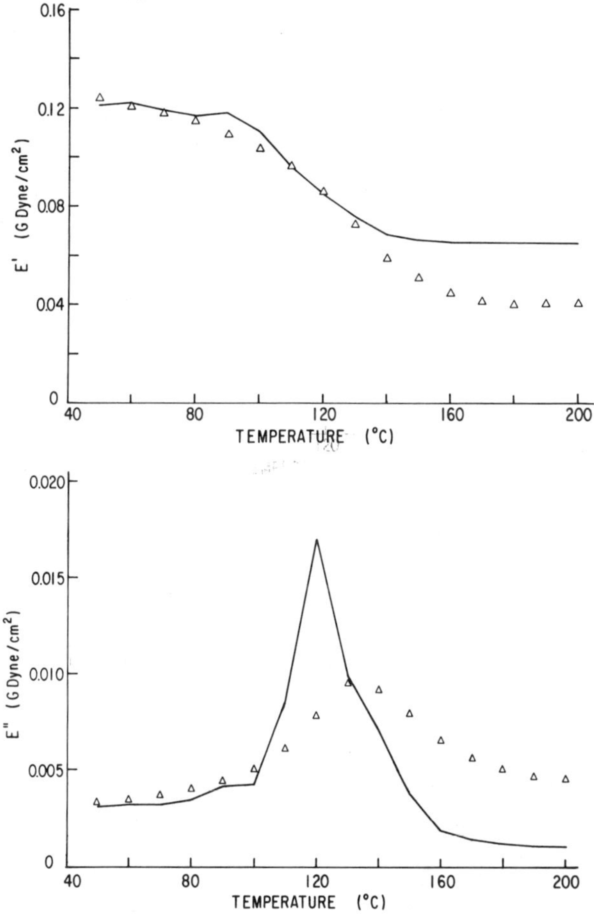

Fig. 5. Takayanagi model fitted to 37%-pHEMA composite with ϕ=0.37 and q
= 0 and 0.1. Points, experimental; dotted line, q = 0; solid line, q =
0.1.

Attempts were made to model these composites by using the Takayanagi model (17). This calculation defines E^* of the composite as a weighted average of those of the components. The weights take into account geometric effects found when the spaces within a non-parallel assembly of fibers are filled with a continuous phase, so that the two components of the composite are arranged spacially neither completely in parallel nor in series. The geometry is accounted for by a parameter q, which would be zero in the case of a parallel assembly. The model as formulated is not appropriate for this system, because it assumes that all the volume of the material is occupied by one or the other component. These composites with amounts of polymer less than 37% contain empty spaces among the fibers. Because of this feature of the fibrous structure the modulus of the unimpregnated leather is orders of magnitude smaller than that of dry collagen or even gelatine. It should be observed, however, that the geometry of the pHEMA phase is the same as that of the collagen, since it shares the same space within the fibers, so an assembly of pHEMA fibers identical to that of the collagen fibers in the leather would be expected to have a modulus reduced by the same amount from that of the neat polymer film. Therefore, to account for the fibrous structure with its free space in the model, we divided the values of E' and E" of the pHEMA (Fig. 3) by 100 and used these values with those of unimpregnated leather to represent the properties of the separate components. This is equivalent to introducing a different geometrical factor into the original calculations of the moduli from the experimental data.

Another difficulty with modeling this system is the large increase in E' of the unimpregnated leather at temperatures above $150\,^\circ C$, which is absent in the samples containing polymer. Since the polymer seems to interact with the collagen to prevent this effect, we artificially suppressed it in the calculations by holding the moduli of collagen, E' and E", constant above $140\,^\circ C$ at their values at this temperature.

The composite containing 37% polymer has the smallest amount of free space within its fibers, while the spaces between the fibers are open, as described above. Its dynamic moduli are compared with those calculated by means of the Takayanagi series-parallel scheme in Fig. 5. Our working equations were shown previously (15), but here the parameter ϕ is the volume fraction of the polymer, instead of the ratio of polymer to collagen. The best fit was obtained with q = 0.1, a rather small value, and

even q = 0 was acceptable. Deviations are the lower values of E' for the
sample at temperatures above the transition (where we held the moduli of
the leather constant) and its higher transition temperature than found in
the model. The low values of the moduli suggest a failure of the model
to account for plasticization by the softened polymer, as suggested
above. The elevated transition temperature of the sample may not be
significant; both 120° C and 140° C are within the transition ranges of the
sample and of the polymer, respectively. Aside from these two reservations,
it must be concluded that the system behaves as an assembly of fibers of
pHEMA and collagen, the two types of fibers lying almost parallel to each
other with little mechanical interaction. This behavior reflects a
geometry of collagen fibrils lying parallel to each other in fibers, with
pHEMA lying in the long interstices among and parallel to the fibrils.

ACKNOWLEDGEMENTS

We thank Dr. Philip E. Pfeffer for helpful guidance and discussion
in NMR, John Lin for supervising the γ-ray irradiations, and Mathew Dahms
for microscopy.

REFERENCES

1. M. Chvapil, R. Holusa, K. Kliment, and M. Stoll, Some chemical
    and biological characteristics of a new collagen-polymer compound
    material, J. Biomed. Mater. Res. 3:315 (1969).

2. P. N. Sawyer, B. Staczewski, and D. Kirschenbaum. The devel-
    opment of polymeric cardiovascular collagen prostheses, Artif.
    Organs 1:83 (1977).

3. K. P. Rao, K. T. Joseph, and Y. Nayudamma, Characterization of
    the collagen-vinyl copolymers prepared by the ceric ion method.
    I. Solution properties, J. Polym. Sci., Part A1, 9:3199 (1971).

4. E. F. Jordan, Jr., B. Artymyshyn, and S. H. Feairheller, Polymer-
    leather composites. IV. Mechanical properties of selected acrylic
    polymer-leather composites, J. Appl. Polym. Sci., 26:463 (1981).

5. R. E. Baier and W. A. Zisman, Wetting properties of collagen and
    gelatin surfaces, in Applied Chemistry and Protein Interfaces, R. E.
    Baier, Ed., Advances in Chemistry Series 145, 155-174 (1975).

6. M. M. Taylor, E. J. Diefendorf, M. V. Hannigan, B. Artymyshyn, J. G.
    Phillips, S. H. Feairheller, and D. G. Bailey, Wet process technology.
    III. Development of a standard process, J. Amer. Leather
    Chem. Assn. (In Press).

7. W. V. Gerasimowicz, K. B. Hicks, and P. E. Pfeffer, Macromole-
    cules (In press).

8. P. E. Pfeffer, D. G. Bailey, and E. F. Jordan, Evidence for limited
    phase homogeneity in PMMA "grafted" chrome-tanned leather by ^{13}C
    CP-MAS, NMR, 25th Experimental NMR Conference, Wilmington, DE 8-12
    April, 1984.

9. M. J. Sullivan, and G. E. Maciel, Anal. Chem. 54:1601 (1982).

10. A. Pines, M. G. Gibby, and J. S. Waugh, J. Chem. Phys., 59:569-590
    (1973).

11. S. J. Opella and M. H. Frey, J. Am. Chem. Soc. 101:5854
    (1979).

12. M. Takayanagi, Viscoelastic properties of crystalline polymers,
    Mem. Fac. Eng. Kyushu Univ. 23:41 (1963).

13. O. Wichterle and D. Lim, Hydrophilic gels for biological use,
    Nature, 185:117 (1960).

14. O. Wichterle, Hydrogels, in "Encylopedia of Polymer Science and
    Technology," N. M. Bikales, ed., Wiley, New York (1971).

15. P.L. Kronick, P. Buechler, Frank Scholnick, and Bohdan Artymyshyn,
    Dynamic mechanical properties of polymer-leather composites, J.
    Appl. Polym Sci. 30:3095 (1985).

16. J. A. Lowell and P. B. Buechler, Studies on impregnation. Determi-
    nation of the depth of polymer penetration, J. Am. Leather Chem.
    Assoc. 60:519 (1965).

17. M. Takayanagi, S. Uemura, and S. Minimi, Application of the equivalent-
    model method to dynamic rheo-optical properties of crystalline
    polymer, J. Polym. Sci. Part C5, 113 (1964).

RADIATION POLYMERIZED GRAFT COPOLYMERS FOR LEATHER

Peter R. Buechler, Paul L. Kronick, and Frank Scholnick

US Department of Agriculture
ARS, Eastern Regional Research Center
600 East Mermaid Lane
Philadelphia, PA   19118

Radiation cured coatings for leather have been discussed in a series
of publications (1-10). Somewhat less attention has been focused on the
use of radiation polymerization of vinyl monomers within the leather to
modify its properties in spite of an early suggestion by Buechler (1, 3)
that this should also be possible and should be considered for full grain
(unbuffed) leathers.

The reasons for this suggestion arose from the trend in the shoe upper
leather industry to replace lightly buffed leathers split from cattle hides
to a desired thickness with unbuffed full grain leather split to a similar
thickness. The ratio of the corium minor (grain layer) which contains fine
collagen fiber bundles to the deeper corium major layer varies with the
thickness to which the leather has been split but is usually less than 1.0.
In multiple creasing of the leather, as occurs in the vamp of a shoe, the
coarse fiber structure in the corium major frequently dominates the fine
fiber structure in the corium minor and leads to unsightly folds. Buechler
(11) has shown that the impregnation of the buffed grain layer with pre-
formed polymer improved both the aesthetics and durability of the leather.
Scuff resistance of the leather is greatly improved by the impregnation
treatment. The polymer impregnant should have a low glass transition temperature, high molecular weight and compatibility with the fibrous component of the leather, the collagen fibers. Polymers employed have included
alkaline-solubilized acrylic polymers in water (12), soluble acrylic polymers (13), soluble polyurethanes (14) and co-reacted polyurethane and
acrylic polymers (15). Due to greater ease in handling, lower cost and
better creasing improvement the acrylics have been favored. The approach
described in this paper has been to use a low molecular weight oligomer
dissolved in monomers to penetrate into the corium minor of full-grain
leathers and then radiation polymerized in situ using either electron beams
(EB) or ultraviolet light (UV). When ultraviolet irradiation was employed
a photosensitive initiator had to be employed to produce free radicals.

EXPERIMENTAL

The oligomer/monomer solutions were applied to the grain surface of
the leather with a thoroughly wet plush swab. The solutions were rapidly
absorbed. The swab was then wiped semi-dry and used to wipe off excess
from the surface.

Radiation with electron beams was carried out using an Energy Sciences Corp. EB unit. Most impregnation work utilized 200 KEV electrons. Dosages were varied from 0.5 to 6 megarads during the course of the experiments. Polymerization occurred in seconds.

Polymer content was determined by changes in hydroxyproline and extractables after impregnation and polymerization. Bound (unextractable) polymer may not be truly grafted (16) but was reported as such in accord with custom. As will be seen below there is evidence of at least mechanical interaction between polymer and leather.

Depth of polymer penetration was determined histochemically using special adaptation of method using Sandocryl Brilliant Red B-4G and an Auramine O counterstain. The microscope employed was a Nikon Optiphot equipped with an ocular micrometer.

Dynamic mechanical testing was carried out with a Rheovibron II dynamic viscoelastometer (Toyo Baldwin Co. Ltd.) (17). The real and imaginary parts of the complex modulus were determined at three frequencies (3.5, 11, and 110 Hz) at an oscillating relative deformation of 3.8×10^{-4}. This instrument measures the amplitudes of the harmonic force and amplitude signals and determines their phase, δ, with a high precision. These data give the real and imaginary parts E' and E'' of the complex modulus: $\hat{E} = E' + iE''$, with $\tan \delta = E''/E'$ where E' is the elastic modulus and E'' is the loss modulus.

RESULTS

Cross-sections of impregnated and unimpregnated leather are compared in Figure 1. Figure 1A shows a sample of the impregnated grain layer of cattlehide leather. The dark stain shows the location of radiation polymerized polymer. Figure 1B shows the control area similarly treated histochemically but polymer is absent.

Table I, shows that epoxy oligomer (Celrad 3500) shows less penetration and a lesser amount of bound polymer (lower graft efficiency) than the polyurethane oligomer (Celrad 6700) mixtures. At higher dosage it seemed to benefit more than the others. This may be due to the fact that N-vinyl pyrrolidone (NVP) accelerates cure more than most monomers, and at low dosages some of the somewhat volatile methyl methacrylate may be lost. With NVP present cure is usually fairly complete at absorbed doses in excess of 2.5 Mrad.

Table II, shows the effect of adding 10% (based on the solution) of crosslinker trimethylolpropanetriacrylate (TMPTA). It was expected to increase the amount of unextractable polymer and hence apparent graft efficiency. However, the percent of bonded polymer was already so great that its effect was insignificant except for a little deeper penetration due to decreasing the viscosity of the oligomer solution to which it was added.

Figure 2 shows the dynamic mechanical properties of a film of the mixture described in Table II which had been drawn on a glass plate and cured in an electron beam. There is evidence of two transitions, neither of them sharp, at -35°C and 30°C, respectively. This can be compared with a similar film containing also 2% diethoxyacetophenone as sensitizer, polymerized with ultraviolet light from a Fusion Systems Corporation (Rockville, Md.) conveyorized Model F440 system. The UV light came from a 300 W/in

---

Reference to brand or firm name does not constitute endorsement by the U.S. Department of Agriculture over others of a similar nature not mentioned.

lamp with principal radiation at 210-270 nm; the radiation time was about
1 s. Only one transition is evident in this film, at 25°C, which is very
sharp (Fig. 3). Thus the elastic modulus of the electron beam-cured film
is orders of magnitude greater than that of the UV-cured film at room temperature but does not abruptly increase between room temperature and 0°C.

1A.   Impregnated

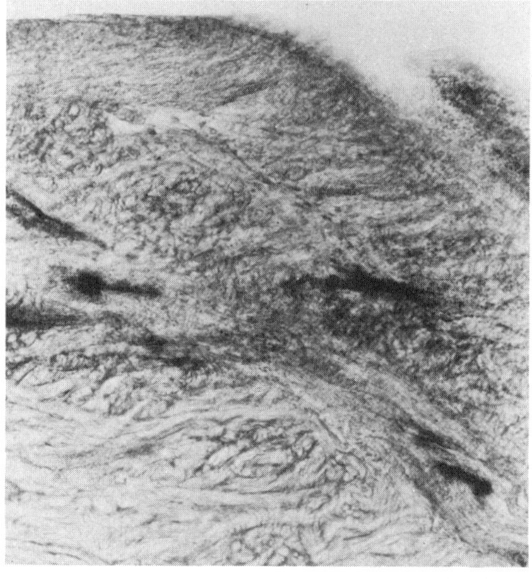

1B.   Control

Fig. 1.   Leather Cross-Sections

TABLE I
Depth of Penetration and Graft Efficiency after EB Cure

Impregnation System	EB Dosage (Mrad)	Depth of Penetration (mm)	Graft Efficiency (%)
50% Epoxy Olig./25% MMA/25% EHA	1.5	0.034	8.61
	3.0	0.050	24.4
	6.0	0.083	63.1
50% Urethane Olig./25% NVP/25% EHA	1.5	0.234	52.5
	3.0	0.246	75.4
	6.0	0.244	87.3
33% Urethane Olig./33% NVP/33% EHA	1.5	0.211	85.0
	3.0	0.176	95.7
	6.0	0.163	75.6

TABLE II
EB CURE (3 Mrad) after Impregnation of Commercial
Leathers with 33% Urethane Oligomer/33% NVP/33% EHA

Leather Type	Crosslinking Agent	Depth of Penetration (mm)	Graft Efficiency (%)
Side Leather A	None	0.254	100
	TMPTA	0.432	73.3
Side Leather B	None	0.254	97.9
	TMPTA	0.330	87.9
Pigskin C	None	0.559	71.9
	TMPTA	0.584	76.7

The polymer impregnated layer from an electron beam-cured composite
was separated and subjected to dynamic mechanical testing using the
Rheovibron instrumentation. The effect of the impregnation on the dynamic
mechanical properties of the leather grain layer can be seen by comparing
the untreated leather (Figure 4) with the impregnated sample data plotted
in Figure 5. The mechanical response of the leather is less sensitive to
temperature than that of the composite near room temperature, which shows
a vague trace of the underlying glass transition of the polymer. We attempted to model the composite by means of the series-parallel scheme of
Takayanagi (18), with his coupling parameter, q, of 0 or 0.3, using the
experimentally determined volume fraction of polymer of 0.6. A parameter
of q=0 refers to infinite parallel fibers of leather and polymer; that of
0.3, shorter fibers in various directions, which optimizes the fit.

It is clear from the poor fit in both cases that the geometry alone
cannot reconcile the very large shift in transition temperature from 30°C
in the film down to about 10°C or lower in the impregnated sample. The
shapes of the q=0 calculated curves in Fig. 5 resemble those of the sample
curves, although shifted to higher temperatures, but the q=0.3 curves represent the data better. In fact, from the dynamic mechanical properties
alone one would not discern that the same polymer mixture was used in the

film and in the leather grain. It appears that ingredients of the leather
have altered the character of the polymer, such as with shorter polymer
chains initiated or terminated by radicals formed in the collagen or a
lower degree of crosslinking.

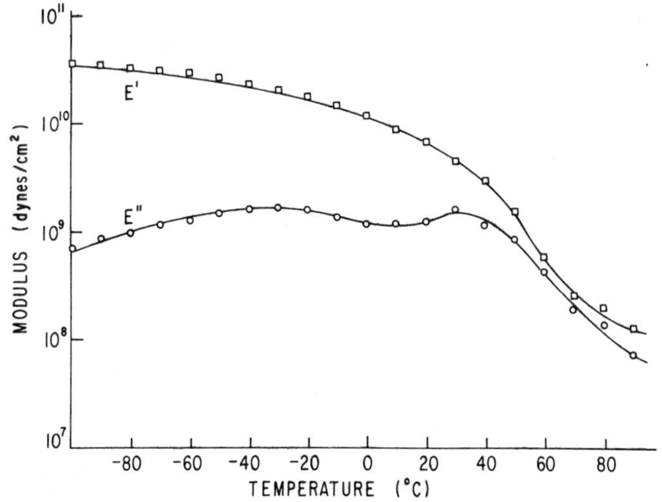

Fig. 2.   EB Cured Film (PU/NVP/EHA = 1/1/1)

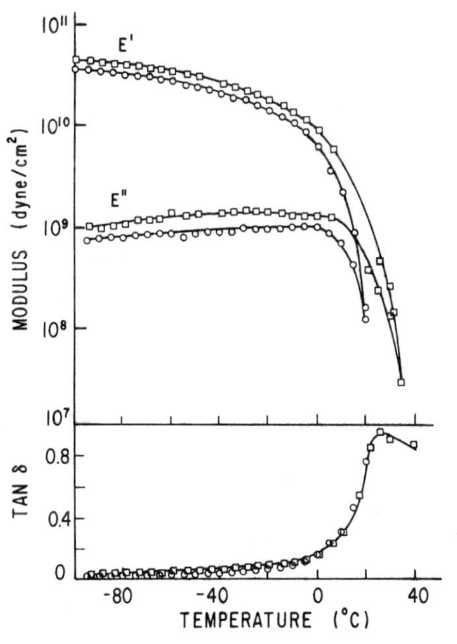

Fig. 3.   UV Cured Film (PU/NVP/EHA = 1/1/1)

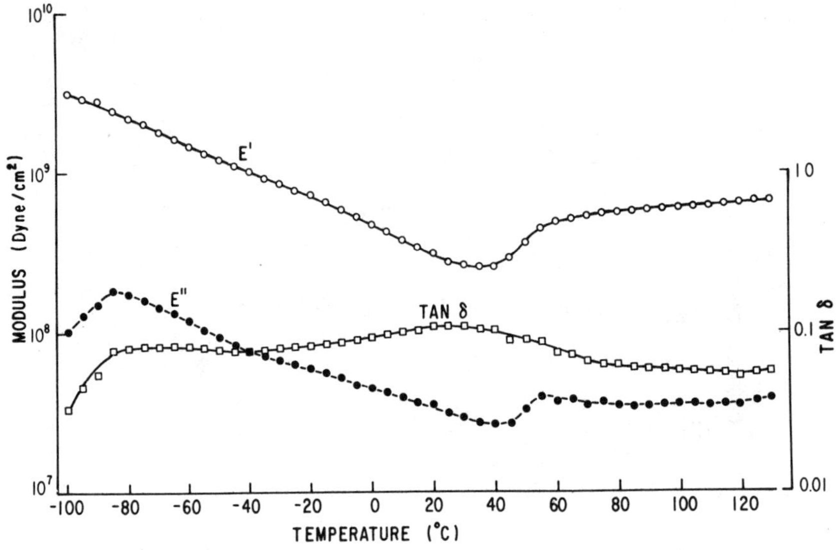

Fig. 4.   Untreated Grain Layer

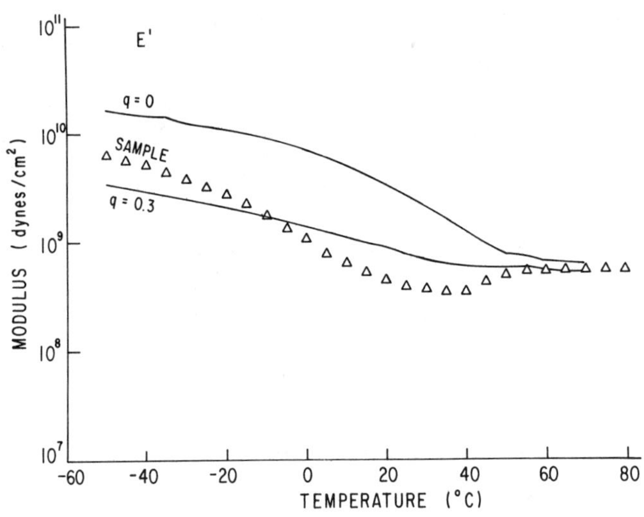

Fig. 5A.   Comparison of Takayanagi Models with EB-Cured Composites

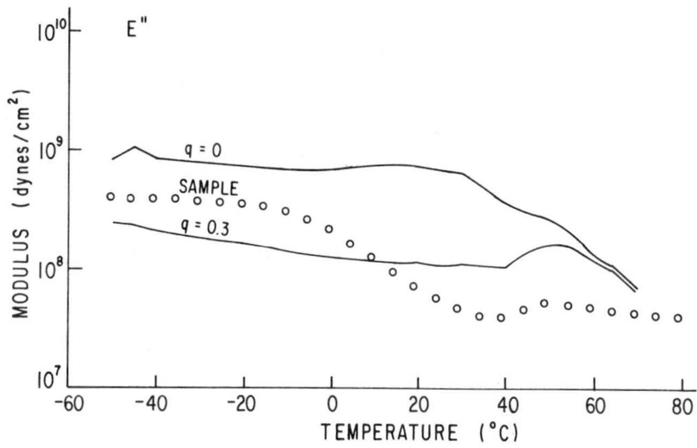

Fig. 5B. Comparison of Takayanagi Models with EB-Cured Composites

References

1. P. R. Buechler, Radiation Curing and Its Potential for Leather
    Coatings, J. Am. Leather Chem. Assoc., 72:193 (1977).
2. M. A. Knight and A. G. Marriott, UV Curing Polymers in Leather
    Finishing, XV Kongress der Internationalen Union der Leder-Techniker-
    und-Chemiker-Verbände, Hamburg, Germany, Texts of Papers VI/5 (September
    1977).
3. P. R. Buechler, Radiation Curing & Its Potential for Leather Coatings,
    J. Am. Leather Chem. Assoc., 73:56 (1978).
4. M. A. Knight and A. G. Marriott, UV Curing Polymers in Leather Finishing,
    J. Soc. Leather Technol. Chem., 62:14 (1978).
5. L. Rasmussen, Prospects for Use of UV Curing in Automotive Upholstery
    Leather Coatings, Radiation Curing IV, Chicago, Illinois, September
    1978, sponsored by the Association for Finishing Processes, SME.
6. F. Scholnick, E. H. Harris, and P. R. Buechler, Radiation Cured Coatings
    for Leather. I, Topcoats for Side Leather, J. Am. Leather Chem. Assoc.,
    77:93 (1982).
7. P. R. Buechler, F. Scholnick, and E. J. Diefendorf, Radiation Cured
    Coatings for Leather. II, Color Coats for Side Leather, J. Am. Leather
    Chem. Assoc., 77:269 (1982).
8. F. Scholnick, E. H. Harris, and P. R. Buechler, Radiation Cured Coatings
    for Leather. III, A Review of Research at the Eastern Regional Research
    Center, Chapter in: "Organic Coatings Science and Technology," Volume 6,
    G. D. Parfitt and A. V. Patsis, ed., Marcel Dekker, Inc., New York (1984).
9. M. A. Knight and A. G. Marriott, UV Drying of Polymers in the Process of
    Leather Finishing, Rev. Tech. Inds. Cuir, 74:58 (1982).
10. J. D. Rock and J. L. Garnett, Radiation Cured Coating for Leather, U.S.
    Pat. 4,268,580, May 19, 1981.
11. P. R. Buechler, Improvement of Leather With Aqueous Impregnating Agents,
    Leather Manufacturer, 78:19 (1961).
12. J. A. Lowell, H. L. Hatton, F. J. Glavis, and P. R. Buechler, Leather
    and Method for Producing It, U.S. Pat. 3,103,447 (September 10, 1963).

13. J. A. Lowell, E. H. Kroeker and P. R. Buechler, Process for Treating Leather and Leathers Obtained, U.S. Pat, 3,231,420 (January 25, 1966).

14. M. B. Neher and V. G. Vely, Leather Treatment Process and Composition, U.S. Pat. 3,066,997 (December 14, 1962).

15. J. A. Lowell and P. R. Buechler, Process and Compositions for Treating Leather and Leathers Obtained, U.S. Pat. 3,441,365 (April 29, 1969).

16. E. F. Jordan, Jr., Polymer - Leather Composites V in "Polymer Applications of Renewable Resource Materials," C. E. Carraher, Jr. and L. H. Sperling, ed., Plenum Press, New York (1983).

17. M. Takayanagi, Viscoelastic Properties of Crystalline Polymers, Mem. Fac. Eng. Kyushu Univ., 23:41 (1963).

18. M. Takayanagi, S. Uemura, and S. Minami, Application of Equivalent Model Method to Dynamic Rheo-Optical Properties of Crystalline Polymers, J. Pol. Sci., Part C5, 113 (1964).

SECTION VI - RUBBER, LIGNIN, AND TANNIN

CHEMICAL MODIFICATION OF NATURAL RUBBER AS A ROUTE TO RENEWABLE
RESOURCE ELASTOMERS

Ian R. Gelling

The Malaysian Rubber Producer's Research Association
Tun Abdul Razak Laboratory, Brickendonbury
Hertford SG13 8NL, England

INTRODUCTION

Natural rubber (NR) is commonly obtained from the latex of the
Hevea brasiliensis tree, which is indigenous to South America. The
latex is contained within a system of anastomosing articulated vessels
which occur in the region of the phloem as sheaths concentric with the
outer bark. There are limited anastomoses between each ring of vessels
forming a continuous network and thus allowing the withdrawal of latex
to be made by repetitive tapping without serious damage to the tree.
NR can also be obtained from the Guayle plant (Parthenium argentatum)
but in this case the latex is present in ordinary parenchyma cells and
the whole plant has to be sacrificed to obtain the rubber. A Hevea
tree has a useful productive life of between fifteen and twenty years
and is by far the most efficient source of natural rubber.

Hevea latex consists of rubber particles ranging in diameter from
0.04 to 2.0 microns dispersed in an aqueous phase. The rubber content
of the latex varies between approximately 25 and 35 w/w % depending on
climatic conditions and other variables. Also present in the latex are
proteins (1.0%), lipids (0.9%), quebtachitol (1.0%), inorganic salts
(0.5%) and many other minor components. The vast majority of NR is
utilized as a dry product which consists of 95% rubber hydrocarbon and
5% non-rubbers.

The chemical structure of NR hydrocarbon was first elucidated by
Harries[2] and Pummerer[3] as a 1,4-polyisoprenoid with the monomer units
C_5H_8 joined "head to tail." Modern analytical techniques[4] have
confirmed this basic structure and shown that NR is the stereoregular
all _cis_ 1,4-polyisoprene (1) with a weight-average molecular weight
($\bar{M}_w$) of one to two million.

$$\underset{\text{(1)}}{\cdots\cdots}$$

(1)

The growth of NR as a large scale commercial polymer commenced with the invention of the pneumatic tyre and increased with the growth of the automotive industry. Production of natural rubber is centred in S.E. Asia, particularly in Malaysia and Indonesia. The growth of NR compared to synthetic rubbers, which are petroleum based products, is illustrated in Fig. 1 with the historic and economic events clearly evident. The severe disruption of supplies during World War II stimulated the development of synthetic rubbers especially in the U.S.A. Cheap and abundant supplies of crude oil in the 1950's and 1960's resulted in a dramatic growth in the production of synthetic rubber, which was briefly checked by the "oil crisis" in the early seventies. Today total rubber consumption is approximately twelve million tonnes per annum of which two thirds is synthetic rubber, mainly styrene butadiene copolymer (SBR) and the remainder NR.

Synthetic rubbers are based on non-renewable crude oil while natural rubber is a renewable resource polymer. The total energy content of synthetic rubbers, from crude oil to finished polymer, can readily be calculated and some values are recorded[5] in Table 1 together

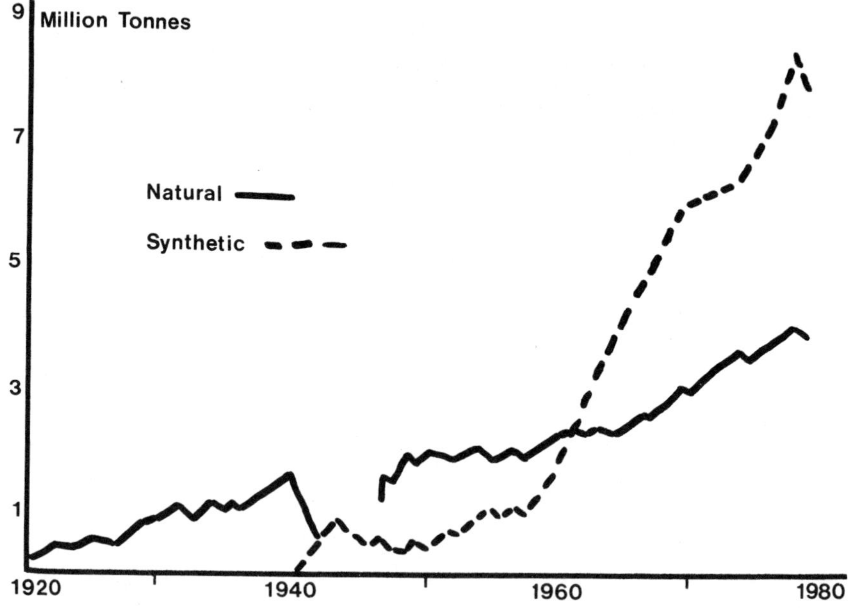

Fig. 1. Annual world production of Natural and Synthetic rubbers.

262

with the non-renewable energy content of NR eg. fertilizers, processing
and transport.   Thus the synthetic rubber industry consumes 24 million
tonnes of oil per annum (the energy content of a tonne of crude oil
is approximately 50GJ) or 450,000 barrels of oil per day.

In certain applications the choice between NR and a general purpose
synthetic rubber is based on financial considerations, but in other
areas technical considerations will be the determining factor.   For
example, NR cannot compete with some of the speciality synthetics with
regard to oil resistance and gas permeability properties.   It has been
recognised for many years that chemical modification can change the
physical properties of NR, and thus extend its applicational area.
Provided the modification is energy efficient, this approach could
result in considerable savings of non-renewable energy.

The chemical modification of NR is reviewed with the emphasis on
recent work, which has both the technical and economic potential to
replace some oil-based synthetic rubbers.

Table 1

Non-renewable Energy Content of Rubbers

	GJ/tonne
SBR	156
Butyl rubber	209
Polychloroprene	144
EPDM	170
Natural Rubber	15

CHEMICAL MODIFICATION OF NATURAL RUBBER

In theory natural rubber can be treated as a simple olefin and
should therefore undergo the many chemical reactions of this species.
However, in practice the non-rubbers can compete or interfere with
many reactions and the polymeric nature can cause special problems.
A large number of chemical modifications have been investigated in
the past with some success.   Chlorinated[6,7],hydrochlorinated (2),
and cyclized natural rubber[9] were all produced as commercial materials.
Chlorinated rubber, a white thermoplastic powder, was principally
employed in protective paints and coatings where resistance to
chemical or corrosive atmospheres was required.   The hydrochlorinated
derivative was utilized as a packaging film (Pliofilm, Goodyear
Tire and Rubber Co.) and cyclized rubber as a thermoplastic resin.
Today these materials have been largely superseded by oil based
products.

Thiols and related compounds add to the double bond of NR (1) by
a free radical mechanism[10].

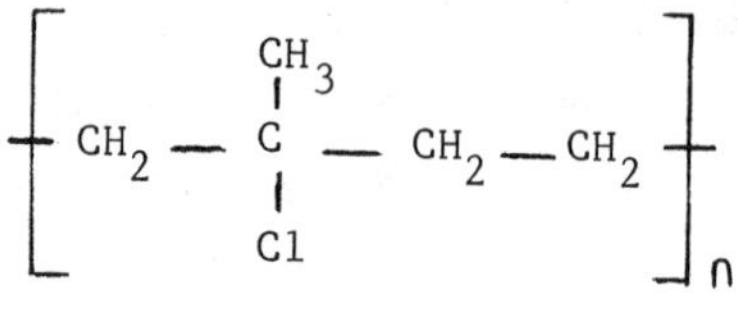

$$\left[CH_2 - \overset{\overset{\displaystyle CH_3}{|}}{\underset{\underset{\displaystyle Cl}{|}}{C}} - CH_2 - CH_2 \right]_n$$

$$(2)$$

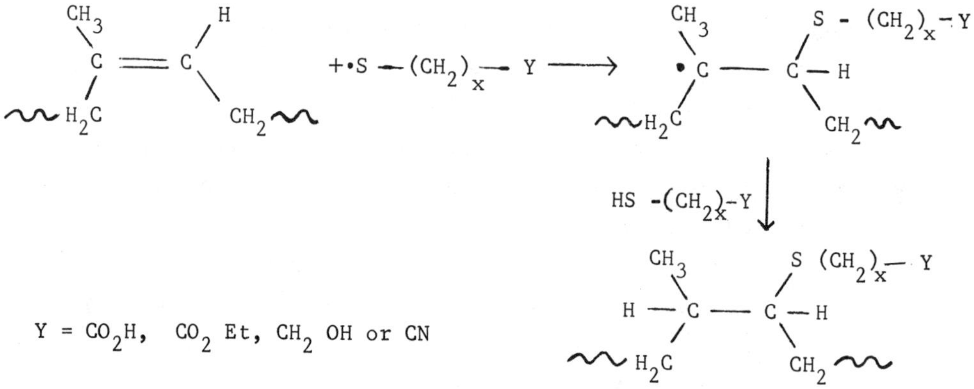

$$Y = CO_2H, \quad CO_2 Et, \quad CH_2 OH \text{ or } CN$$

However, this type of addition has the disadvantage that
cis-trans isomerisation can be induced by the intermediate thiyl
radical. Nitrenes and carbenes[11], maleic anhydride and maleimides[12],
nitrones[13], and aldehydes[14], are among the other classes of chemicals
that have been employed to modify NR, but all have technical or
economic disadvantages.

It is now recognized that any chemical modification of NR
must obey the following criteria if it is to be technologically
and commercially viable.

1. Reagents should be specific and not induce changes in the
natural rubber molecule (ie degradation, cyclization, crosslinking
etc) other than that intended.

2. No catalysts that are poisoned by non-rubbers should be
employed.

3. Reactions should be capable of being carried out with
high efficiency in latex or during conventional dry rubber mixing
or curing.

4. Reagents and processes should be cheap.

During the last few years a number of reactions have been
studied which fit the above criteria and have the potential to
yield commercially viable materials.

<u>'Ene' Reactions</u>

The general 'ene' reaction is illustrated below and early work
with nitroso 'ene' reagents lead to rubber-bound antioxidants[15]
and a novel di-urethane (Novor) crosslinking system[16]. Subsequently,
azodicarboxylates have been examined in some detail as 'ene' reagents

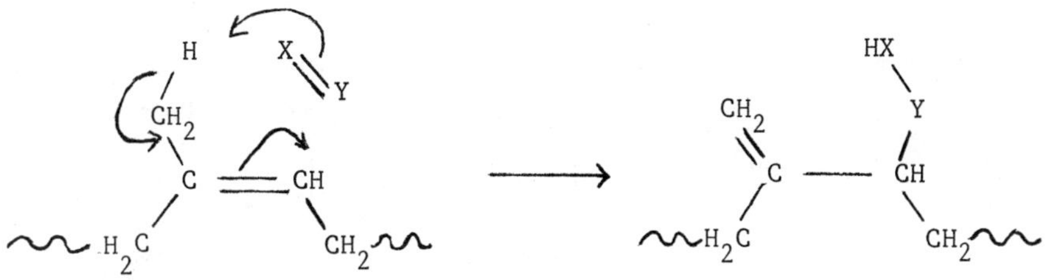

X = Y can be $-N = O$, $- N=N$, $> C = S$, $> C = O$, $> C = C<$
as it is known that their reaction with most olefins is insensitive
to radical initiators or scavengers and to solvent type[17].

Ethyl N-phenylcarbamoylazoformate (ENPCAF) (3) readily reacts
with dry NR (1) and modification can be carried out in an internal
mixer as part a of standard rubber compounding exercise.
Substantial changes in physical properties can be achieved and a
variety of functional groups introduced onto the rubber backbone[18]
by substitution of the phenyl group.

The effect of ENPCAF modification on the gas permeability
and solvent resistance properties is illustrated in Table 2.

Table 2

Effect of ENPCAF modification on the gas permeability
and solvent resistance of Natural Rubber

Modification level (mole %)	Permeability const, p/N_2 gas $(cm^2 s^{-1} atm^{-1}) \times 10^8$	Linear Swelling in Petroleum ether (60-80) $1/1_o$
0	3.5	1.51
5	2.04	1.3
10	0.98	1.2

1_o is the initial length of rubber sample and 1 length after
swelling to equilibrium

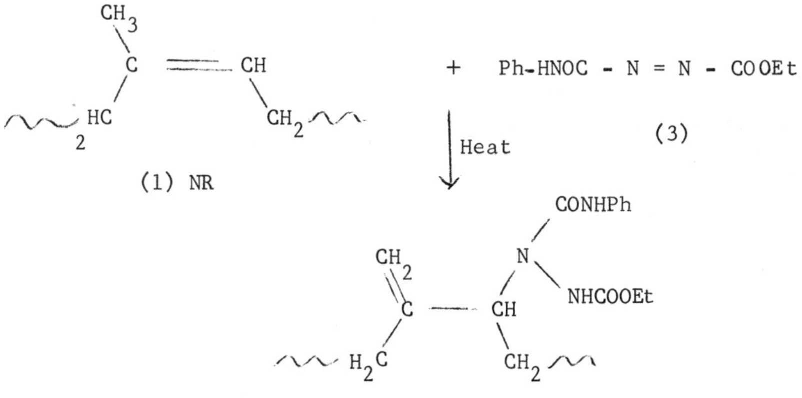

ENPCAF MODIFIED NR

The introduction of a trialkoxysilyl group $\left[-CH_2CH_2CH_2Si(OR)_3\right]$
into the ENPCAF molecule results in a modification that also
functions as a silane coupling agent[19], which enhances the degree of
reinforcement by silica fillers.

The hydrazo-ester pendent groups, introduced by ENPCAF
modification, are very polar and readily undergo hydrogen bonding
with one another. This interaction retards the rate of
crystallization of NR. A modification level of one mole %
has been estimated to be sufficient to delay the crystallization
hardening that occurs when NR is stored or transported at low
temperatures.

Thermoplastic rubbers, ie materials which behave as if
vulcanized at ambient temperatures but which can be processed
like plastics at elevated temperatures, are becoming increasingly
important, as significant savings in both production and energy
costs can be achieved. Thermoplastic NR's have been developed
by employing the azo 'ene' reaction to attach polystyrene chains
to the NR backbone in the form of a comb effect. Anionically

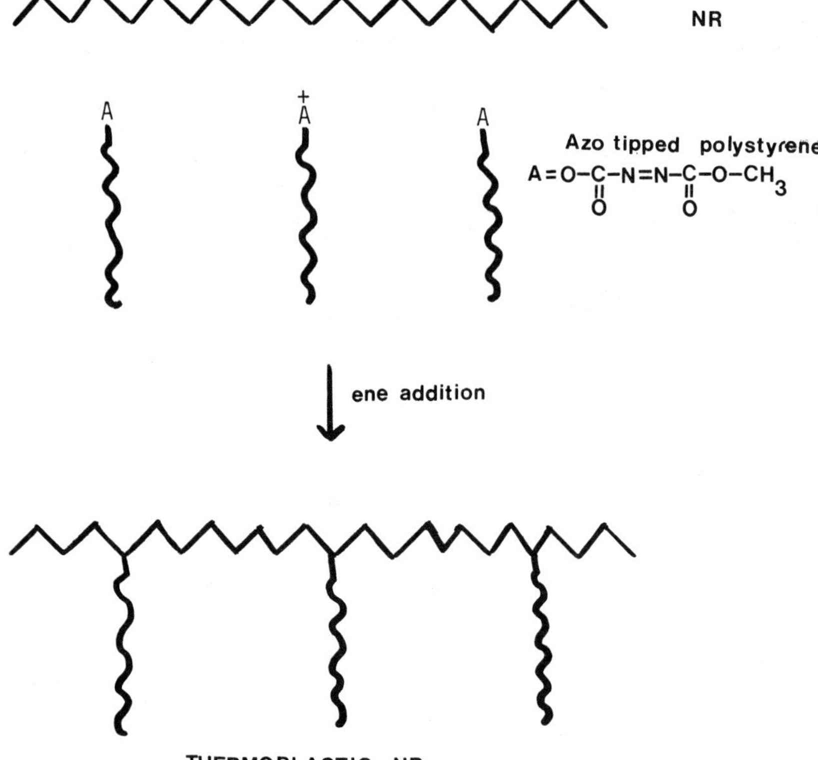

Fig. 2. The grafting of azo tipped polystyrene onto NR.

polymerised polystyrene with a terminal hydroxyl group is modified
to yield a terminal azo group. At high shear and at temperatures
above the softening point of polystyrene the mixing of NR and azo
tipped material yield a graft copolymer (Fig.2)[20,26]. Grafting
efficiencies of over 70% can be obtained. Within well defined
limits these graft materials show thermoplastic rubber properties
very similar to the styrene butadiene styrene (SBS) block copolymers.
Optimum strength properties are obtained with polystyrene molecular
weights of around 8000 (Fig. 3) and at a polystyrene content of 40%
w/w (Fig. 4).

At present these NR graft materials are not cost competitive
with the all synthetic SBS block copolymers in general applications,
as the azo tipping chemistry is relatively expensive and the
natural rubber based materials cannot be oil extended to the same
extent as the SBS copolymers. However, they have potential as
adhesives and blending aids. A variety of polymers, including
blends of different materials can be grafted onto NR by this
technique.

Although not a chemical modification in the true sense,
thermoplastic NR can also be obtained by blending NR with polyolefins
at high shear rates and temperatures in excess of the melting
point of the polyolefin phase[22]. A range of materials based on

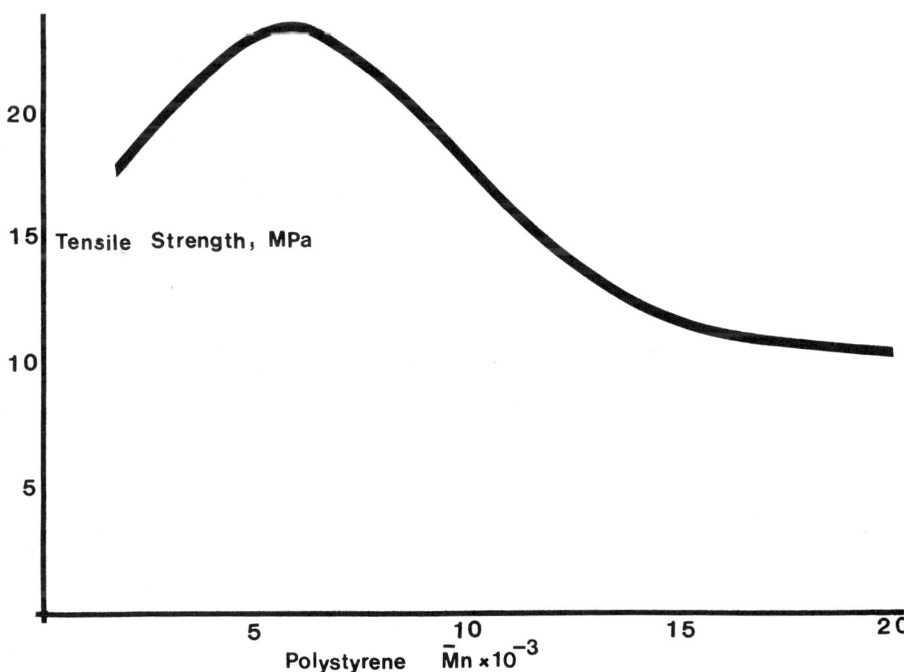

Fig. 3. Tensile strength as a function of polystyrene Molecular
weight for azo graft polystyrene 40/NR 60 thermoplastic rubbers.

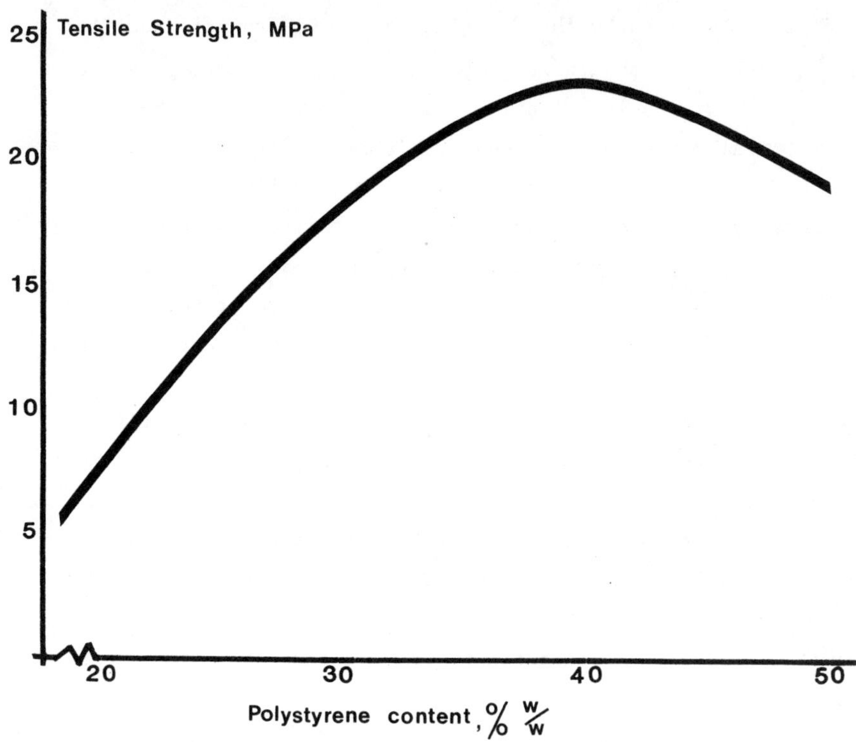

Fig. 4. Tensile strength as a function of polystyrene ($\bar{M}n$ 8200) content for NR thermoplastic rubbers.

polypropylene and polyethylene have been produced and the properties vary with the polyolefin composition. At high polyolefin concentrations the NR basically acts as a modifier to improve impact strength, but at lower levels of polyolefin the products are thermoplastic rubbers. The physical properties of these blends are very similar to the EPDM/polyolefin materials. These NR based materials are currently undergoing commercial development and are expected to be economically attractive, especially in NR producing countries.

Epoxidized Natural Rubber

Of the recent chemical modifications of NR, epoxidized natural rubber (ENR) is the most attractive from both a technical and economic viewpoint. Under controlled conditions the reaction of NR latex with either a 35% w/w solution of peroxyacetic acid[23] or peroxyformic acid, formed 'in-situ' from hydrogen peroxide and formic acid[24], yields epoxidized natural rubbers (Fig. 5). The latter method is more attractive from an economic standpoint.

A range of ENR's, 25-75 mole % epoxidized, have been prepared and within the limits of detection of the analytical techniques employed no other modifications were observed.

Infrared spectra showed absorptions at 870 and 1240 cm^{-1} (epoxide),
but absorptions due to hydroxyl or carbonyl groups which are
characteristic of secondary epoxide ring-opened structures were
absent. The ^{1}H nuclear magnetic reasonance (n.m.r.) spectra[25] were
consistent with published data on epoxidized synthetic _cis_, 1,4-
polyisoprene[26]. The signals at 2.7 or 5.05 ppm were used to determine
the degree of epoxidation and good agreement was observed between these
results and elemental oxygen data. The properties of these materials
will depend to a significant extent on the epoxide unit sequence
distribution. Although these materials were prepared from latex
and the physical constraints of this heterogeneous system could
well control the epoxide unit distribution, [13]C n.m.r. showed the
epoxide groups to be randomly dispersed along the polymer backbone[27,28].

The epoxidation of NR results in a systematic change in many
properties. Every mole % epoxidation raises the glass transition
temperature (Tg), as measured by differential scanning calorimetry,
by $1^{o}C$, and this and other changes are reflected in the physical
properties of the materials.

ENR can be crosslinked using any of the standard sulphur
formations normally employed for NR[29] or by a peroxide system,
although in the latter case the efficiency decreases with increasing
degree of epoxidation. The comparative vulcanization
characteristics of ENR's and NR are illustrated in Fig. 6.

The tensile strengths of ENR gum vulcanizates (Table 3)
are high and characteristic of polymers that undergo strain
crystallization, and this can be attributed to the stereospecificity
of the epoxidation reaction[31] together with the relatively small
size of the oxygen atom.

The ability of ENR's to strain crystallise is also reflected
in the non-relaxing fatigue properties (Table 3). Although the
number of cycles to failure decreases with increasing epoxide
content, the fatigue resistance of ENR-50 is still an order of
magnitude greater than a comparable non-crystallizing rubber
such as acrylonitrile (31%)butadiene (NBR) copolymer (Table 3).

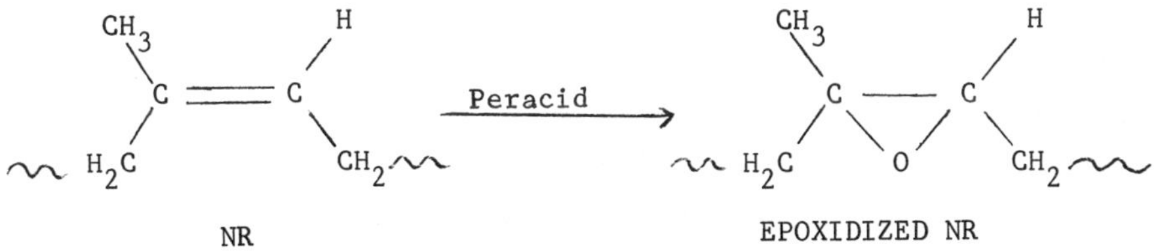

Fig. 5. Epoxidation of NR

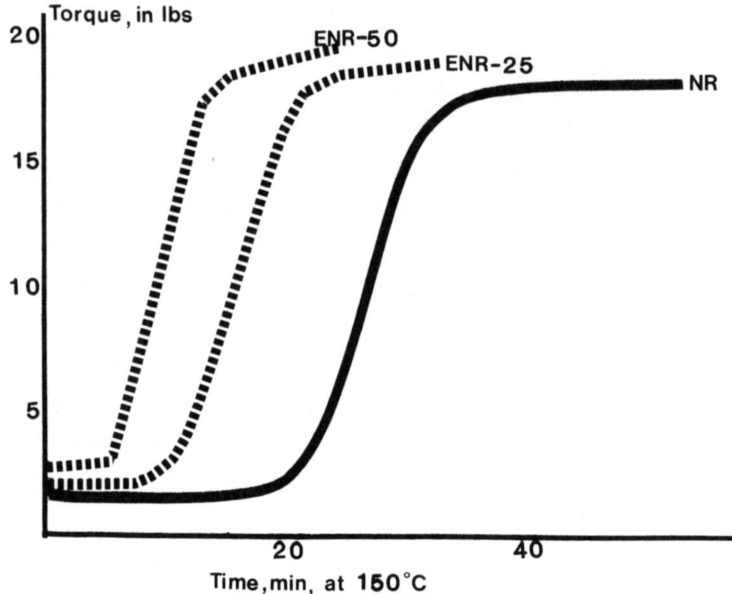

Fig. 6. Vulcanization characteristics of ENR-25 and ENR-50 compared to NR in a semi EV formulation, S 1.5, MOR 1.5.

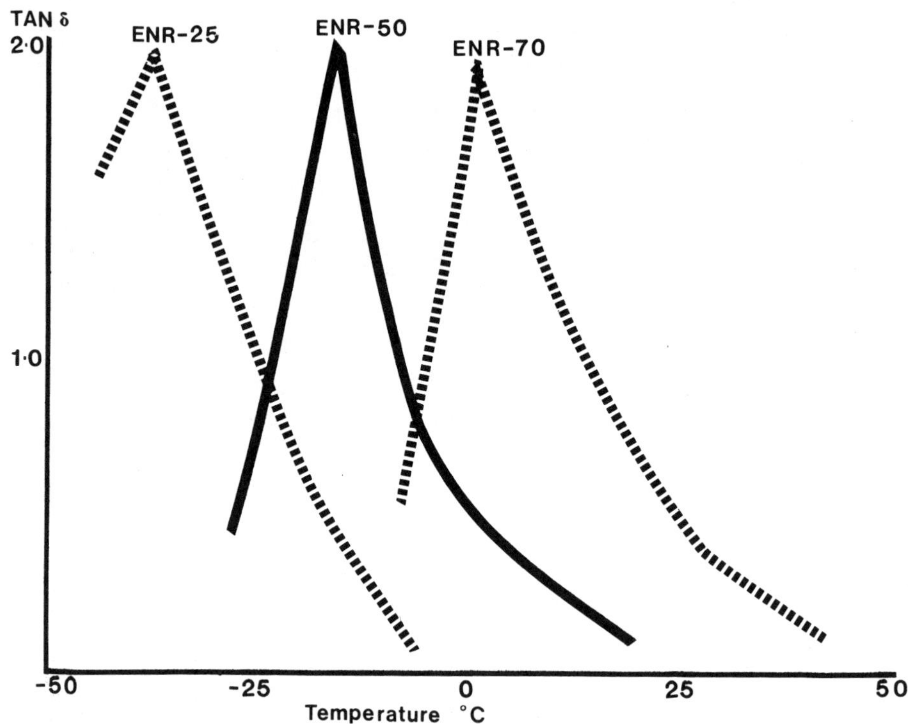

Fig. 7. Damping properties of ENR gum vulcanizates.

270

Table 3

Physical properties of ENR vulcanizates

Gum vulcanizates[a]	NR	ENR-25	ENR-50	ENR-75
Modulus at 100% extension (MPa)	0.68	0.69	0.74	0.95
Tensile strength (MPa)	25.7	24.3	28.3	27.9
Elongation at break (%)	760	770	770	650
Degree of crystallinity %[b]	11	11	10	4
Wet skid Resistance[c] (Concrete)	100	130	132	-

Black filled vulcanizates[a] (30 phr N220)	NR	ENR-25	ENR-50	NBR
Modulus at 100% extension (MPa)	1.5	1.8	1.9	1.6
Tensile strength (MPa)	32.6	28.4	28.3	17.4
Elongation at break (%)	660	590	580	704
Hardness (IRHD)	57	59	61	58
Dunlop Resilience (23°)	78	67	21	31
Ring Fatigue (KCS)				
0-100% extension	103	165	230	41
50-150% extension	1560	1206	550	39
Oil resistance				
% Volume change 70h/100°C				
ASTM No 1 oil	87	14	-0.5	-0.2
ASTM No 2 oil	141	69	12	14
ASTM No 3 oil	225	137	38	29

[a]Semi EV formulation, S 1.5/MBS 1.5 phr cured to opt. at 150°C.

[b]Determined from the variation of peak intensity of the amorphous halo in the X-ray diffraction patterns of samples strained to 400%.

[c]The NR control was taken as 100 and the other results noted accordingly.

The increase in the Tg of ENR with the degree of epoxidation is reflected in the damping properties (Fig. 7) and reduction in gas permeability (Table 4).

Table 4

Comparative air permeabilities at 23°C

Permeation Constant $\times 10^8$ (cm^2/sec/atmos)

NR	6.1
ENR-25	1.5
ENR-50	0.51
ENR-70	0.15
IIR	0.41
NBR	0.30

The change in hysteresis also results in an increase in the
wet coefficient of friction.   Data obtained with the portable wet
skid tester (Stanley, London) are recorded in Table 3.

Increasing levels of epoxidation significantly improve the
resistance to hydrocarbon oils (Table 3) and solvents, although
the reverse in time for more polar liquids such as hydraulic
brake fluids.

Most fillers require the use of a coupling agent to maximise
the reinforcement with silica fillers, however, with ENR a high
degree of reinforcement is obtained in the absence of coupling
agent.   This effect is illustrated in Table 5, similar properties
are obtained from black and silica filled ENR's.

Table 5

Comparison of Black and Silica filled vulcanizates[a]

	NR		ENR-25		ENR-50	
	Black	Silica	Black	Silica	Black	Silica
Hardness (IRHD)	65	69	69	67	73	68
Modulus at 300% (MPa)	11.9	5.8	12.4	12.8	13.5	12.6
Tensile strength (MPa)	29.4	23.2	25.4	21.0	24.5	22.4
Elongation at break (%)	495	720	435	405	500	435
Akron Abrasion (mm^3/500 rev)	21	63	14	15	11	14
Compression Set (%)	18	32	17	18	21	22
Ring Fatigue (0-100%) Kcs.	70	51	65	52	93	58

[a] 50 phr silica or N330 black in Semi EV formulation (S 1.5 phr/MBS
1.5 phr) cured to optimum at 150°C.

The epoxidation of NR results in significant property changes,
which open up new market areas.   Potential applications include
oil resistant products, adhesives, high wet grip and low rolling
resistant tyre tread compounds, inner tubes and tyre liners and
belting.   Current estimates from a one tonne pilot plant indicate
that on a commercial basis ENR should be price competitive with
the synthetic materials.

CONCLUSIONS

Chemical modification of NR can be used to produce new
materials, which can compete both technically and economically
with certain oil based synthetic elastomers.   In the longer term
other forms of modified NR should become cost competitive as oil
based products become more expensive.

Every tonne of synthetic rubber that can be replaced by a
naturally derived product will save in the region of three tonnes

of crude oil. This assumes that the modification chemistry is not
based on crude oil or energy intensive, which is true in the case
of ENR and the other modifications described.

Any increased demand for NR can readily be met. Currently the
NR yield in Western Malaysia is 1300 Kg/hectare/year, but new clones
have been developed which can produce up to 5000 Kg/hectare/year
and yield stimulants can further increase the immediate supply of NR.

REFERENCES

1. B. L. Archer, D. Barnard, E. G. Cockbain, P. B. Dickenson and
    A. I. McMullen, Composition of Hevea Latex, in: "The Chemistry
    and Physics of Rubber-Like Substances", L. Bateman, ed.,
    Maclaren & Sons Ltd., London (1963).

2. C. D. Harries, "Untersuchungen uber die Naturlichen und
    kunstlechen Kaukschukarten", Springer, Berlin (1919).

3. R. Pummerer, G. Ebermeyer and K. Gerlach, Ber, 64, 809(1931).
    R. Pummerer, G. Matthews and L. Socias-Vinals, Ber, 69, 170
    (1936).

4. M. A. Golub, M.S Hsu and L. A. Wilson, Rubber Chem. Tech.,
    48, 953 (1975).

5. F. Burton, Product energy in the rubber industry, Elastomerics,
    22 (1978).

6. J. Le Bras and A. Delande, "Les Derives Chemiques du Cautchouc
    Natural", Dunod, Paris (1950).

7. G. F. Bloomfield and E. H. Farmer, J. Soc. Chem. Ind. 53,
    43T, 47T (1934).

8. G. J. Van Veersen, Proceedings Second Rubber Technical Conference,
    London (1948).

9. H. L. Fisher, Ind. Eng. Chem., 19, 1325 (1927).

10. J. I. Cunneen, C. G. Moore and B. R. Shephard,
    J. Appl. Polymer Sci., 3 (7), 11 (1960).

11. M. L. Kaplan, P.G. Bebbington and R. L. Hartless, J. Polym. Sci.
    Polym. Lett. Ed., 11, 357 (1973)

12. C. Pinnazzi, J. C. Danjard and R. Pautrat, Proc. Nat. Rubber
    Conference, Kuala Lumpur, 555, (1960).

13. A. C. Udding, British Patent 1,343,554 (1972).

14. C. Pinazzi, R. Pautrat and R. Cheritat. Makromolek. Chem., 70,
    260 (1964).

15. M. E. Cain, K. F. Gazeley, I. R. Gelling and P. M. Lewis,
    Rubber Chem. Technol, 45 204 (1972).

16. C.S.L. Baker, _Kautschuk und Gummi,_ **36**, 677 (1983).

17. H.M.R. Hoffmann, 'The ene reaction', _Angew. Chem. Int. Ed._ **8** (8) 556 (1969).

18. D. Barnard, K. Dawes and P. G. Mente, 'Chemical Modification of Natural Rubber', Proc. Int. Rubber Conf., Kuala Lumpur, 4,215 (1975).

19. K. Dawes and R. J. Rowley, Chemical Modification of NR - a new silane coupling agent.   Rubbercon 77, Paper 18, Brighton, England (1971).

20. D. S. Campbell and A. J. Tinker, _Polymer,_ **25**, 1146 (1984).

21. D. S. Campbell, P. G. Mente, A. J. Tinker, _Kautschuk und Gummi Kunststoffe,_ **34**, 636 (1981).

22. D. S. Campbell, D. J. Elliott and M. A. Wheelans, NR Technology, **9**, 21 (1978).

23. I. R. Gelling and J. F. Smith, Proceedings of Int. Rubber Conf., Venice, 1, 140 (1979).

24. British Patent Appl. 2, 113, 692 (1983).

25. C.S.L. Baker, I. R. Gelling and R. Newell, _Rubber Chem. Technol.,_ **58** (1) 67, (1985)

26. H. V. Gemmer and M. A. Golub, _J. Polymer Sci. (Chem. Ed)_ **16**, 2985 (1978).

27. I. R. Gelling, _Rubber Chem. Tech_ , **58** (1), 86 (1985).

28. J. E. Davey and M. J. R. Loadman, _British Polym. J.,_ **16**, 134 (1984).

29. I. R. Gelling and N. J. Morrison, _Rubber Chem. Tech._ 58(2), 243 (1985).

30. C. Davies, S. Wolfe, I. R. Gelling and A. G. Thomas, _Polymer_, 24, 107 (1983).

31. L. P. Witnauer and D. Swern, _J. Amer. Chem. Soc.,_ **72** 3364 (1950).

SHORT FIBRE-RUBBER COMPOSITES

Dipak K. Setua

Defence Materials & Stores Research & Development Estt.
Post Box No. 320
Kanpur 208 013
India

INTRODUCTION

Short fibres have found a variety of applications in rubbers because of
the ease of mixing, processing advantages and improvement in many mechanical properties. Design of a short fibre-rubber composite depends on several
factors : preservation of high aspect ratio (average length to diameter ratio
of the fibre), control of fibre orientation, generation of a strong fibre-rubber
interface, establishment of a high state of dispersion and optimal formulation
of the rubber compound itself to accommodate processing and facilitate
stress transfer. In addition, short fibres provide high green strength and high
dimensional stability during fabrication, improved creep resistance, good
ageing resistance, damping, improved tear and impact strengths and anisotropy in mechanical properties. The manufacture of complex shaped engineering articles is impractical from elastomers reinforced with continuous
fibres but is easily accomplished with short fibres. Short fibres can be incorporated directly into the rubber compound along with other additives, and
the compounds are amenable to the conventional standard rubber processing
operations such as extrusion, calendering and compression, injection or transfer mouldings. Economic advantages are thus readily apparent since dipping,
wrapping, laying and placing of the fibres generally associated with continuous cord reinforcement are avoided.

The term 'short fibre' means that the fibres in the composites have a
critical length which is neither too high to allow individual fibres to entangle
with each other, nor too low for the fibres to lose their fibrous characteristics. The term 'composite' signifies that the two main constituents i.e., the
short fibres and the rubber matrix remain recognisable in the designed material.

COMPONENT MATERIALS

Types of Fibre Reinforcement

Glass fibres. Although a high initial aspect ratio can be obtained with
glass fibres, their brittleness causes breakage of the fibres during processing.
Many investigators have considered short glass fibres for reinforcing rubbers
because of their high modulus, high resilience and low creep[1-3]. Murty and
De[4,5] have studied the extent of fibre-matrix adhesion and physical properties of short glass fibre reinforced NR and SBR composites. The advantages

of using small diameter fibre glass in SBR and NBR have been patented by
Heitmann[6]. Monceau[7] has reported that glass fibres have a markedly lower
reinforcing capability than cellulose fibres but can undergo higher elongation. In a patent[8] to PPG Industries Inc., it was reported that pre-impregnated
glass fibres give better reinforcement to rubber.

    <u>Cellulose fibres</u>.   Bonded composites of discontinuous cellulose fibres
and vulcanizable elastomers having modulus and strength sufficiently high
for use as replacement for composites from continuous fibre have been suggested by Boustany and Coran[9]. Two major advantages associated with short
cellulose fibre reinforcement are its resistance to breakage during mixing
and its rough surface which allows good fibre-rubber adhesion. Anthoine
et al.[10] and Coran and Hamed[11] have reviewed the reinforcement of elastomers with discontinuous cellulose fibres while Goettler and Shen[12] have discussed the properties of different types of cellulose fibres and their mechanism of reinforcement. Various applications and the technical details of
Santoweb[R] fibre reinforcement of rubber are given in technical reports of
Monsanto Co.[13,14]. Rahman and Hepburn[15] have reported the technical
advantages of using Santoweb H[R] fibres in oil-extended ethylene propylene
diene (EPDM) rubbers.

    <u>Asbestos fibres</u>.   Asbestos fibre is mostly used where working conditions
are severe e.g., brake linings and gaskets. In contrast to other fibre-rubber
composites, the proportion of fibre to rubber in these applications is normally
high and the rubber acts only as a binder. Bohmhamel[16] quotes rubber usage
as low as 8 per cent. Bament[17] described the use of asbestos fibre bonded
with polychloroprene to improve the dimensional stability of roofing sheet
of unvulcanized chlorosulphonated polyethylene. Brokenbrow et al.[18] have
studied asbestos fibre reinforcement of rubbers. In general, however, asbestos
has little potential as reinforcing element in rubber because of its poor properties and the health hazards associated with its usage.

    <u>Miscellaneous fibres</u>.   The use of fibres derived from natural materials
as reinforcing elements for rubber compounds has been investigated. These
include jute[4,19-21], bagasse[22] and others[23]. Zuev et al.[24] have reviewed the
use of asbestos flax and cotton fibres for reinforcing different rubbers.
Setua et al.[25-28] have introduced short silk fibre as a reinforcing agent in
short fibre-rubber composites and examined, in detail, the advantages associated with the usage of silk fibre over other conventional reinforcing fibres.
Recently carbon fibres are drawing considerable attention. However, their
use is restricted because of their high cost. Carbon fibres reinforce the composites by contributing to their abrasion resistance. Grinblat et al.[29] have
observed that carbon fibres enhance the strength and resistance to ageing
and compression set when present in fluorine containing rubbers. Improvement
in tensile properties achieved by the addition of short carbon fibres in fluoro-
elastomers has been reported by Sieron[30]. The effect of carbon fibres on
the properties of rubber vulcanizates has also been reported by Lewitt[31].

    Nowadays synthetic fibres such as polyester, kevlar, nylon, rayon and
acrylic are widely used as reinforcing materials for rubbers. Good bonding
between fibre and rubber is, however, difficult to achieve because the fibre
surface is not too reactive. Aramid fibres have also been used for reinforcing
rubber[32]. The breakage resistance of aramid fibres is found to be better
than that of glass fibres and the high strength of the fibre itself is responsible for the high strength of the composite[33].

    Another method of reinforcement involves in-situ generation of short
fibres. Getson and Lewis[34,35] have reported a process in which they produced
short fibres in-situ by grafting excess olefinic monomers (such as styrene
and butylacrylate) onto the polymer (polymethyl siloxane fluid) in the form
of agglomerates. It is also perhaps permissible to include the preparation

of composite by a reverse process where the polymer is formed in-situ and not the fibre. Brokenbrow et al.[18] studied nylon fibres incorporated in a low molecular weight non-terminally reactive liquid SBR. Recently, Coran and Patel[36] have used a similar technology to reinforce chlorinated polyethylene with nylon fibrils. Investigations have also been made on the possibility of using rubber-fibre compositions, obtained by comminuting waste from rubberised textile materials and cord, in the development of waste free production of V-belts[37-38]. The effect of a mixture of two or more fibres at low volume fractions on the composite modulus has been discussed by Moghe[39]. He has studied NR, CR and SBR compounds reinforced with flexible cellulose and high strength aramid fibres. Boustany and Coran[40] have reported that a combination of cellulose fibres and chopped textile fibres in a hybrid composite showed improved performance.

Types of Elastomers

Short fibres find application in essentially all conventional rubber compounds. NR and EPDM are used most often[1-3,10,20,21,41-45] but SBR, CR, NBR and XNBR have also received much attention[6,19,33,41-43]. Setua et al.[25-28] have studied the reinforcement characteristics of short silk fibre in NR, NBR, CR and SBR. Boustany and Hamed[43] discussed the advantages of using short cellulosic fibres (Santoweb[R]) in NR, SBR, SBR-NR, EPDM, CR, NBR and polyurethane. Dzyura and Serebro[46-47] described the effects of adhesion and orientation of chopped nylon and steel fibres on the tensile strength of isoprene rubber.

Various speciality elastomers have also found utility as composite matrices. Reinforcement of fluoro-rubbers by polyamide and other fibres have been discussed by Novikova et al.[48] and Grinblat et al.[29] respectively. Sheeler[49] studied urethane, EPDM and ethylene propylene rubbers as matrices for chopped glass reinforcing fibres. A new urethane rubber that can be reinforced by glass fibres has been described by Turner et al.[50].

Types of Bonding Systems

The tricomponent dry bonding system, HRH (consisting of hexamethylenetetramine (hexamine), resorcinol and precipitated silica) compounded into the rubber stock has been widely used to secure a high level of adhesion between the fibre and the rubber matrix[19,25,26,28] Derringer[2] has evaluated the HRH system for various fibres in NBR and NR. O'Connor[33] studied the effect of three different bonding systems e.g., HRH system, RH system without silica, and a resin bonding agent on NR composites containing 17 volume per cent fibres (e.g., glass, carbon, aramid, cellulose and nylon). In case of short silk fibre-CR composites[27] 'cohedur RK - cohedur A-silica' bonding system was found to provide better processing safety and superior fibre-matrix adhesion than the HRH system. A viable alternative is fibre pretreatment with either isocyanate-based resins or RFL (resorcinol-formaldehyde-latex) dips. Two patents to Owens-Corning fibre glass corporation[51,52] suggested a resorcinol-formaldehyde treatment for glass fibres for improved fibre-rubber adhesion. Esser[53] reviewed the use of CR latices in bonding glass and asbestos. For a commercially available treated cellulose fibre containing resorcinol alone or resorcinol-based resins and hexamethoxymethylenetetramine was found to be satisfactory for reinforcing EPDM or NR-SBR blends[54]. Boustany and Hamed[43] have developed a special polymeric bonding agent for cellulose fibres. The effect of modifying the surface of cotton fibres with a styrene graft in reinforcing SBR and NR has been discussed by Zuev et al.[55]. The use of silane coupling agents in establishing proper bonding between short fibres and ethylene-vinyl acetate rubber has been discussed by Fetterman[56].

PREPARATION OF COMPOSITES

Mixing

Conventional mixers such as open mixing mill and Banbury can be utilised for mixing short fibres with rubber. The mixing procedure (distributive or dispersive) adopted depends on the type of fibre. Distributive mixing increases the randomness of spatial distribution of the minor constituent within the major base material without further size reduction, while dispersive mixing serves to reduce the agglomerate size. Thus, brittle fibres such as glass or carbon which break severely during mixing require more distributive mixing, but for organic fibres such as nylon, silk, jute and cellulose which tend to agglomerate during mixing, dispersive mixing is called for.

Fibre dispersion

An essential requisite for high performance composites is good dispersion of the fibres. Two major factors which contribute towards good fibre dispersion are (a) level of fibre-fibre interaction and (b) fibre length. It is found, for example, that naturally occurring fibres such as cellulose tend to agglomerate during mixing as a result of hydrogen bonding. A pretreatment of fibres is at times necessary to reduce fibre-fibre interaction. Such treatments include making of predispersions and formation of a soft film on the surface. Leo and Johansson[57] have described predispersions of chopped polyester, glass and rayon fibres in a CR latex for better mixing into CR or SBR rubber. Goettler[58] has reported that cellulose pulp may be dispersed directly into a concentrated rubber masterbatch or into the final compound, if it is sufficiently wetted to reduce fibre-to-fibre hydrogen bonding. Secondly, the fibre length should be small enough to facilitate better dispersion. According to Derringer[2] the commercially available fibres such as nylon, rayon, polyester and acrylic floc must be cut into smaller lengths of approximately 0.4 mm for better dispersion. The dispersion of fibres can be improved by adding fibres first in the Banbury. Shen and Rains[59] have shown that a dimensionless dispersion number N_{RS} which is a function of rotor length, rotor diameter, rotor tip clearance, mixing chamber volume, rotor speed and mixing time is reliable scale-up parameter for short fibre mixing.

Fibre Breakage

The length of fibre in composite is a critical parameter. Many investigators[1,2,42,8] have studied the importance of fibre length and its influence on the properties of the composite but a detailed study on the effect of fibre length on dispersion and ultimate properties of the composite is still lacking probably because of the difficulties associated with controlling the fibre length during mixing. O'Connor[33] has studied the extent of fibre breakage after both processing and vulcanization and concluded that fibre breakage and distribution of fibre length occur only in the uncured stock during processing and not in the cured vulcanizate. Setua et al.[19,25-27,60] have determined with the help of optical microscopy the extent of fibre breakage in short jute fibre-XNBR system[19] due to mixing in open mill and in short silk fibre-NR, NBR and CR systems[25-27,60] due to mixing both in the open mixing mill and in the Brabender plasti-corder. Glass and carbon fibres being brittle, possess low bending strength and suffer severe damage during mixing unlike silk, cellulose and nylon fibres which are flexible and have high resistance to bending.

Processing Characteristics

Setua et al[19,25-27] extensively studied the processing characteristics of fibre-rubber compositions. In short jute fibre-XNBR system[19], addition of fibres to the mixes increases Mooney viscosity, mill shrinkage and green strength and reduces the Mooney scorch time. Whereas in the case of short silk fibre-NR, NBR and CR systems,[25-27] although the improvement in Mooney viscosity, mill shrinkage and green strength are similar to that in the jute fibre-XNBR system, the Mooney scorch time increases with the addition of fibres to the mixes due to the presence of acidic (carboxyl) group in the fibre. Similar results have also been reported in other fibre-rubber systems[1,20,42].

RHEOLOGICAL AND EXTRUSION CHARACTERISTICS

Shear Viscosity. The effect of short glass fibre reinforcement on the rheology of various elastomers has been described by Lutskii and Fridman[61]. Goettler et al[62] have also studied viscosity changes for various EPDM compounds reinforced with treated cellulose fibre. The effect of fibre concentration and type of base polymer on reheological behaviour of short silk fibre-filled NR, NBR and CR compounds have been studied by Setua[60]. The shear viscosity-shear rate relationships of fibre-rubber compounds are reported to obey power law model for fluids and are similar to those of fibre-filled polymer melts. The effect of fibre concentration on the shear viscosity is more prominent at low shear rates e.g., at 30 sec^{-1}. Rheological characteristics such as extrusion and flow of asbestos reinforced rubber composites have been studied by Vershchev et al[63,64].

Die swell and extrudate distortion. As a consequence of the reduction in the elastic recovery when short fibres are present in the compounds short fibre-filled rubber mixes exhibit low or negligible die swell, as observed by Chan et al[65]. They have suggested that the normal stress and inability of oriented fibres to disorient are responsible for this reduction. According to Setua[60], mill shrinkage may be the consequence of the effect of fibre on the elasticity of the compound. Both mill shrinkage and die swell reduce considerably on the addition of fibres to the mixes and decrease progressively with increase in the fibre concentration. Extrudate distortion decreases considerably in the presence of short fibres in the rubber compounds. Goettler et al[54] have reported in detail the extrudability of fibre-filled rubber compounds.

FIBRE ORIENTATION

During processing and subsequent fabrication of short fibre-rubber compounds the fibres orient preferentially in a particular direction depending on the nature of the flow e.g., convergent, divergent, shear or elongational as explained by Goettler et al[66]. If the flow is of convergent type, the fibres align themselves in the direction of flow. The divergent type of flow causes an alignment of fibres away from the direction of flow. In the case of shear flow, the fibre alignment can be from random to unidirectional depending on the shear rate and if the flow is of elongational type the fibres orient themselves in the direction of the applied stress. All the conventional rubber processing techniques viz., milling, extrusion, calendering etc. are applicable to short fibre composites as well. Since the direction parallel to the fibre alignment shows the highest reinforcement, the utility of different processing equipment lies in controlling fibre orientation in the preferred direction to meet the anticipated loads on the fabricated product. A detailed review of short fibre orientation is given by Mcnally[67].

FIBRE-MATRIX ADHESION

Adhesion between fibre and rubber matrix in short fibre-rubber composites is extremely important. When the fibres are not properly bonded with the matrix, they slide past each other under tension resulting in low strength properties. According to Lee[68], the development of strong adhesive forces between the rubber and various substrates through the interaction of silica, resorcinol and formaldehyde donor involves a complex mechanism that is not yet clearly understood. Based on the proposed explanation of the function of RFL dip process and present adhesion theories it has been postulated that adsorption is the only significant mechanism for the adhesion of rubber to fibre (such as nylon). The effect of silica and silicates either alone or in combination with resin-formers to improve fibre-rubber adhesion has also been discussed by Creasey et al[69]. The increase in adhesion by the addition of silica is due to improved wetting and hydrogen-bonding between the rubber compound and the substrate to which it adheres. The evaluation of rough guidelines for the manipulation of the compounding and processing factors e.g., the resin-former ratio, the amount of resin-formers, the amount and type of silica, the amount of zinc oxide in the course of designing an optimum adhesion compound has been discussed by Dunnom[70]. He suggested that resorcinol and hexamine should be dispersed in the rubber at temperatures low enough to prevent resin formation prematurely. During curing of the composite, polymerization of resorcinol and formaldehyde donor is initiated. As vulcanization of the rubber proceeds, the low molecular weight polymer species are able to diffuse to the interfacial region between the rubber and the solid substrate. Silica helps in reducing the extent of reaction of resorcinol and formaldehyde, thus providing low molecular weight species which are able to diffuse through the rubber matrix. A boundary layer at the rubber surface high in resin formation results, which hydrogen-bonds with the solid substrates. The importance of zinc oxide as a rate determining factor, and the role of stearic acid in the 'resorcinol-hexamine-silica' bonding system to promote fibre-matrix adhesion has been demonstrated by Hewitt[81]. Since in certain cases the resorcinol-hexamine system reduces the scorch safety of the compound, Nicholas and Ohm[72,73] have studied alternatives. Setua and Dutta[27] have utilised dry bonding 'cohedur RK - cohedur A - silica' system in the case of short silk fibre-CR composites to avoid scorch safety problems.

DESIGN PROPERTIES

The physical properties of short fibre composites are intermediate between those of composites containing continuous cords and particulate-filled materials. The mechanics of short fibre reinforcement of elastomers have been discussed by many authors[74,75-77]. Paipetis and Grootenhuis[78,79] have studied the dynamic properties of composites having either particulate or fibre reinforcements. The effect of bonded and unbonded fibres on properties such as heat build-up, static and dynamic compression, permanent set, rupture elongation and low extension moduli has been discussed by Das[80] and Setua et al[19,23-27]. Some of the important physical properties manifested on the rubber composites as a result of short fibre reinforcement are discussed below.

<u>Tensile strength</u>. Classical theories to explain the mechanism of stress-strain properties developed for continuous and discontinuous fibre reinforced plastics are applicable to short fibre reinforced rubber composites, subject to certain modification and the theories applicable to particulate filler reinforced rubbers[81] may also be extrapolated to low aspect ratio fibre composites. Broutman and Krock[75] have developed theories for polymer composites where elastomer matrices can be considered as a special case. Rosen[82] has discussed the effect of fibre length on tensile properties and used shear-log analysis to explain the mechanism of stress transfer. The response of tensile strength to a variation in the volume loading of fibre is a complex one. For strain crystallizing rubbers (e.g., NR and CR), the tensile strength first decreases upto a certain volume fraction of fibre as a result of the dilution effect, even when the fibres are properly bonded to the rubber matrix[21,25,27]. The minimum fibre loading value depends upon the nature of the fibre, nature of the rubber, bonding level and state of dispersion and is different for different fibre-elastomer systems. For non-strain crystallizing rubbers where the strength of unfilled matrix is poor (e.g., NBR and SBR), the presence of even a small fraction of fibre increases the overall

strength of the composite[4,26]. Dzyura[83] and Setua[28] have reported that the tensile strength does not drop in the case of non-strain hardening SBR. But if the matrix strength is increased with the help of reinforcing carbon black the tensile strength is found to decrease[4].

The above mentioned theoretical consideration holds good for unidirectional composites and for randomly oriented composites when the load is applied along the direction of principal fibre orientation. But when the fibres are aligned transversely to the direction of application of stress, the fracture of the composites takes place mainly through the matrix and the fibres do not affect the strength properties significantly. Variation of physical properties of the composites with the direction of fibre orientation has been reported by Moghe[84]. Dzyura has proposed that the strength of a rubber-fibrous composition may be described by the additivity rule provided the adhesion and orientation coefficients are introduced and the true influence of the matrix is considered.

<u>Tear strength.</u> The tear resistance of composites reinforced with short fibres is considerably higher than that for other rubber compounds. Beatty et al.[85,86] have reported that incorporation of low loadings (< 5 per cent) of short fibres causes an increase in tear strength of the composite above that of the non-reinforced rubber matrix. Murty and De[4,20,21] have reported that in the case of composites of short jute fibre with NR and SBR systems a sharp increase in tear strength occurs upto a certain fibre concentration beyond which it remains almost constant. However, Sheeler[49] has observed that fibres have little effect on the tear strength of the composite which is mainly dependent on the strength of the matrix. Setua et al.[19,25-28] have extensively studied the improvement in tear strength on the addition of short jute fibre in XNBR and the short silk fibre in NR, SBR, CR and NBR systems. The anisotropy in tear property due to variation in the direction of fibre orientation and dependence of tear on the type and extent of crosslinks of rubber matrix has also been reported. In another report, Setua[87] has studied the temperature dependence of the tear strength of short silk fibre-filled NR, CR and NBR composites. High temperature (150°C) causes substantial deterioration of the fibre-matrix adhesion in fibre-filled composites which has a pronounced effect on the tear strength values. Composites with longitudinally oriented fibres show higher retention of tear properties at elevated temperatures than composites with transversely oriented fibres.

<u>Fatigue and hysteresis properties.</u> Generally, short fibre reinforcement particularly at high fibre loadings and high strain has an adverse effect on flex fatigue. Fatigue failure is associated with crack generation and its propagation in the matrix, followed by dewetting and destruction of the fibre-matrix bond. It has been reported that the flex cracking resistance is slightly more when the fibres are oriented transversely than when they are oriented longitudinally[21,27]. Derringer[3] pointed out that composites containing 9 phr rayon exhibit lower heat build-up and permanent set than carbon black (FEF, 50 phr) reinforced vulcanizates. Many investigators have explained that mechanical damping near the fibre-matrix interface at high frequencies accounts for the higher heat build-up and is in part responsible for the low fatigue life of these composites[4,19-21].

<u>Creep.</u> Addition of short fibres to an elastomer reduces the creep substantially[88]. Coran et al[41] have reported their results on the creep behaviour of short cellulose fibre reinforced NR composites. Derringer[3] discussed the advantages of short glass fibre composites over FEF black-filled composites with reference to their creep behaviour. The time dependent failure of fibre reinforced elastomers under cyclic strain conditions has been discussed by Moghe[89].

<u>Modulus and elongation at break.</u> Addition of short fibres to rubber

compounds always increases the modulus[25-27,90]. Young's modulus is frequently used and its estimation at low strain has been described[10]. O'Connor[33] studied a range of fibres at 16-17 volume per cent concentration in the presence of a bonding system. He showed how the elongation at break originally at 620 per cent can be reduced e.g., to 63 per cent with glass, to 96 per cent with carbon, to 13 per cent with kevlar and cellulose and to 40 per cent with nylon. Derringer[3] suggested that the rapid loss of elongation with increased fibre loading is due to good fibre-matrix adhesion and ultimate elongation is a good index of fibre-matrix adhesion especially at higher fibre loadings.

Effect of short fibres on critical cut length in tensile failure of short fibre-rubber composites. So far only Setua and De[91] have studied the critical cut length (l_c) phenomenon in the case of short silk fibre-filled NR, NBR and CR composites. For all types of rubbers, addition of short fibres causes a significant improvement in the l_c values which show a gradual increase with increase in fibre concentration. l_c exists in the composites wherein the fibres are oriented along the direction of application of tensile stress rather than across it, in which case the decrease in tensile strength is marginal at initial stages and is followed by a sharp fall with increasing size of cut length. l_c values of the composites increase with ageing but remain unchanged with increase in the test temperature.

APPLICATION

Generally, short fibres can find application wherever continuous fibres are being used now. If factors such as aspect ratio and adhesion of fibres to rubber can be suitably controlled short fibres can conveniently replace continuous cord as they offer flexibility in both design and processing. Some major applications of short fibre reinforced rubber composites are presented below:

V-belts. The power transmission device commonly known as a V-belt represents an extreme dynamic application for short fibre-rubber composites. As a belt runs over the pulleys, it is bent or flexed in the direction of rotation while at the same time as it wedges in and out of each pulley, it is subjected to transverse forces on the side wall. Here the anisotropy of short fibre-rubber composites which exhibit high modulus in the transverse direction and low modulus coupled with high flexibility in the axial direction was found to be very useful. Rogers[92] and Yagnyantinskaya et al[93] have studied the use of short cellulose fibre along with polyester fibre as reinforcement for V-belt compounds. Rogers and Carison[94] have studied the performance of four fibres viz., cotton staple, adhesive treated polyester, cotton floc and treated unregenerated cellulose in CR or SBR-NR matrix.

Hoses. The major advantages associated with short fibre reinforcement in the area of hoses are easy processing, economy and higher production rates and exclusion of the braiding operation without affecting the physical properties adversely. Goettler et al[62,66,95,96] have reported extensively on the manufacture and application of such hoses. They have designed specially developed extrusion dies to align the fibres into a predominantly circumferential disposition within the tube wall to provide the necessary burst strength specified for heater, radiator and fuel hoses. Extrusion shaping of curved hoses to produce hoses with bends have also been reviewed[62]. Iddon[97] has discussed an optimum screw design and extruder head construction in hose manufacturing.

Tires. The short fibres can be used in practically all parts of the tire but because of their high green strength, they find particular application in the construction of the tire inners. They are popular in tire treads because of the high chipping and cutting rersistance[85]. Inoue et al[98] have reported

282

improvement in modulus and cut/crack resistance of urethane rubber compositions with chopped organic fibres. Boustany and Coran[99] have recommended other tire applications. Dzyura et al[44] have also discussed extrusion of a bead filler stock containing short glass fibres to increase stiffness. Goettler et al[54] have studied the extrusion of treated cellulose fibre reinforced rubber profiles with controlled fibre orientation and their use as tire components such as OTR tire treads, tire chafer, shoulder inserts and bead filler. The advantages of using rubber-fibre composite (RFC) in extending service life of tractor tires have been described by Dzyura et al[100].

<u>Other applications</u>.   The application of cotton or other cellulosic fibre reinforced thermoplastic polyisoprene[101] as sheeting in shoe constructions is given by Georgieva and Vinogradova. Some applications of cellulose fibre-EPDM composites for automotive applications have been reviewed[62]. Application of Santoweb[R] fibre in rubber goods such as diaphragm, roofing, sheeting, mouldings and sealants have been described[13]. The high degree of anisotropy of fibre-rubber composites is useful in designing applications such as tubing, where swell can be minimised without decreasing elasticity[5].

<u>References</u>

1. A.P. Foldi, Rubber Chem. Technol. 49:379 (1976).
2. G.C. Derringer, J. Elastoplast. 3:230 (1971).
3. Idem, Rubber World 165:45 (1971).
4. V.M. Murty and S.K. De, J. Appl. Polymer Sci. 29:1355 (1984).
5. V.M. Murty, Int. J. Polymeric Mater. 10:149 (1983).
6. G.A. Heitmann, US Patent No. 4048137 (September 1977).
7. F. Monceau, Rev. Gen. Caoutch. Plast. 95:542 (1979).
8. D.D. Dunnom, M.P. Wagner and G.C. Derringer, Chemical Division, PPG Industries Inc., US Patent No. 3746669 (July 1973).
9. K. Boustany and A.Y. Coran, US Patent No. 3697364 (October 1972).
10. G. Anthoine, R.L. Arnold, K. Boustany and J.M. Campbell, Eur. Rubber J. 157:28 (1975).
11. A.Y. Coran and P. Hamed, <u>in</u> : "Additives for Plastics", R.B. Seymour ed., Academic Press, New York (1978).
12. L.A. Goettler and K.S. Shen, Rubber Chem. Technol. 56:619 (1983).
13. Technical Report No. 34, Rubber Chemicals Div., Monsanto Co., Louvian-La Neuve, Belgium.
14. Technical Report No. 31, Rubber Chemicals Div., Monsanto Co., Louvian-La Neuve, Belgium.
15. B.K.I. Ku Abd Rahman and C. Hepburn, Paper presented at the Int. Rubber Conference on 'Structure Property Relations of Rubber', Indian Institute of Technology, Kharagpur, India (December 1980).
16. H. Bohmhamel, Gummi Asbest. Kunst. 26:924 (1973).
17. J. Bament, Eur. Rubber J. 158:24 (1976).
18. B.E. Brokenbrow, D. Simes and A.L. Stokoe, Rubber J. 151:61 (1969).
19. S.K. Chakraborty, D.K. Setua and S.K. De, Rubber Chem. Technol. 55:1286 (1982).
20. V.M. Murty and S.K. De, J. Appl. Polymer Sci. 27:4611 (1982).
21. Idem, Rubber Chem. Technol. 55:287 (1982).
22. A.M. Usmani, I.O. Salyer, G.L. Ball III and J.L. Schwendeman, J. Elastoplast. 13:46 (1981).
23. Anonymous, Rubber World 171:42 (1974).
24. Y.S. Zuev, T.I. Karpovich and M.F. Bukhina, Kauch. Rezina 6:28 (1978).
25. D.K. Setua and S.K. De, Rubber Chem. Technol. 56:808 (1983).
26. Idem, J. Mater. Sci. 19:983 (1984).
27. D.K. Setua and B. Dutta, J. Appl. Polymer Sci. 29:3097 (1984).
28. D.K. Setua, Kauch. Gummi Kunst. (in press).
29. M.P. Grinblat, A.M. Lundstrem, R.M. Levit and N.M. Veselinova, Kauch. Rezina 33:15 (1974).
30. J.K. Sieron, Rubber World 148:50 (1963).

31. R.M. Lewitt, Kauch. Rezina 8:27 (1981).
32. L. Bergomi, Ger. Offen Patent No. 2115444 (November 1971).
33. J.E. O'Connor, Rubber Chem. Technol. 50:945 (1977).
34. J.C. Getson and R.N. Lewis, Rubber Chem. Technol. 49:402 (1976).
35. Idem, Paper presented at the Meeting of Rubber Division, ACS, Minneapolis, Minnesota (April 1976).
36. A.Y. Coran and R. Patel, Paper presented at the Meeting of Rubber Division, ACS, Chicago (October 1982).
37. A. Kuznetsova, E.M. Solov'ev, N.D. Zakharov,Y.N. Gorodnichev and G.L. Malakhova, Proizvo. Shin Rezino. Asbest. Izdelii 12:15 (1979).
38. E.M. Solov'ev, I.A. Kuznetsova, N.M. Levkina and N.D. Zakharov, ibid. 11:4 (1970).
39. S.R. Moghe, Rubber World 187:16 (1983).
40. K. Boustany and A.Y. Coran, US Patent No. 3709845 (January 1973).
41. A.Y. Coran, K. Boustany and P. Hamed, Rubber Chem. Technol. 47:396 (1974).
42. K. Boustany and R.L. Arnold, J. Elastoplast. 8:160 (1976).
43. K. Boustany and P. Hamed, Rubber World 171:39 (1974).
44. E.A. Dzyura, V.L. Mamon, A.M. Krivonos and K.S. Putankin, Int. Polymer Sci. and Technol. 4:101 (1977).
45. P. Hamed and P.C. Li, J. Elastomers Plast. 9:395 (1977).
46. E.A. Dzyura and A.L. Serebro, Kauch. Rezina 7:32 (1978).
47. Idem, Int. Polymer Sci. and Technol. 4:77 (1977).
48. L.A. Novikova, N.N. Kolesnikova and F.S. Tolstukhina, Kauch. Rezina 6:19 (1978).
49. J.W. Sheeler, J. Elastomers Plast. 9:267 (1977).
50. R.B. Turner, R.E. Morgan and J.H. Waibel, ibid. 12:155 (1980).
51. Owens - Corning Fibre Glass Corporation, Neitherland Patent No. 6512528 (March 1966).
52. Owens - Corning Fibre Glass Corporation, Neitherland Patent No. 6606254 (November 1966).
53. H. Esser, Gummi Asbest. Kunst. 26:574 (1973).
54. L.A. Goettler, J.A. Sezna and P.J. Dimauro, Rubber World 187:33 (1982).
55. Y.S. Zuev, T.I. Karpovich and M.F. Bukhina, Kauch. Rezina 9:26 (1978).
56. M.Q. Fetterman, J. Elastomers Plast. 9:226 (1977).
57. T.J. Leo and A.M. Johansson, US Patent No. 4263184 (April 1981).
58. L.A. Goettler, US Patent No. 4248743 (February 1981).
59. K.S. Shen and R.K. Rains, Rubber Chem. Technol. 52:764 (1979).
60. D.K. Setua, Int. J. Polymeric Mater. (in press).
61. M.S. Lutskii and I.D. Fridman, Kauch. Rezina 1:10 (1978).
62. L.A. Goettler, A.J. Lambright, R.I. Leib and P.J. Dimauro, Rubber Chem. Technol. 54:273 (1981).
63. A.A. Vershchev and N.P. Shanin, Soviet Rubber Technol. 30:20 (1971).
64. A.A. Vershchev, N.P. Shanin and Y.A. Kolbovskii, ibid. 30:11 (1971).
65. Y. Chan, J.L. White and Y. Oyanagi, Polymer Eng. Sci. 18:268 (1978).
66. L.A. Goettler, R.I. Leib and A.J. Lambright, Rubber Chem. Technol. 52:838 (1979).
67. D.L. Mcnally, Polymer Plast. Technol. Eng. 8:101 (1977).
68. L.H. Lee, J. Polymer Sci. 5:751 (1967).
69. J.R. Creasey, D.B. Russel and M.P. Wagner, Rubber Chem. Technol. 41:1300 (1968).
70. D.D. Dunnom, Hi-Sil Bulletin No. 40, Chemical Division, PPG Industries Inc. Pittsburgh (August 1969).
71. N.L. Hewitt, Rubber Age 104:59 (1972).
72. M.J. Nicholas and R.F. Ohm , Adhes. Age 9:25 (1976).
73. Idem, ibid. 10:31 (1976).
74. J.M. Campbell, Prog. Rubber Technol. 41:43 (1978).
75. L.J. Broutman and R.H. Krock, "Modern Composite Materials", Addison-Wesley Publishing Co., Reading, MA (1967).
76. T.S. Chow, J. Mater. Sci. 15:1873 (1980).
77. J.L. Kardos, Paper presented at the Meeting of Rubber Division on Orga-

nic coatings and Plastics Chem., ACS (March/April 1981).

78. S.A. Paipetis and P. Grootenhuis, Fiber Sci. Technol. 12:377 (1979).
79. Idem, ibid. 12:353 (1979).
80. B. Das, J. Appl. Polymer Sci. 17:1019 (1973).
81. B.B. Boonstra, Polymer 20:691 (1979).
82. B.W. Rosen, "Fiber Composite Materials", American Society for Metals, Metals Park, Ohio (1965).
83. E.A. Dzyura, Int. J. Polymeric Mater. 8:165 (1980).
84. S.R. Moghe, Rubber Chem. Technol. 47:5 (1974).
85. J.R. Beatty and P. Hamed, Elastomerics 110:27 (1978).
86. J.R. Beatty and B.J. Miksch, Rubber Chem. Technol. 55:1531 (1982).
87. D.K. Setua, Polymer 25:345 (1984).
88. S. Turner, British Plastics 38:44 (1965).
89. S.R. Moghe, Rubber World 187:16 (1983).
90. D. Mclean and B.E. Read, J. Mater. Sci. 10:481 (1975).
91. D.K. Setua and S.K. De, J. Mater. Sci. (in press).
92. J.W. Rogers, Rubber World, 183:27 (1981).
93. S.M. Yagnyantinskaya, B.B. Goldberg, E.M. Dubinker and C.V. Pozdnya-kava, Kauch. Rezina 32:28 (1973).
94. J.W. Rogers and D.W. Carlson, Paper presented at the Symposium of Rubber Division, ACS, Minneapolis, Minnesota (April 1976).
95. L.A. Goettler and A.J. Lambright, US Patent No. 4056591 (November 1977).
96. L.A. Goettler, R.I. Leib, P.J. Dimauro and K.E. Kear, Paper presented at the Meeting of Detroit Rubber Group, ACS (October 1979).
97. M.I. Iddon, Paper presented at the Scandanavian Rubber Conference, Ronneby, Sweden (May 1980).
98. S. Inoue, T. Nishi, S. Shibata, T. Matsunaga and Y. Kaneko, US Patent No. 3968182 (July 1976).
99. K. Boustany and A.Y. Coran, US Patent No. 3802478 (April 1974).
100. E.A. Dzyura, A.V. Kuz'min, L.G. Klimenko and V.N. Belkovskii, Kauch. Rezina 12:27 (1982).
101. V.S. Georgieva and G.G. Vinogradova, Kozh-obuvn Prom-st 22:45 (1980).

<u>Glossary of the terms used</u>

BR	: Polybutadiene rubber
Cohedur A	: Methoxymethylmelamine
Cohedur RK	: Condensation product of resorcinol and formaldehyde
CR	: Polychloroprene rubber
FEF	: Fine extrusion furnace black
NBR	: Nitrile rubber
NR	: Natural rubber
Santoweb[R]	: Discontinuous cellulose fibre
SBR	: Styrene-butadiene rubber
Silk	: Mulberry type of silk fibre
XNBR	: Carboxylated nitrile rubber

KRAFT LIGNINS: A NEW PERSPECTIVE*

Theodore M. Garver, Jr., and Simo Sarkanen

Department of Forest Products
University of Minnesota
St. Paul, Minnesota 55108

KRAFT LIGNINS: CURRENT OPINION

Among the naturally occurring polymers, lignins are second only to cellulose in abundance, and in terms of energy content they might actually be the most abundant. They are found as cell-wall components in all dry-land arborescent and herbaceous plants. Native lignins are generally conceived as being cross-linked macromolecules of "infinite" extent that are consituted from (p-hydroxyphenyl)propane units by an essentially *random* proportionate distribution of ten different linkages (Figure 1);[1] about half of these are, remarkably, of the same arylglycerol β-aryl ether type (C in Figure 1). The prominent interunit linkages in softwood lignins may be assembled to fashion formal molecular fragments such as the one exemplified by Figure 2. Despite the considerable effort expended in lignin related studies, however, satisfactory structural representations for lignin macromolecules have continued to elude adequate definition. This is in no small measure due to the fact that over 90% of the intermonomer linkages in lignins are much more resistant towards degradation than those found in almost all other biopolymers.

The escalation in oil prices during the past decade has fostered a great deal of interest in the potential of lignins as renewable alternative chemical resources. Unfortunately, the separation of lignins from polysaccharide biomass components is not a straightforward process; in practice, the lignins currently available are those formed during the production of cellulosic fibers. The quantity of such byproduct lignins is very large: annually about 40 million tons of lignin are produced by the worldwide kraft pulping industry. Although they represent by far the most plentiful group of aromatic chemical components available from renewable resources, less than 0.1% of these kraft lignins are consumed each year in applications other than their use as fuel.[2] At first sight this might cause little surprise. Having been recovered under quite severe conditions (typically at 170°C for 2 h in aqueous solution containing 45 gL^{-1} NaOH and 12 gL^{-1} Na$_2$S), kraft lignins are thought to have undergone major structural modifications compared with the native polymer.[3]

---

*Paper No. 14,503 of the Scientific Journal Series of the Minnesota Agricultural Experiment Station funded through Minnesota Agricultural Experiment Station Project No. 43-68, supported by Hatch funds.

The various chemical transformations of lignins under kraft pulping conditions have been understood to arise from competition between degradation reactions resulting in fragmentation and "condensation" reactions forming additional interunit covalent bonds.[3] Although they lead to opposing effects, these two types of reactions often bear a close relationship to one another: extensive studies with appropriate model compounds suggest that they may proceed through common intermediates.[3]

Degradation Reactions

Alkyl aryl ether linkages (A-D, Figure 1) in lignins may undergo cleavage during kraft pulping, but diaryl ethers (E, Figure 1) are largely unaffected; alkyl aryl and alkyl alkyl C-C bonds (G-I, Figure 1) can be cleaved to some extent, while diaryl C-C bonds (J, Figure 1) are quite stable. Of these transformations, the abbreviated description that follows is limited to the most frequent interunit ether linkages which are predominantly responsible for lignin degradation under kraft pulping conditions: more comprehensive summaries may be found in a number of other accounts.[3,4]

The alkyl aryl ethers react through pathways that depend on whether these structures are attached to phenolic or nonphenolic aromatic residues.

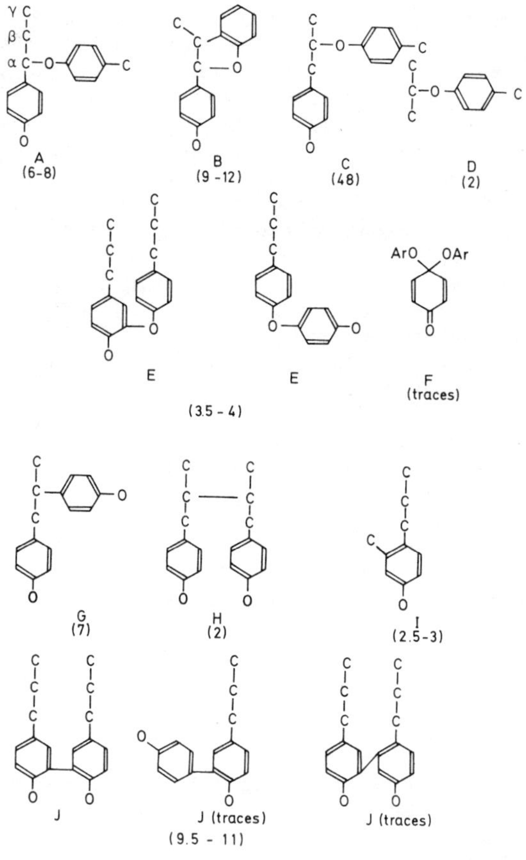

Fig. 1. Estimated frequencies (%) for different types of linkages between arylpropane units in softwood lignin.[1] Reprinted with permission from reference 3. Copyright 1977 Springer-Verlag.

Accordingly, when linked to units with a hydroxyl group *para* to the
3-carbon side chain, α-aryl ethers (A & B, Figure 1) are readily cleaved by
a mechanism involving a transient methylene quinone intermediate (Figure
3).[5]   On the other hand, α-aryl ethers attached to units with an alkoxy
group *para* to the side chain exhibit much greater stability.

Phenolic aromatic residues linked to the lignin macromolecule through
β-aryl ether structures (C, Figure 1) may similarly be converted to methylene quinone intermediates providing that the α-substituent is an effective
leaving group (Figure 4).   Nucleophilic addition of HS⁻ can then give the
corresponding benzyl mercaptide which facilitates elimination of the
β-aroxy substituent through anchimeric assistance (Figure 4).[6]   In the
absence of HS⁻, however, the corresponding β-aroxy styrene becomes the predominant (~70%) reaction product (Figure 4).[7]   On the other hand, methylene
quinone intermediates cannot be formed from nonphenolic arylglycerol β-aryl
ethers which are therefore cleaved relatively slowly; the mechanism may
involve nucleophilic displacement of the β-aroxy group by the conjugate

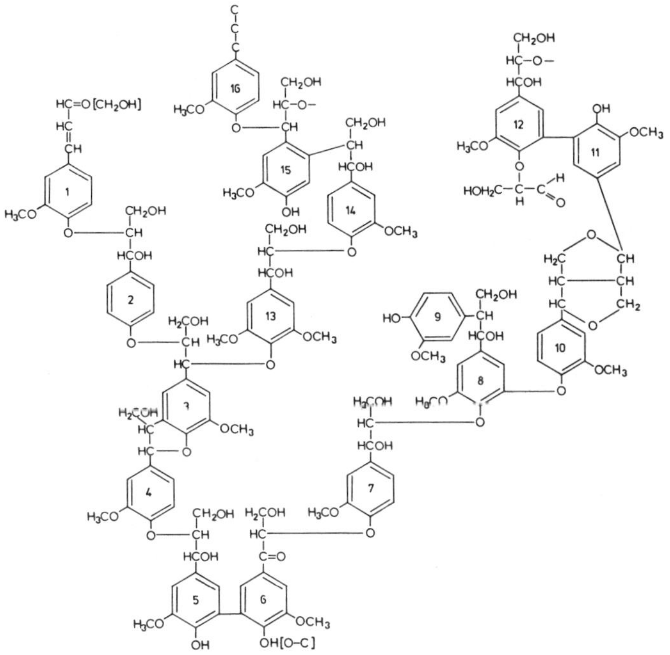

Fig. 2. Structural characteristics of softwood lignin represented in a
molecular fragment comprising 16 arylpropane units.[1]   Reprinted
with permission from E. Sjöström, "Wood Chemistry: Fundamentals
and Applications," Academic Press, New York, (1981).   Copyright
1977 Springer-Verlag.

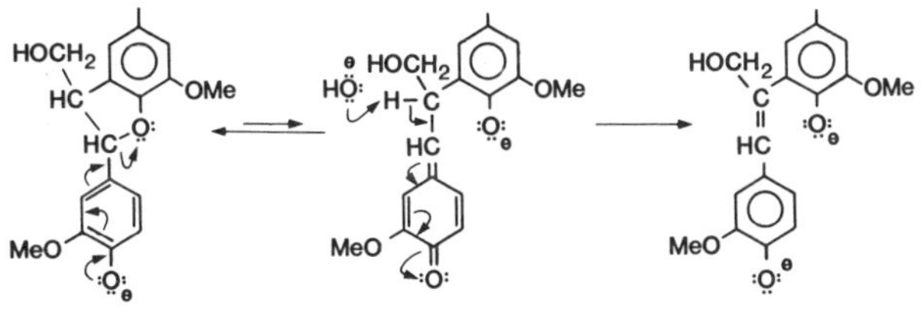

Fig. 3. Alkaline cleavage of α-aryl ether linkages in phenolic
phenylcoumaran structures.

base of a vicinal hydroxyl on the glycerol side chain (Figure 5).[8]
Consequently cleavage of the preponderant interunit linkage in lignins
during kraft pulping is thought to involve neighboring group participation
in both cases.

Condensation Reactions

The transient methylene quinone intermediates formed during cleavage
of phenolic α- and β-aryl ether linkages in lignins (Figures 3 & 4) may be
susceptible to nucleophilic addition by phenolic aromatic units acting as
quasi-carbanions (Figure 6); removal of a proton or proton and formaldehyde
from the resulting adduct can establish a stable C–C bond between two residues that were originally independent.[9] Phenolic aromatic units may also
exercise their quasi-carbanion character by condensing with liberated formaldehyde to yield diaryl methane derivatives.[9] These condensation
reactions have often been invoked as factors contributing to incomplete
delignification during kraft pulping.[10]

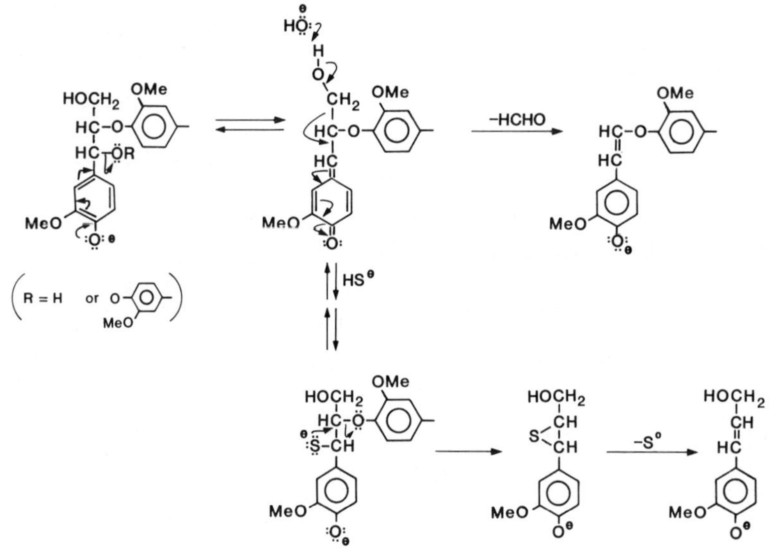

Fig. 4. Cleavage of β-aryl ether linkages in phenolic arylpropane units
under kraft pulping conditions.

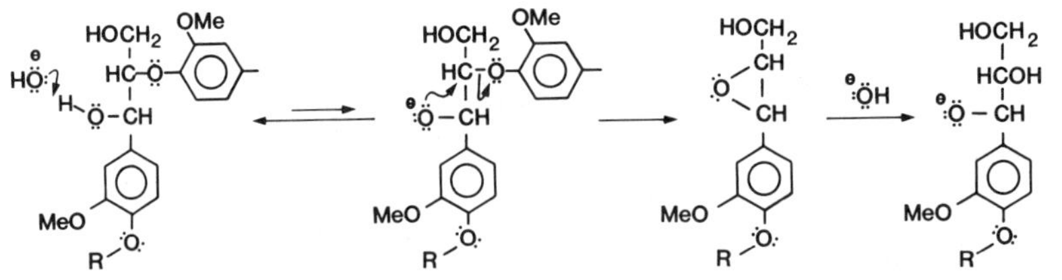

Fig. 5. Cleavage of β-aryl ether linkages in nonphenolic arylpropane units
under alkaline conditions.

It is not unreasonable to inquire if the proposed occurrence of methylene quinone intermediates during the transformations of phenolic β-aryl ether linkages in lignins under alkaline conditions has always been correctly rationalized. For example, 1-(4'-hydroxy-3'-methoxyphenyl)-1-(4"-hydroxy-3"-methylphenyl)ethane 2-O-(2'''-methoxyphenyl) ether undergoes a base-catalyzed reaction[11] involving 2-methoxyphenyl group displacement to form a spiro cyclohexadienone intermediate[12,13] which rapidly rearranges to the corresponding stilbene; the net anionotropic [1,2] aryl shift favors preferential migration of the phenolic benzene ring possessing the more electron-donating substituents.[14] The question arises as to whether a similar mechanism could be operative for the alkaline cleavage of arylglycerol β-aryl ether linkages to phenolic lignin units, a process which in the absence of sulfide occurs to the extent of about 30% while the predominant reaction product is the β-aroxy styrene (cf. Figure 4).[7] However, no reaction has been observed with 1-(4'-hydroxy-3'-methoxyphenyl)ethane 2-O-(2"-methoxyphenyl) ether during a 2 h period at 170°C in aqueous 9% ethyleneglycol monomethyl ether containing 0.9 M NaOH (the reaction conditions adopted for these studies).[14] Such a finding suggests that aryl participation in the reaction may be restricted to substituted 1,1-bis(4'-hydroxyphenyl)ethane 2-O-aryl ether compounds. The unreactivity of 1-(4'-hydroxy-3'-methoxyphenyl)ethane 2-O-(2"-methoxyphenyl) ether relative to 2-(p-hydroxyphenyl)ethyl bromide[13,14] can be accounted for by the fact that the 2-methoxyphenoxy group is poorer than bromine as a leaving group by two or three orders of magnitude.[15]

The foregoing results have been taken to support the contention that the 30% cleavage of 1-(4'-hydroxy-3'-methoxyphenyl)glycol 2-(2"-methoxyphenyl) ether encountered in the absence of sulfide under comparable alkaline conditions does not involve aryl group migration.[14] It has been claimed, rather, that the reaction proceeds through neighboring participation of the benzoxide moiety acting as a nucleophile (cf. the behavior of nonphenolic arylglycerol β-aryl ether linkages illustrated in Figure 5).[14] Nevertheless, it would be prudent to allow for the possibility that some transformations of arylglycerol β-aryl ether linkages under aqueous alkaline conditions may occur through mechanisms which are nontrivially different from those of the corresponding glycol derivatives so often adopted for model compound studies.

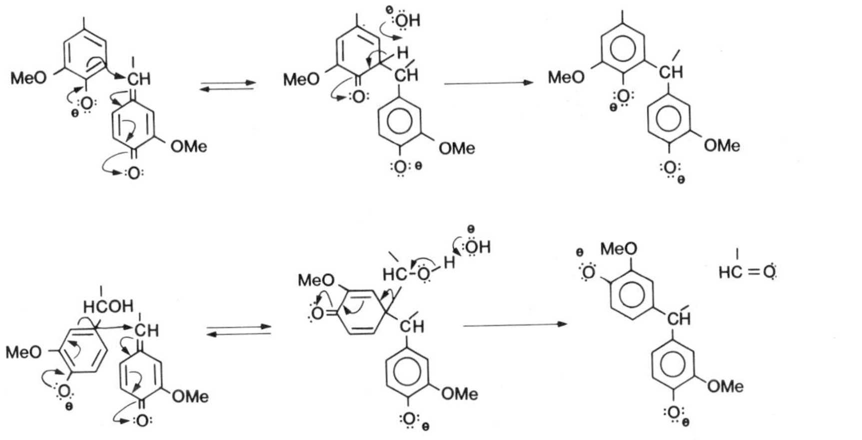

Fig. 6. Examples of condensation reactions occurring through nucleophilic addition of delocalized aryl carbanions to methylene quinone intermediates under alkaline conditions.

<u>Net Effect of Kraft Pulping</u>

When all the possible transformations of lignins during kraft pulping are contemplated, it would not be unreasonable to assume that the reaction products should extend to almost hopelessly complicated mixtures of components. The situation would certainly be beyond salvation if a random distribution of interunit linkages were actually intrinsic to the configuration of native lignin macromolecules. Indeed, no fewer than 22 monomers and 16 dimers from spruce kraft lignin have been unambiguously characterized,[16] and on this basis alone it would seem that the higher molecular weight fractions will contain an enormous variety of species. Such considerations have prompted assertions that "a 'formula' for kraft lignin ... would be quite impossible to devise" and that "this makes the assignment of a definite structure to these 'lignins' quite irrelevant".[4] One reservation should, however, be introduced at this juncture: among the many different reactions that can alter lignin structure during kraft pulping,[3] only a few have been partially characterized in kinetic terms through studies of suitable model compounds.[5,8,17-19] Consequently the relative importance of each type to the overall transformation of lignins is, at present, unclear.

<u>Phases of Kraft Pulping</u>

The rate of lignin dissolution varies in such a way during kraft pulping that the overall process can be divided into three phases.[20] The initial phase is usually complete when the temperature has risen past 150°C as the wood chips and aqueous NaOH-NaSH solution are first being heated; between 20 and 25% of the lignin has dissolved at a relatively rapid rate which is first order with respect to lignin concentration but essentially independent of hydroxide and bisulfide ion concentration. The facile cleavage of α- and β-aryl ether structures linked to phenolic aromatic units (Figures 3 and 4) are thought to be responsible for delignification during this first phase.[3]

The subsequent bulk phase results in dissolution of about 60% of the lignin from the wood through a process which is approximately first order with respect to both lignin and hydroxide ion concentration but is only slightly dependent on the concentration of bisulfide.[21] Cleavage of β-aryl ether linkages attached to nonphenolic aromatic units (Figure 5) has been invoked as the rate-determining reaction during this second phase of delignification.[3,22]

Of the lignin originally present in the wood, 10-15% dissolves during the residual delignification phase. Here the process might be partly facilitated by alkaline cleavage of interunit C-C bonds while competing condensation reactions could retard lignin dissolution which has now become quite slow.[3]

The results from kraft pulping of selectively premethylated pine wood shavings have been claimed to support the foregoing correlations between lignin degradation reactions and the respective phases of delignification.[23] Premethylation of phenolic hydroxyl groups with diazomethane caused the apparent elimination of the initial phase but left the bulk phase delignification rate unaffected. On the other hand, premethylation of the benzylic hydroxyl groups with methanolic hydrochloric acid had little effect on the initial phase but appreciably retarded the bulk phase. Finally, exhaustive methylation of all accessible aliphatic and aromatic hydroxyl groups with dimethyl sulfate in aqueous hydroxide solution largely suppressed *both* the initial and bulk phases of delignification.[23] It has, however, been pointed out[19] that the absence of the initial rapid delignification phase during kraft pulping of wood shavings pretreated with diazomethane could arise from the poorer swelling and dissolution behavior of the methylated lignin.[24]

The pseudo first order rate for aqueous alkaline cleavage of veratrylglycol β-guaiacyl ether[25] can be very similar in magnitude to the phenomenologically first order net delignification rate[21] encountered during the bulk phase of kraft pulping under comparable conditions.[22] This coincidence could be appropriated to identify the cleavage of nonphenolic arylglycerol β-aryl ether linkages as the rate-determining chemical transformation that governs bulk phase delignification — even though the comparison remains one between a homogeneous and a complicated heterogeneous system.[22]

In any case, veratrylglycerol β-aryl ethers are more appropriate model compounds for such correlations since they more closely resemble the actual structural units in native lignins than do the corresponding glycol homologues. *Erythro* and *threo* veratrylglycerol β-guaiacyl ethers exhibit markedly different rates of alkaline cleavage.[8] The *erythro* isomer is cleaved about 4.6 times faster than the *threo* in 0.25 M aqueous NaOH at 170°C so that the rate of alkaline cleavage for an approximately equimolar (3:2) mixture of the two does not display first order behavior.[19] Although *erythro* and *threo* arylglycerol β-aryl ether linkages occur in native lignin with roughly equal frequencies,[26] the bulk delignification phase during kraft pulping clearly exhibits phenomenologically first order kinetics.[21] Consequently the similarity between the pseudo first-order rates of nonphenolic model β-aryl ether cleavage and bulk phase delignification under kraft pulping conditions should not be regarded as indicative of the actual mechanism of lignin dissolution.[19]

It has been suggested that delignification is probably facilitated by the cleavage of certain "key" β-aryl ether linkages (to either nonphenolic or phenolic units) which are so positioned in the native macromolecule as to allow significant lignin dissolution upon depolymerization.[24] Subsidiary effects can be anticipated from β-aryl ether cleavages that may enhance delignification merely by increasing the phenoxide group content of the lignin and consequently its solubility.[19] The cleavage of other β-aryl ethers may minimally assist the process through sequential removal of superficial end-groups or oligomers from the macromolecule (a kind of lignin "peeling").[22] The latter could be largely responsible for generating significant concentrations of coniferyl alcohol during the initial phase of kraft pulping.[27]

Coniferyl Alcohol and the Incidence of Condensation

Although it is formed during delignification (Figure 4), coniferyl alcohol (4-hydroxy-3-methoxycinnamyl alcohol) appears only transiently as a result of kraft pulping owing to its instability under the prevailing conditions. Significant levels of the compound have been detected in the early stages of the process, the highest concentration being attained at the end of the initial delignification phase.[27] Kinetic considerations have suggested that little or no more coniferyl alcohol is formed once this point has been reached if the temperature is not increased further.[27] After the alkaline cleavage of nonphenolic β-aryl ether linkages during bulk phase delignification, more of the compound could be formed[22] although its subsequently rapid chemical transformation would render it difficult to detect. Not surprisingly,[19,23,24] premethylation of black spruce wood meal with diazomethane has been found to reduce coniferyl alcohol formation substantially (— by 95% during the initial delignification phase).[27] The maximum yield for coniferyl alcohol of 3.5% (w/w lignin) was actually observed after 1 h at 125°C in 1 M aqueous NaOH containing 0.8% anthraquinone[28,29] and 20% glucose (w/w wood). This accounts for only one third

of the estimated phenolic aromatic units present in black spruce
lignin[30] -- some of which evidently either are not attached through β-aryl
ether linkages or are bound through C5 to the native macromolecule (Figure
2).

     Part of the coniferyl alcohol liberated from the lignin seems to be
converted to vinyl guaiacol and isoeugenol in aqueous alkaline solution at
170°C.[27]   However, by exercising its carbanion character at either the β-
or an appropriate ring position, the compound could participate nucleo-
philically in condensation reactions (Figure 6). Alternatively, through
elimination of the γ-hydroxyl group, coniferyl alcohol could form a conjugated methylene quinone intermediate that might be susceptible to attack
by quasi-carbanions.[10]   The consequences of such potential condensation
reactions have been explored through the behavior of 5.6 mM coniferyl alcohol with 2 gL^{-1} of a low molecular weight lignin fraction in aqueous 21
gL^{-1} NaOH containing 8.6 gL^{-1} Na$_2$S at 170°C.[31]   (Ranging between 200 and
4800 in molecular weight, the lignin sample had been isolated by treating
Western Hemlock wood platelets for 1 h with a similar solution at 100°C in
the absence of coniferyl alcohol.) Under these simulated kraft pulping
conditions at 170°C, the proportion of components with molecular weights
above 1000 was substantially reduced.[31]   To the extent that they might
occur, therefore, condensation reactions in solution do not engender higher
molecular weight kraft lignin species.

     These findings have identified an important feature in the constitu-
tion of complete kraft lignin preparations -- which usually possess molecular weight distributions extending from 200 to ~100,000 (*vide infra*).
Taken together, the lower molecular weight oligomers[16] should not reflect
the predominant structures present in the kraft lignin sample as a whole.

THE PHYSICOCHEMICAL PROPERTIES OF KRAFT LIGNIN COMPONENTS

     Little has emerged from the preceding account to dispel the notion
that the lignin derivatives formed during the kraft pulping process should
harbor a considerable complexity. Yet recent studies have disclosed the
distinct possibility that kraft lignins may be much more tractable than
current opinion would allow. These investigations[32-35] have established
that, with the possible exception of the lignosulfonates, lignin components
generally exhibit a pronounced tendency to associate reversibly in solution. Indeed, the field of lignin chemistry has been beset by a conceptual
deficiency at a relatively fundamental level: the underlying associative
interactions, although among the most prominent physicochemical characteristics of lignin samples, have received little attention from wood chemists.
Such effects collectively represent perhaps the single most profound
obstacle confronting attempts to determine the structures of the individual
molecular species comprising byproduct lignin preparations.

     Association, a reversible phenomenon between lignin components, should
always be distinguished from non-reversible aggregation between the resulting complexes.[36,37]   The first indication that lignin components tend to
associate with one another appeared more than twenty years ago,[38] but the
publication three years later of a second article[39] confirming the existence of such effects prompted no further studies of the phenomenon for
more than a decade. Consequently reliable methods for determining the
molecular weight distributions (m.w.d.'s) of lignin samples were only
recognized quite recently.[32]

     In nonaqueous DMSO and DMF solutions without dissolved electrolytes,
the apparent m.w.d.'s of gymnosperm kraft lignins extend to molecular
weights 10,000 times larger than those of the individual components.[32]

Rather than being continuous, these distributions are multimodal in form;
the associated kraft lignin complexes therefore must somehow be well-
defined. Discrete (or individual) kraft lignin components behave as appreciably expanded random coil homologues in aqueous alkaline solution.[34] The
associated complexes, on the other hand, appear to have a flexible disklike
configuration with a thickness of about 1.6 nm and hydrodynamic properties
intermediate between those expected for a non-freedraining coil and an
Einstein sphere.[34,40]

Exhaustive acetylation or methylation of lignin samples does not
appreciably affect the relative proportions of high molecular weight
species observed in nonaqueous DMF solutions,[41] so hydrogen bonding cannot
provide the driving force for association. Yet hydrophobic interactions
alone would not be sufficient to overcome the electrostatic repulsion
between polyionic lignin components associating in aqueous alkaline solution. Presumably the stability of the associated complexes arises from
nonbonded orbital interactions between the aromatic moieties of the par-
ticipating components. The consequences of these effects probably recall
the configuration of native lignin macromolecules in the plant cell wall
where the benzene rings of the (p-hydroxyphenyl)propane units are parallel
to one another.[42] Semi-empirical molecular orbital calculations[43] have
hinted that the structures of the associated complexes could be based upon
a head-to-tail orientation between the aromatic moieties of the interacting
molecular species. However, the conformational analyses reported in this
connection were based upon a method using atom-atom potential functions
which leads to results that have not been reliably consistent with experiment.[44] It is anyway unfortunate that lignin model compounds with appropriate substitution patterns around the benzene rings have not so far been
selected for such inferential theoretical studies.[43,44]

In aqueous alkaline solutions, where kraft lignin components are polyionic, the maximum attainable degree of association is inevitably smaller
than in nonaqueous solvents such as DMSO or DMF. Nevertheless, an important facet of the associative process is more readily apparent under
aqueous alkaline conditions: the components representing the oligomeric
subset with molecular weights below 3500 associate reversibly in constant
(roughly 1:1) stoichiometric ratios with respect to each other.[35] The
associated complexes evidently are not only well-defined, therefore, but
also behave as though they are topologically regular. Many of the foregoing claims support new perspectives on the nature of kraft lignins that
differ quite radically from the prevailing view in the field. Examples of
recently published observations which argue in their favor are summarized
in the two sections that follow.

Associative Behavior of Kraft Lignin Components

The apparent m.w.d. of gymnosperm kraft lignin is affected markedly by
the method employed for isolating the sample and subsequent incubation conditions in solution. The straightforward acidification of Weyerhaeuser Co.
black liquor from Douglas fir (*Pseudotsuga menziesii* (Mirb.) Franco) is
cited here as an illustration. After the stock solution had been carefully
acidified to pH 2.5, centrifugation of the kraft lignin precipitate
followed by freeze-drying from aqueous solution at pH 8.5 yielded a sample
for which the effective m.w.d. is described by Sephadex G100/0.10 M aqueous
NaOH elution profile no. 1 in Figure 7. After incubation at 20 gL^{-1} in
0.10 M aqueous NaOH for 24 h, the proportion of higher molecular weight
species in the same preparation increased significantly (profile 2, Figure
7). On the other hand, when the original precipitate was instead filtered
from acidified black liquor and air-dried, only a portion (~80%) of the
resulting solid could be redissolved in aqueous solution at pH 8.5 and
freeze-dried therefrom; the apparent m.w.d. for this portion after incuba-

tion at 20 gL^{-1} in 0.10 M aqueous NaOH for 48 h is represented by profile 3, Figure 7. The other 20% of the filtered sample remained in the form of granular particles and was presumably derived from locally moister regions in the *originally homogeneous* precipitate; upon dissolution and subsequent 72 h incubation at 150 gL^{-1} in 1.0 M aqueous NaOH, this fraction displayed the component m.w.d. depicted by profile 4, Figure 7.

Despite their common origins, these three kraft lignin preparations exhibit substantial differences from one another in the contribution that the higher molecular weight species make to their respective m.w.d.'s. There is consequently no reason to expect that the m.w.d. of a kraft lignin preparation directly isolated from black liquor should represent that for the discrete molecular components. Indeed incubation of the sample at concentrations of 1.0 gL^{-1} and 0.5 gL^{-1} in 0.10 M aqueous NaOH was found to elicit marked dissociation of the higher molecular weight complexes (Figure 8) during a process which reached equilibrium within 1700 h. In as far as the equilibrium m.w.d. in 0.10 M aqueous NaOH at a 0.5 gL^{-1} sample concentration (profile 4, Figure 8) approximates that for the discrete components, the formal degree of association for the kraft lignin in the black liquor (profile 1, Figure 8) was 0.45, while the maximum degree of association

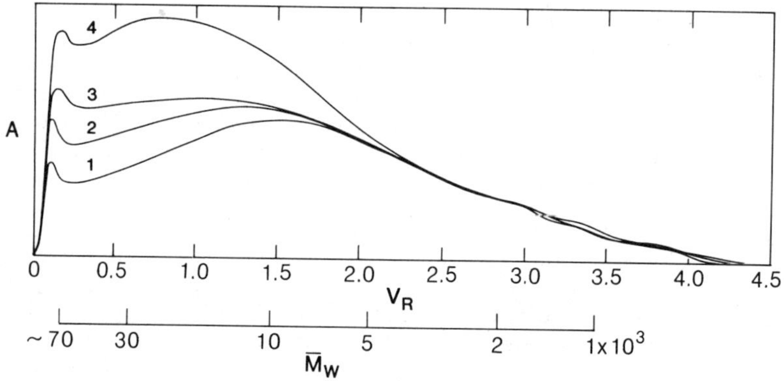

Fig. 7. Combined effects on kraft lignin molecular weight distribution arising from isolation method and incubation conditions prior to size-exclusion chromatography (Sephadex G100/0.10 M aqueous NaOH, monitored at 254 nm). Reprinted with permission from reference 35. Copyright 1984 American Chemical Society.

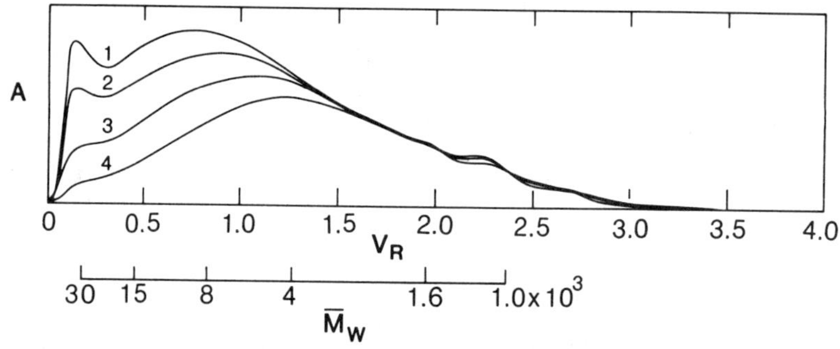

Fig. 8. Dissociation of kraft lignin components in 0.10 M aqueous NaOH after (1) 0.8 h at 1.0 gL^{-1}, (2) 98 h at 1.0 gL^{-1}, (3) 122 h at 1.0 gL^{-1}, then 287 h at 0.50 gL^{-1}, and (4) 122 h at 1.0 gL^{-1}, then 1610 h at 0.50 gL^{-1} (Sephadex G75/0.10 M aqueous NaOH elution profiles monitored at 280 nm). Reprinted with permission from reference 35. Copyright 1984 American Chemical Society.

observed for the sample in aqueous alkaline solution was 0.70 (profile 4, Figure 7).

The elution profiles in Figures 7 and 8 have been scaled so as to confer the same value upon the areas under the respective segments of these curves below $\overline{M}_w$ = 3500. It has hereby become evident that while the proportions of the higher molecular weight species exhibit considerable variations, the oligomeric components of the subset with molecular weights below ~3500 remain, within experimental error, in the same relative ratios. Such behavior would, *a priori*, be surprising for a subset of oligomers which have degrees of polymerization greater than the critical chain length for associated complex formation: the equilibrium constant for association would be expected to increase rapidly with molecular weight so that preferential association of the higher molecular weight species should be strongly favored.[45] The constraints that act upon the associative processes involving kraft lignin components obviously demand careful examination.

Mode of Association between Kraft Lignin Species

The variation of weight-average molecular weight, $\overline{M}_w$, with number-average molecular weight, $\overline{M}_n$, for a lignin sample during the course of association can provide quite specific information about the mode of the process. The relationship between the two parameters is given by:[33]

$$\partial \overline{M}_w / \partial \overline{M}_n = 2 \langle m_i m_j \rangle (1 / \overline{M}_n{}^2)$$

where $\langle m_i m_j \rangle$ at any particular point is the appropriate ensemble average product of the molecular weights of associating components. Expressions deduced for $\langle m_i m_j \rangle$ can be evaluated numerically from the populations of lignin species empirically observed during the associative/dissociative process. Direct comparison with experiment is facilitated by considering a plot of $\overline{M}_w$ vs. $1/\overline{M}_n$, the slope of which at any point is equal to $-2\langle m_i m_j \rangle$.

The dependence of $\overline{M}_w$ on $1/\overline{M}_n$ that characterizes the dissociation of the gymnosperm kraft lignin sample in 0.10 M aqueous NaOH (Figure 8) is illustrated in Figure 9. The plot indicates that $\langle m_i m_j \rangle$ remains effectively constant in magnitude ($2.22 \times 10^7 \pm 3.7\%$) while the formal degree of association for the kraft lignin components varies from 0.45 to 0. Consequently the species whose relative concentrations change during the course of dissociation (namely those with molecular weights greater than 3500) must exhibit zero order kinetic behavior.

The minimal scheme which can account for the associative/dissociative process is therefore one incorporating two kinetically distinguishable steps. This can be envisaged in the following terms. Discrete components L_{di}, which cannot participate directly in association, undergo a slow conformational transformation to species L_{ci} that may then interact productively with complexes C_j. Inclusion of the respective rate coefficients allows the overall situation to be summarized as

$$L_{di} \underset{k_{-ci}}{\overset{k_{ci}}{\rightleftarrows}} L_{ci} \underset{k_{-aij}}{\overset{k_{aij}[C_j]}{\rightleftarrows}} L_i C_j$$

There is one constraint alone which will meet the condition that $\langle m_i m_j \rangle$ remain constant during association:[35] each complex C_j must associate *exclusively* with a particular component L_i so that k_{aij} is nonzero *only* for a given i and j. Hereby the kinetic scheme may, within experimental error, account for the numerical value of $\langle m_i m_j \rangle$ when the components L_i are identified as the oligomers with molecular weights below 3500 and the species

C_j are taken to be the remaining higher molecular weight entities (Figure 8). Agreement with experiment requires that productive associative encounters occur when the *same* minimum number of (effectively independent) chain segments in each L_i are conformationally compatible with part of the corresponding locus on C_j; subsequent conformational rearrangements can then lead to interactions between all potentially complementary segments on L_i and C_j.

The available data therefore suggest that the mode of association between gymnosperm kraft lignin species is quite specific: not only do the higher molecular weight complexes interact selectively with lower molecular weight oligomers, but complexes of a given type are evidently compatible with only one kind of component. Simply stated, the associated complexes appear to be, in some sense, topologically regular so that the associative process is stoichiometrically constrained. It is difficult to understand how this could be the case unless the individual molecules within each increment of the kraft lignin m.w.d. were to possess a limited number of dominant structural features. The isolation and characterization of discrete components with molecular weights spanning the entire range of a kraft lignin sample thus becomes particularly meaningful propositions.

Analytical Studies of Kraft Lignins

Any approach to the chemical characterization of kraft lignins should be facilitated by working with preparations for which the degree of association is small. Failure to observe this precept may foster conclusions

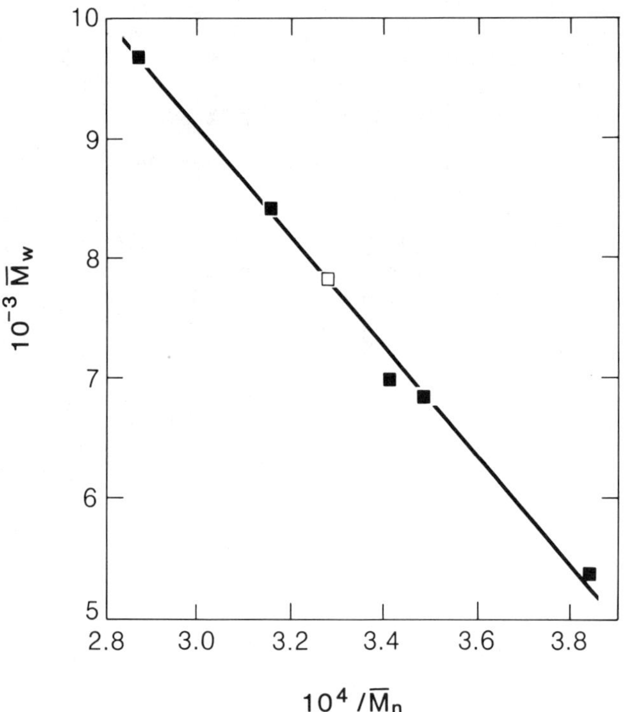

Fig. 9. (■) Relationship between $\overline{M}_w$ and $\overline{M}_n$ during dissociation of kraft lignin components present at 1.0 gL^{-1} and 0.5 gL^{-1} in 0.10 M aqueous NaOH. (□) Effect of 0.10 M betaine on $\overline{M}_w$ and $\overline{M}_n$ of kraft lignin in 0.10 M aqueous NaOH. Reprinted with permission from reference 35. Copyright 1984 American Chemical Society.

that are not consistent with one aspect or another of the analytical data subsequently acquired. A recently published investigation[46] illustrates the problem.

A series of kraft lignin fractions isolated at successive stages during the delignification of pine (*Pinus silvestris*) wood chips were subjected to analyses of their elemental compositions, methoxyl group contents, aromatic hydroxyl group frequencies and ^{13}C NMR spectra (obtained with gated proton noise decoupling).[46] It should be pointed out that the empirical formulae for these fractions may have been influenced by the presence of bound water in the samples which are exceptionally difficult to dry completely.

An important finding from the study was contained in the observation that kraft lignins apparently undergo little further chemical modification after their dissolution during the delignification process. This is presumably a consequence of the protective shielding enjoyed by the molecular components incorporated in associated complexes.

The frequencies of phenolic units in the kraft lignin fractions were estimated by selective aminolysis with pyrrolidine[47] of the aromatic acetoxy groups in the acetylated derivatives. The method could be seriously affected by the presence of acetylated polysaccharide reducing end-groups,[48] although the kraft lignin preparations here were not appreciably contaminated with such impurities.[46] Nevertheless, the reported phenolic hydroxyl group frequencies were rather high (~0.65). They were estimated by assuming that each aromatic unit contained ~9.2 carbon atoms, a value previously determined through the specific incorporation ratio of radioactivity in a pine kraft lignin derived from the appropriately ^{14}C labeled native biopolymer in wood meal.[49]

The number of aromatic carbon atoms calculated on the foregoing basis became so large, however, that very few vinylic carbon atoms could be accommodated within the corresponding regions of the ^{13}C NMR spectra. Nevertheless the presence of β-aroxy styrene structures (cf. Figure 4) had already been implicated at least in those kraft lignins entering solution around the beginning of bulk phase delignification;[50] their existence had been inferred from the detection of homovanillin among the degradation products formed by subjecting the kraft lignin fractions to "acidolysis" (hydrolysis at 100°C in aqueous 90% dioxane containing 0.2 M HCl).

Ultimately, in order to reconcile the empirical formulae per aromatic unit with the ^{13}C NMR spectra for the respective kraft lignin samples, a substantial proportion of the side chain carbon atoms were relegated to alkyl and carboxylic acid groups -- even though the corresponding ^{13}C NMR signals were not in evidence. The dilemma could be resolved if the frequency of phenolic hydroxyl groups were somewhat higher than originally estimated. This supposition would both be consistent with the reported analytical data and allow for contributions from reasonable numbers of vinylic carbon atoms to the ^{13}C NMR spectra. The requisite concession is probably granted by association between lower molecular weight oligomers and the other kraft lignin components.

Recent observations by the authors of this article have confirmed that such interactions are present even under circumstances where they might not be anticipated. In work directed towards isolating individual kraft lignin components, the parent samples were allowed to dissociate fully at a 0.5 gL^{-1} concentration in aqueous 0.10 M NaOH (Figure 8). The dissociated kraft lignin preparations isolated hereupon by precipitation through acidification to pH 3.0 represented about 60% of the original sample. Following elution from Sephadex G100 with aqueous 0.10 M NaOH, each paucidisperse

kraft lignin fraction was freeze-dried after neutralization to pH 7.5.
Subsequent elution through Sephadex G25 with water effected the clean
separation of every fraction into two subsets of components B and C.

The component subsets B contribute primarily to the higher molecular
weight region, while the lower molecular weight components are predominantly composed of subsets C (Figure 10). Ultracentrifuge sedimentation
equilibrium measurements have indicated that the species C initially co-
eluting with the higher molecular weight components B include a significant
proportion of oligomers (Figure 11). Even after prolonged incubation under
conditions favoring dissociation, therefore, an appreciable number of low
molecular weight kraft lignin species remain tenaciously attached to the
higher molecular weight components.

It would hardly be an exaggeration to declare that such pronounced
associative interactions represent the principal obstacle to the unambig-
uous chemical characterization of the individual molecular constituents in
kraft lignin samples. Yet this obstacle cannot be avoided if the most
abundant byproduct lignins are ever to realize their full potential as
renewable-resource materials.

FUTURE DEVELOPMENTS

It remains to be elucidated *specifically* to what extent the structural
features of native lignins are reflected in their kraft derivatives. The
molecular determinants of the associative interactions between kraft lignin
species requires clarification so that the topological regularity evident
in the resulting macromolecular complexes[35] may be understood. Consequently, the isolation and characterization of representative *individual*
components from throughout the molecular weight range encompassed by kraft
lignin samples are particularly important concerns. The middle and higher
molecular weight (acetylated) kraft lignin components have all too often
been found to exhibit rate-limiting desorption under chromatographic conditions that would otherwise facilitate peak resolution. The isolation of
these species can thus become quite laborious although absolutely necessary
for subsequent characterization studies.

Such work should be capable of unravelling the basis for an intriguing
aspect of macromolecular lignin structure that seems to be reflected in
partially degraded preparations. Three independent lines of argument have
now appeared which, although not mutually consistent, could be based upon a
common denominator. From elemental analyses of an apparently homologous

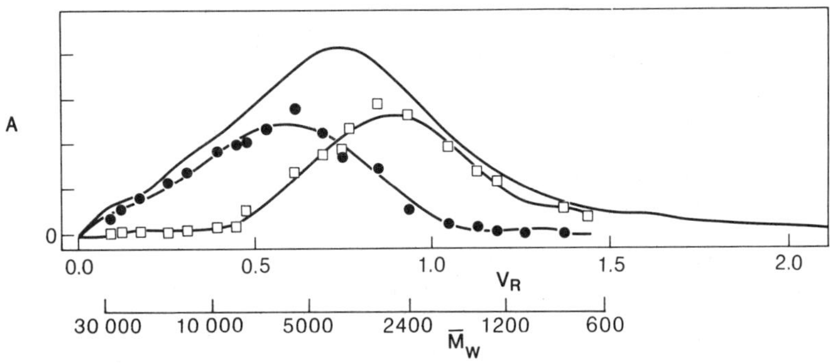

Fig. 10. Contributions of component subsets B (●) and C (□) to dissociated
kraft lignin elution profile from Sephadex G100/0.10 M aqueous
NaOH.

series of low molecular weight lignosulfonates, it was suggested over twenty years ago[51] that the structure of (spruce) lignin is based upon an 18-mer repeating unit.[52] Much more recently, the kinetics of spruce wood delignification with 0.1 M HCl-dioxane solution have been interpreted as indicating a cross-linking density of 1/19 in the native lignin macro-molecule[53] -- assumed here to be constituted from *randomly* distributed interunit linkages. From apparent m.w.d.'s in DMSO (a solvent facilitating marked association between lignin components[32]), intermediate "modules" with degrees of polymerization ~20 have been implicated in the assembly of "model" lignins during the *in vitro* polymerization of coniferyl alcohol by horseradish peroxidase.[54] Corresponding intermediates have similarly appeared during the degradation (at 160°C in aqueous dioxane *and* alkaline aqueous solution with and without sulfide) of a spruce "milled wood lignin" sample (the fraction soluble in aqueous 90% dioxane after cellulase treat-ment of ground wood).[54] It may not be entirely coincidental that a ~20-mer represents the upper bound to the lower molecular weight component subset which associates with the complementary group of higher molecular weight species in Douglas fir kraft lignin (*vide supra*).[35] Appropriate structural determinations will presumably lead to the identification of the underlying molecular features which account for these observations.

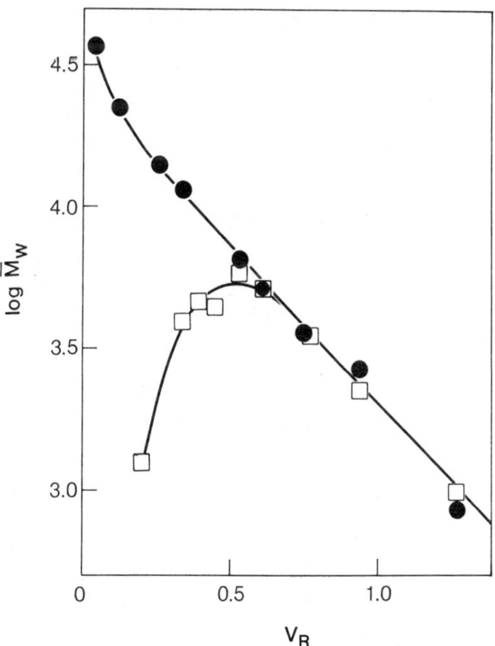

Fig. 11. Semilogarithmic plots of weight-average molecular weight vs. relative retention volume for component subsets B (●) and C (□) in paucidisperse kraft lignin fractions isolated by elution from Sephadex G100 with 0.10 M aqueous NaOH.

301

ACKNOWLEDGMENTS

Acknowledgment for partial support of this research is made to the
donors of The Petroleum Research Fund, administered by the ACS, and to the
National Science Foundation (Grant CBT 8412604), the Blandin Foundation,
the Graduate School of the University of Minnesota, the Minnesota
Agricultural Experiment Station, and the University of Minnesota Computer
Center. Advice and assistance from Ms. Lynda L. Tucker is also much appre-
ciated.

REFERENCES

1. E. Adler, Wood Sci. Technol., 11: 169 (1977).
2. S. Y. Lin, in: "Progress in Biomass Conversion," D. A. Tillman and E. C.
   Jahn, Eds., Academic Press, New York (1983), Vol. 4, p. 31.
3. J. Gierer, Wood Sci. Technol., 14: 241 (1980).
4. G. Brunow and G. E. Miksche, Appl. Polym. Symp., 28: 1155 (1976).
5. G. E. Miksche, Acta Chem. Scand., 26: 3269 (1972).
6. J. Gierer and L. Å. Smedman, Acta Chem. Scand., 19: 1103 (1965).
7. J. Gierer and I. Norén, Acta Chem. Scand., 16: 1713 (1962).
8. G. E. Miksche, Acta Chem. Scand., 26: 3275 (1972).
9. J. Gierer, Sven. Papperstidn., 73: 571 (1970).
10. J. Gierer, F. Imsgard and I. Pettersson, Appl. Polym. Symp., 28: 1195
    (1976).
11. J. Gierer and I. Pettersson, Can. J. Chem., 55: 593 (1977).
12. S. Winstein and R. Baird, J. Am. Chem. Soc., 79: 756 (1957).
13. R. Baird and S. Winstein, J. Am. Chem. Soc., 79: 4238 (1957).
14. J. Gierer and C. R. Nelson, J. Org. Chem., 43: 4028 (1978).
15. D. R. Marshall, P. J. Thomas and C. J. M. Stirling, J. Chem. Soc.,
    Perkin Trans., 2: 1898 (1977).
16. J. Gierer and O. Lindeberg, Acta Chem. Scand., Ser. B, 34: 161 (1980).
17. G. E. Miksche, Acta Chem. Scand., 26: 4137 (1972).
18. J. Gierer and S. Ljunggren, Sven. Papperstidn., 82: 503 (1979).
19. J. R. Obst, Holzforschung, 37: 23 (1983).
20. P. Axegård, S. Norden and A. Teder, Sven. Papperstidn., 81: 97 (1978).
21. P. Axegård and J.-E. Wikén, International Symposium on Wood and Pulping
    Chemistry, Stockholm (1981), Preprints Vol. II, p. 22.
22. S. Ljunggren, Sven. Papperstidn., 83: 363 (1980).
23. J. Gierer and I. Norén, Holzforschung, 34: 197 (1980).
24. J. R. Obst and N. Sanyer, Tappi, 63, No. 7: 111 (1980).
25. J. Gierer and S. Ljunggren, Sven. Papperstidn., 82: 71 (1979).
26. H. Nimz and H.-D. Lüdemann, Holzforschung, 30: 33 (1976).
27. R. D. Mortimer, J. Wood Chem. Technol., 2: 383 (1982).
28. H. Holton, Pulp Pap. Can., 78: T 218 (1977).
29. L. L. Landucci, Tappi, 63, No. 7: 95 (1980).
30. J.-M. Yang and D. A. I. Goring, Can. J. Chem., 58: 2411 (1980).
31. R. Kondo and J. L. McCarthy, J. Wood Chem. Technol., 5: 37 (1985).
32. W. J. Connors, S. Sarkanen and J. L. McCarthy, Holzforschung, 34: 80
    (1980).
33. S. Sarkanen, D. C. Teller, J. Hall and J. L. McCarthy, Macromolecules,
    14: 426 (1981).
34. S. Sarkanen, D. C. Teller, E. Abramowski and J. L. McCarthy,
    Macromolecules, 15: 1098 (1982).
35. S. Sarkanen, D. C. Teller, C. R. Stevens and J. L. McCarthy,
    Macromolecules, 17: 2588 (1984).
36. N. S. Yaropolov and D. V. Tishchenko, Zh. Prikl. Khim. (Leningrad), 43:
    1120 and 1351 (1970).
37. T. Lindström, Colloid Polym. Sci., 257: 277 (1979).
38. J. Benko, Tappi, 47: 508 (1964).
39. W. Brown, J. Appl. Polym. Sci., 11: 2381 (1967).

40. D. A. I. Goring, <u>ACS Symp. Ser.</u>, No. 48: 273 (1977).
41. M. A. Ivanov, P. P. Nefedov, A. E. Rusakov, L. D. Sherbakova, M. A. Lazareva and V. I. Zakharov, <u>Khim. Drev.</u>, No. 6: 108 (1979).
42. R. H. Atalla and U. P. Agarwal, <u>Science</u>, 227: 636 (1985).
43. J. Gravitis and P. Erins, <u>Appl. Polym. Symp.</u>, 37: 421 (1983).
44. M. Remko, <u>Cellulose Chem. Technol.</u>, 19: 47 (1985).
45. E. Tsuchida and K. Abe, <u>Adv. Polym. Sci.</u> 45: 77 (1982).
46. D. R. Robert, M. Bardet, G. Gellerstedt and E. L. Lindfors, <u>J. Wood Chem. Technol.</u>, 4: 239 (1984).
47. P. Månsson, <u>Holzforschung</u>, 37: 143 (1983).
48. G. Gellerstedt and E. L. Lindfors, <u>Sven. Papperstidn.</u>, 87: R 115 (1984).
49. N. Terashima, H. Araki and N. Suganuma, <u>Mokuzai Gakkaishi</u>, 23: 343 (1977).
50. G. Gellerstedt, E. L. Lindfors, C. Lapierre and B. Monties, <u>Sven. Papperstidn.</u>, 87: R 61 (1984).
51. K. Forss and K.-E. Fremer, <u>Paperi ja Puu</u>, 47: 443 (1965).
52. K. Forss and K.-E. Fremer, <u>Appl. Polym. Symp.</u>, 37: 531 (1983).
53. J. F. Yan, F. Pla, R. Kondo, M. Dolk and J. L. McCarthy, <u>Macromolecules</u>, 17: 2137 (1984).
54. M. Wayman and T. I. Obiaga, <u>Can. J. Chem.</u>, 52: 2102 (1974).

REVIEW OF THE SYNTHESIS, CHARACTERIZATION, AND TESTING OF GRAFT
COPOLYMERS OF LIGNIN

John J. Meister

Department of Chemistry
Southern Methodist University
Dallas, TX. 75275

INTRODUCTION

Lignin [8068-00-6] is a natural product produced by all woody
plants. It is second only to cellulose in mass of polymer formed per
annum.[1] Lignin constitutes between 15 and 40 percent of the dry weight
of wood with variation in lignin content being caused by growing conditions, species type, the parts of the plant tested, and numerous other
factors[2]. Plants use lignin to 1. control fluid flow, 2. add strength,
and 3. protect against attack by microorganisms[3].Each cell of the plant
grows its own lignin. The cell undergoes "lignification" in response to
an internally-orchestrated series of reactions which take place all
during cell differentiation[3]. Lignin appears first in the primary
(exterior) wall of the cell "corners". As the cell grows, lignin
deposits throughout the primary wall and then appears in the secondary,
interior wall of the cell. During this growth period, lignin deposits
develop in the intercellular region, also. Lignin appears to be
attached to the crystalline microfibrils of cellulose by phenylpropane
linkages to carboxyl groups. Such a bond structure would be a uronic
acid ester linkage.[3]

Recent work by Atalla[61] supports the idea that lignin is at least a
semi-ordered substance in wood with the plane of the aromatic ring
parallel to the cell wall surface. Woody plants synthesize lignin from
trans-confieryl alcohol 1 (pines), Trans-sinapyl alcohol 2 (deciduous),
and Trans-4-coumaryl alcohol 3 by free radical crosslinking initiated by
enzymatic dehydrogenation[4]. Structures of these alcohols are given in
Figure 1.

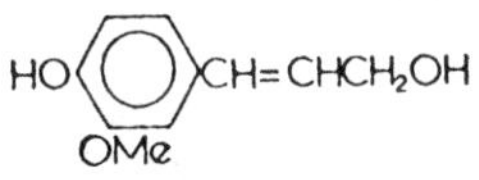

HO—⟨O⟩—CH=CHCH₂OH
OMe

OMe
HO—⟨O⟩—CH=CHCH₂OH
OMe

HO—⟨O⟩—CH=CHCH₂OH

1

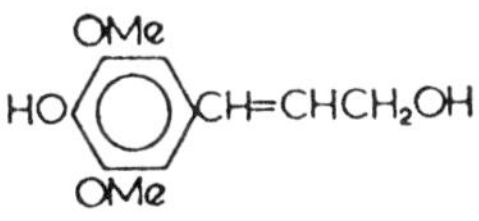

2

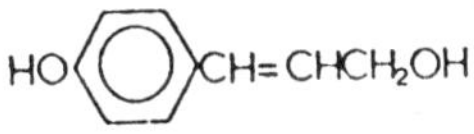

3

Figure 1.

Different ratios of these alcohols are used by different species of plants to form lignin, with the result that lignins from different sources will have different elemental and functional-group compositions. This alone would give lignin an extensive chemical diversity. Lignin recovery processes which extract lignin from wood, change the chemical composition of lignin and make this material extremely heterogeneous.

Methods for recovering lignin are the alkali process, the sulfite process, ball milling, enzymatic release, hydrochloric acid digestion, and organic solvent extraction. Alkali lignins are produced by the kraft and soda methods for wood pulping. They have low sulfur content (< 1.6 wt.%), sulfur contamination present as thioether linkages, and are water-insoluble, nonionic polymers of low (2,000 to 15,000) molecular weight. Approximately 20 million tones of kraft lignin are produced in the United States each year.

The sulfite process for separating lignin from plant biomass produces a class of lignin derivatives called lignosulfonates. Lignosulfonates contain approximately 6.5 weight percent sulfur present as ionic sulfonate groups. These materials have molecular weights up to 150,000 and are very water-soluble. Graft copolymerization of lignosulfonates will not be discussed in this review.

Milled wood lignin (MWL) is produced by grinding wood in a rotary or vibratory ball mill. Lignin can be extracted from the resulting powder using solvents such as methylbenzene or 1,4-dioxacyclohexane[5]. Milling only releases 60 weight percent or less of the lignin in wood, disrupts the morphology of lignin in wood, and may cause the formation of some functional groups on the produced lignin[6]. Despite these limitations, milling appears to be an effective way of recovering lignin from plants with only slight alteration. Enzymes which hydrolyze polysacchardes can be used to digest plant fiber and release lignin. After digestion, the lignin is solubilized in ethanol[7]. Extensive analytical studies support the idea that enzymatically produced lignin has undergone no major modification in removal from plant material[8-12].

Acid hydroylsis of the polysaccharide portion of wood will release lignin but also causes major condensation reactions in the product[13]. These reactions can be minimized by using 41 wt. percent hydrochloric acid in place of other mineral acids but some condensation reactions still occur[14]. This is not an effective method by which to obtain unaltered lignin. On the other hand, lignin can be solvent extracted from wood at temperatures of 175°C using solvent mixtures such as 50/50 by volume water/1,4-dioxacyclohexane[15]. Changes in lignin under these conditions appear to be minor.

Once lignin is separated from other plant products, it can be grafted. Extensive studies on the modification of lignin have been made[16] because of the enormous mass of kraft lignin produced each year by the pulp and paper industry. A major method of forming derivatives is the formation of graft copolymers of lignin, molecules in which a sidechain of synthetic polymer has been grown off of a lignin molecule. Graft copolymerization sharply changes the properties of lignin and allows useful products to be made from this waste biomass[17].

Lignin has been grafted with 1-phenylethylene, 1-(1-oxo-2-oxy-n-propyl)-1-methylethylene, 1-amidoethylene, 1-(2-azoethinyl)ethylene, and urethane derivatives. An index of compounds listing compound number and trivial name is given in Table 8. Two types of methods, radiation or chemical, have been used to attach sidechains of these repeat units to lignin. The radiation methods have used both electromagnetic and particle radiation to produce grafting. Low-energy, electromagnetic irradiation based on visible or ultraviolet light relies on exciting or decomposing a particular bond either in lignin or in an initiator present in the reaction mixture. This method, photoinitiation, has not been used to graft lignin. High energy radiation grafting using either electromagnetic or particle beams proceeds by ionization and excitation reactions that produced anionic, cationic, and free-radical sites. Ionic grafting reactions have not been conducted on lignin and there-fore, only the free radical polymerization is known to contribute to grafts. Lignin is quite stable to ionizing radiation having a G_R value of 0.6 to 0.7[18]. This stability makes lignin a poor candidate for radiation grafting since in the presence of neat monomer or in solution, initiation will occur far more readily to form homopolymer than it will to form graft copolymer.

Some reduction in the amount of homopolymer produced can be achieved by initiating the grafting reaction by chemical methods. Chemical initiation can be applied in two ways. First, a reagent which attacks functional groups on the lignin backbone to produce a grafting site can be used as grafting initiators. Alternatively, a reagent which reacts with lignin to form a reactable functional group is used to derivatize lignin. The added groups are structures such as peroxide or ethene bonds and are then treated or reacted to initiate grafting. The chemical method can be used to initiate all homogeneous polymerization reactions[19] but only step and free radical chain reactions have been conducted on lignin.

In the following sections, efforts to attach a sidechain of a given structure to lignin will be summarized together in sections titled by the name of the polymeric side chain. Radiation-initiated grafting studies will be discussed first and chemical methods discussed second.

Poly(1-phenylethylene)

A significant number of studies have been done on how to attach poly(1-phenylethylene) [I] to lignin. This sytem also provides a good example of the difficulties encountered in forming derivatives of lignin. Zoldners has done extensive studies of making derivatives of birch wood with I and shown that in radiation or chemical-initiated poly-erization 1. extensive amounts of homopolymer are formed,[20-21] 2. poly-saccharide is easier to graft than native lignin[22], 3. grafted I is connected to polysaccharide units, not lignin[23], 4. lignin actually inhibits polymerization of 1-phenylethene [II] by free radical propaga-tion[24,25], 5. this inhibition of propagation by lignin can be reduced by methylating the lignin[22], and 6. grafting does occur on hydrochloric acid, pine lignin but is not seen on hydrochloric acid, hard wood lignin. These results suggest extensive homopolymerization in any radia-ion initiation of lignin grafting conducted in neat monomer where the monomer has $G_R > 0.7^*$ or in solvents with $G_R > 0.7$. In such systems, it is possible that no detectable graft copolymer will be formed. The best method for radiation initiation of lignin grafting is the two step process: 1. irradiate lignin, 2. contact with monomer. This process is aided by the fact that lignin radicals are quite stable and persist for days after formation[26]. The data of Table 1 show lignin free radicals to be very stable with 40 number percent of the originally formed radicals present after 25 to 250 days. This stability is partially the reason that kraft lignin inhibits formation of I during radiation grafting[27]. Irradiation of lignin and II in the absence of oxygen produces graft copolymer but irradiation in the presence of oxygen produces no grafting of either II or 1-amidoethene. Irradiation of lignin in the presence of

---

*G_R is the number of free radicals generated in the substance per 100 ev of energy absorbed.

oxygen produces a polymerization teminator and use of radiation doses of
16 Mrad or more causes partial depolymerization.

TABLE 1.   Free radicals produced from lignin irradiated in air.
Total Dos = 7.6 megarads

Samples	Rad. intensity ($\times 10^{-4}$ rad/hr)	Free radical contents** of peroxidized lignin, $\times 10^{-17}$ per g. of lignin		
		Immediately after irradiation	25 days after irradiation	250 days after irradiation
1	1.57	16.0	6.4	7.9
2	2.46	14.1	6.0	6.4
3	6.70	16.5	6.4	8.2
4	10.6	15.8	6.2	6.0

** Determined by heating peroxidized lignin with diphenylpicryl
hydrazene at 70°C.

Further, more-detailed studies of the grafting of hydrochloric acid
lignin by Koshijima et. al. have shown that radiation grafting by II
proceeds in two stages[28].   This lignin had been methylated before
grafting so these studies were done on lignin lacking the reactive
phenolic hydroxyl groups common to extracted lignin[29].   In the first
stage of the grafting reaction, propagating I chains terminate by chain
transfer at the number 5 or 6 carbon atom of the benzene unit of lignin.
Numbering pattern on the lignin residue is given in Figure 2.   The rate
of termination is such that this process predominates as a grafting
mechanism until a mass of I equal to that of lignin has been added to
the backbone[28].

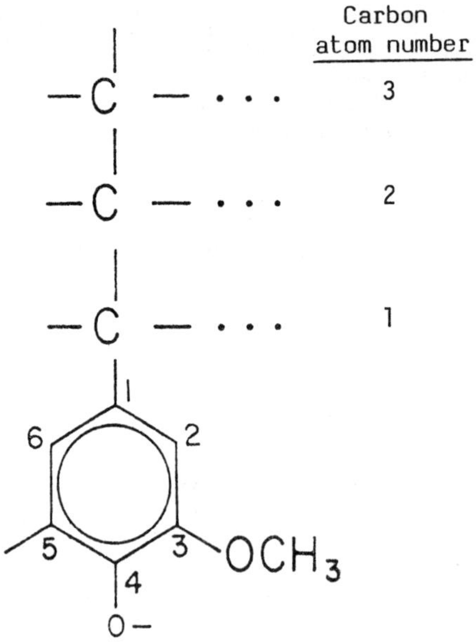

Figure 2.

308

Stage 2 of the grafting reaction occurs when grafting is initiated
by electron abstraction reactions on the propyl group of the lignin
repeat unit. Koshijima et al. propose that these chains propagate to
higher molecular weight than the homopolymers terminating in stage 1.
This occurs because previous termination reactions have removed all of
the lignin residues useful for chain transfer-termination. This work
also shows that when 64 weight percent of the original lignin is
grafted[28], additional initiation of grafting sites begins in the I
sidechain itself and the formation of branched sidechains occurs.

Studies of solvent effects were carried out by the Koshijima
research group. In these experiments, methanol, ethanol, 2-propanol, 1-
butanol, n-hexane, or benzene were added to reaction mixtures before
irradiation under vacuum. A solvent of 2 weight percent methanol in
neat II gave a maximum grafting level of a mass of I added to lignin
equal to 4.3 times the mass of lignin. Figure 3 contains representative
data for solvent effects in grafting.The authors propose that methanol
aids the grafting reaction by formation of methylenehydroxide
radicals[30]. They have also studied the formation of graft copolymer from
hydrochloric acid lignin after radiation-peroxidation in air[31]. This tech-
ique was much less efficient in adding I to lignin since it produced
only 1/6 the mass increase found by direct radiation of lignin in neat
II. In these studies, the oxidation of graft copolymer by nitrobenzene
and the analysis of the resulting I to show it contained lignin repeat
units provide firm proof of grafting. Table 2 contains representative
analytical data from these digestion studies.

Irradiated at 0.1 Mrad for 30 hr.

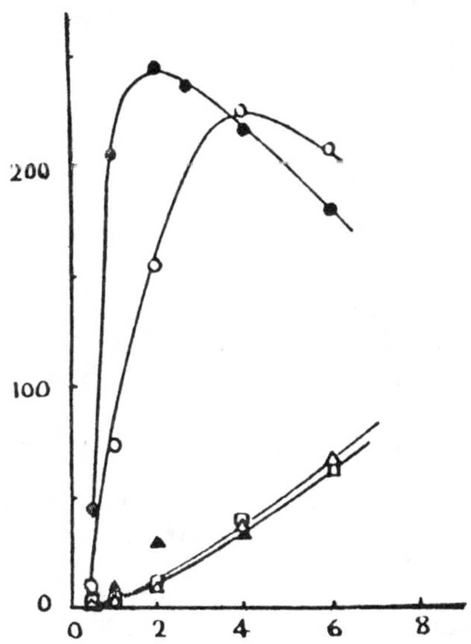

Weight Percent Concentration of Alcohols

Solvents : ●MeOH, ◯EtOH, △2-PrOH,
☐1-BuOH, ▲MeOH (Unmethylated lignin)

Figure 3

These studies were continued and expanded upon by Stannett, et al.
In their studies, they grafted hydrochloric acid lignin as well as
methylated hydrochloric acid lignin[32]. Results of this work show that

for all reactions of lignin in neat II, conversion is high but grafting is slight. The methylation of acid-extracted lignin increases grafting in neat II by swelling the lignin polymer, making more of the lignin coil accessible to monomer and grafting. The addition of 2 weight percent methanol to the reaction mixture also produces a solvent effect and expands lignin coil dimensions. The solvated coil then reacts more readily with II. The role of methanol-derived radicals, proposed by Koshijima, in the grafting reaction is minimal. The termination of propagating I chains onto lignin to form graft copolymer, another idea proposed by Koshijima et al., does not occur. Instead, grafting proceeds by hydrogen or alkyl abstraction reactions which produce quinone-like radicals in acid lignin and benzyl radicals in methylated lignin. Once these radicals are formed, radical reactivity controls the rate and extent of grafting. Quinoid radicals can be far less reactive than benzylic radicals[33], depending on the substituents about the oxygen on the benzene ring. Radicals in acid lignin, then, can have propagation reaction rates ranging from 0, complete inhibition, to that for pure I. Grafting on acid lignin is found to be retarded when compared to methylated lignin. Further, grafting of I on hard wood lignin, where R_5 = -OCH$_3$, is minimal. Both of these experimental findings are coincident with Stannett's mechanism.

TABLE 2. Methoxyl contents of grafted branches separated by nitrobenzene oxidation of lignin-1-phenylethylene graft polymers.

Grafting ratio of original graft polymer (Weight percent increase from grafting)	108	208	265	304	392
Vanillin produced, wt. % of lignin contained in sample	6.00	5.03	5.82	5.60	5.81
Total methoxyl in vanillin, x10^{-3} M	0.28	0.23	0.27	0.26	0.27
Amounts of grafted branches, g. per 700 mg of lignin	0.58	1.68	1.94	2.21	2.84
Methoxyl in grafted branches, wt. %	4.50	3.80	3.60	3.34	2.53
Total methoxyl in grafted branches, x10^{-3} M	0.84	2.1	2.3	2.4	2.3
Degree of Polymerization of grafted branches calculated per one guaiacyl residue	5.2	6.4	6.8	7.5	10.3

Original lignin contained 3.66 x 10^{-3} M. methoxyl.

Simionescu et al. show that the thermal decomposition temperature of I - methylated, hydrochloric-acid-lignin graft copolymer is 290°C by TGA tests[47].

Stannett et al. have also studied radiation-induced grafting of I on kraft pine lignin[34]. Analytical data which support graft copolymerization were obtained by solubility tests and fractionation of reaction products. Methanol, 1,4-dioxacyclohexane, and 1-oxo-2-imino-2-methyl-propane were used as solvents for I. This solubilization of the grafted molecule by I into solvents for I occurs because of the low molecular weight of kraft lignin.

Generally, the largest amount of grafting occurs in reaction mixtures containing 50 or more weight percent solvent. This use of solvents which swelled the lignin molecule but precipitated grafted co-polymer sharply increased the fraction of original lignin grafted. Reactions run in 80 weight percent methanol produced grafting on approxi-

mately 80 weight percent of lignin. Grafting could be performed on both
kraft and methylated kraft lignin. However, degradation of graft copolymer appeared to occur at radiation doses above 8 Mrads. Table 3
contains representative data from these solvent studies.

TABLE 3. Radiation-Induced Graft Copolymerization of Lignin and 1-
          Phenylethene and 1,4-dioxacyclohexane

II/1,4-dioxa-cyclohexane ratio	Dose, Mrads	Conversion, %	Original lignin extracted, %	Graft of inextractable fraction, %
80:20	5	15.6	17.9	9.1
	10	31.3	19.7	13.5
50:50	5	15.6	-21.0[a]	--
	10	30.7	-32.1[a]	--
20:80	5	25.7	28.9	18.6
	10	42.9	32.6	29.4

[a] -- OCH_3 analysis not available; figure given is % weight change
after extraction with benzene.

Graft sidechains of I can be freed from lignin by nitrobenzene
oxidation. They show a polydispersity index of 2.6 and an average
number of grafts per backbone molecule of 1.1. Torsion braid thermal
analysis of fractionated, graft copolymer showed that samples containing
more than 15 weight percent of both lignin and I had two glass transition temperatures. (See ref. 47). The presence of two glass transition
temperatures in a 2-component solid strongly suggests that the
chemically distinct components are phase separating upon solidification.
This would mean that these polymers are surface- or interface-active
materials and spontaneously form composite solids.

Model compound studies done by Stannett et al.[35] ohow that 2-methoxy-
phenol, a structure common to pine lignins, competes slightly with
monomer for the radical growing site during I polymerization in vacuum
but inhibits this reaction when the oxygen needed to form quinone-like
structures is present. A slight change of structure of the lignin model
to 2-hydroxy-1,3-dimethylbenzene, a structure typical of hardwood
lignin, results in a sharp increase in the reactivity of the model
toward radical groups. As a result of this increased reactivity, the
model compound acts as a chain transfer agent. These data do not
clarify the grafting process since the chain transfer behavior of 2-
hydroxy-1,3-dimethyl phenyl groups would constitute a good mechanism for
high grafting on hardwood lignin. Hardwood lignin showed no significant
grafting of I by radiation initiation. Table 4 contains data on chain

TABLE 4. Chain Transfer Constants for Guaiacol and 2,6-Dimethoxyphenol
          in 1-Phenylethene[a]

Compound	$R_p \times 10^{+7}$, M/sec	$\bar{M}_w$	$\bar{M}_n$	$\bar{M}_w/\bar{M}_n$	$C_s \times 10^4$ [b]
1-Phenylethene	72	101,000	51,000	1.98	...
0.05M Guaiacol	88	100,000	50,200	1.99	69
0.05M 2,6-Di-methoxyphenol	66	74,000	38,600	1.91	1,150

[a] Dose rate 41 rad/sec.

[b] Calculated from $\dfrac{1}{P_n} = \dfrac{1}{P_0} + C_s \dfrac{[S]}{[M]}$, P_n = degree of
polymerization of phenol solution reaction; P_0 = degree of
polymerization of pure 1-phenylethene; and $C_s = k_{tr,solv.}/k_p$.

transfer constants and product properties from Stannett's model compound studies[35].

Kraft pine lignin is known to contain a higher concentration of alkyl double bonds conjugated with aromatic groups[36] than other lignins. The effect of such structures on polymerization was determined by adding 1-(3-methoxy-4-hydroxyphenyl)propene to radiation polymerizations of II. These studies[36] showed that these structures inhibit polymerization which is coincident with the reduced rate of polymerization seen when kraft pine lignin is added to polymerizations of II.

Chemical

Hydrochloric acid spruce lignin can be grafted by thermal initiation of polymerization of II in the presence of lignin which has been ozone peroxidated[37]. This chemical grafting reaction is run under an inert atmosphere in a sealed ampoule. The ratio of lignin mass to II mass was 1 to 10. Polymerization was initiated by heating the reaction mixture to between 70 and 100°C. The presence of graft copolymers in the reaction product was suggested but not proven by exhaustive extraction of the reaction products by benzene.

Kinetic studies on these polymerizations show that ozonolysis produces both inhibitors and potential grafting sites on lignin. Methanol can be used as a solvent to both increase polymerization rate and fraction of I grafted to lignin. Both reaction properties reach a maximum when the mole fraction of methanol in the methanol/ II mixture placed in the reaction ampoule is 0.33. This is a significantly higher methanol concentration than found most effective by Stannett during radiation grafting.

To test the role of phenolic hydroxyl groups on graft copolymerization, portions of the hydrochloric acid spruce lignin were methylated with diazomethane before peroxidation. Methylation increased oxidizing unit content after peroxidation by over 60 percent. Further, fewer inhibitors to the polymerization of II are formed during peroxidation of methylated lignin and the propagation and grafting rates of II in the presence of methylated lignin are 8 or more times higher than rates at the same temperature in the presence of underivatized lignin.

Comparing the radiation grafting of lignin by I performed by Stannett et al. and the chemically initiated grafting described above shows that conversion of II to graft copolymer is much higher at any level of monomer polymerization by chemical means that it is by radiation techniques. Chemical initiation of grafting is better than 8 times more effective in converting monomer to graft copolymer than is radiation when both are applied to hydrochloric acid lignin. Stannett's work shows this greater efficiency of grafting by chemical initiation in his own work on lignin - II - benzoyl peroxide reactions. He notes[32] that "chemically initiated grafting at 60° was more effective than the radiation-induced grafting at room temperature." For a system in which grafting would have to occur by chain transfer reactions since the chemical initiation used is decomposition of benzoyl peroxide in 98 weight percent II, 2 weight percent methanol, this is not simply a kinetics effect caused by sharp differences in the production rate of radicals. The polymerization times for chemical and radiation methods are comparable in these experiments. Temperature might be a significant variable promoting grafting in the chemical initiation methods because all chemical initiations were conducted at a temperature 30°C above that for radiation-induced reactions. All of these studies and others in which lignin grafting was not proved[38,39], show lignin to be an inhibitor of polymerization of II.

<u>Poly(1-(1-oxo-2-oxy-n-propyl)-1-methylethylene) (III)</u>

Graft copolymers of lignin with 2-oxy-3-oxo-4-methylpent-4-ene[80-62-6] (methylmethacrylate = IV) or 2-oxy-3-oxopent-4-ene [96-33-3] (methylacrylate = V) have been made. These materials have sharply different synthesis behaviors and physical properties than the 1-phenylethene derivatives of lignin. This is partially due to the differences between the monomers being compared. The ethene acid or ester discussed here is a significantly more polar molecule than 1-phenylethene. Graft copolymers containing these acid or ester units will be more hydroscopic and tend to be more soluble in polar solvents such as alcohols or ketones when compared to copolymers containing phenyl side units. Further, the tendency of these monomers to form free radicals when irradiated is 20 to 40 times higher than that of 1-phenylethene[40]. This is shown by G_R values at approximately 0.1 rads/S of 11 and 22 for IV and V, respectively, but a value of only 0.63 for 1-phenylethene.

These monomers also initiate an immediate gel effect upon polymerization[41] which will tend to promote higher molecular weight graft copolymers and higher conversions of monomer to polymer. All of the polymerizations of the monomers onto lignin have been conducted by free radical mechanisms. Grafting of these monomers to lignin derivatives such as lignosulfonates[42] and nitrolinin[43,44] has been also described. Further discussion will concentrate on the grafting of these monomers to extracted, underivatized lignins.

<u>Radiation</u>

Koshijima et al.[45,46] have also studied the formation of poly(1-(1-oxy-2-oxo-n-propyl)-1-methylethylene) (III) sidechains on hydrochloric-acid-extracted lignin from Pinus densiflora (pine)and Betula Tauschii (hardwood) and found they could be produced by radiation initiation. The mass of III produced on hardwood lignin was only 40 weight percent of that that could be added to pine lignin. These results show that lignin containing 1-alkyl-3,5-dimethoxy-4-oxylbenzene units (syringyl groups) is harder to graft than lignin containing the characteristic pine repeat units, 1-alkyl-3-methoxy-4-oxylbenzene. Also, the mass of III which could be grafted to pine lignin increased 4-fold when the methoxyl content of the lignin added to the grafting reaction was increased from 16 to 23 weight percent by methylation. Continued methylation to give lignin with 38 weight percent methoxyl group content produced a material which only grafted 1.6 times as much III as underivatized pine lignin, however. These results show that there is a degree of methylation which will produce maximum mass of graft sidechain mass on lignin. Further methylation will reduce conversion to graft copolymer. The authors speculate that because of the high G_R value of IV when compared to lignin, most of the grafted sidechains are attached to the surface of undissolved lignin particles in the reaction mixture. This is not necessarily true, however, since IV, as a more polar monomer, would swell lignin more and allow internal grafting in swollen but undissolved lignin. A balance between monomer coil permeation from lignin swelling and the inhibiting behavior of phenolic groups may explain the maximum seen in the effect of methylation on pine lignin grafting. When phenol groups are methylated, free radical traps are removed and inhibition decreases. As a result, graft copolymer mass increases. However, methylated lignin is less polar than pine lignin and the capacity of the polar monomer, IV, to behave like methanol for II and swell methylated lignin decreases. A maximum appears in the mass of III added to methylated lignin when further methylation causes coil collapse and loss of IV access to the radical chain end that fails to be offset by the reduction in inhibitor sites produced by the methylation.

Koshijima et al[46] also showed that this graft copolymer is radiation sensitive and degrades by sidechain cleavage. Careful characterization of the parts of this copolymer and determination of some of its physical properties have been done by Simionescu et al.[47] The graft copolymer begins to undergo extensive thermal decomposition at or above 260°C.

Chemical

Cerium ion has been used to graft III to unbleached, pine pulp containing lignin[48]. No digestion tests were done on the resulting product and grafting is assumed to occur on the basis of failure to extract some III from the pulp after the reaction. Since cerium (+IV) is known to react with polysaccharides at the 2-3 carbon bond of the anhydroglucose unit, it is very probable that if any grafting did occur in these samples, it occured on cellulose and not lignin. Lignin had a pronounced effect on the grafting reaction, however. As lignin content of the pulp increased from 0 weight percent to 20 weight percent, the weight fraction of III retained by the pulp after the reaction decreased from 3.30 to 0.40. Further, as lignin content of the pulp increases from 0 to 20 weight percent, the concentration of cerium (+IV) which can be added to the reaction but fail to produce retention of III by the pulp increases. This implies that lignin is preferentially reacting with Ce(+IV) but not producing conditions which result in the permanent retention of III by the pulp. See Figure 4 for Kubota's[48] data showing these changes.

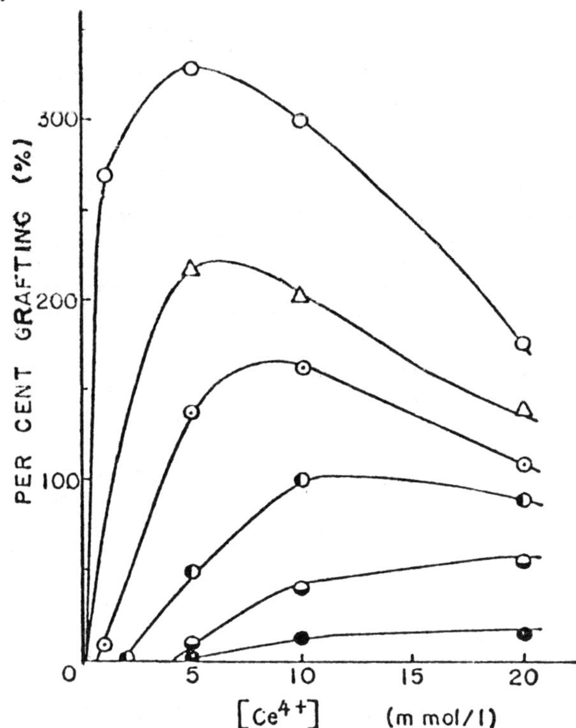

Relation between percent grafting and ceric ion concentration: polymerization temperature, 45°C; polymerization time, 60 min. Lignin content (%): (O) 0.19; (⊙) 0.77; (◑) 2.50; (◓) 6.10; (●) 21.5; (△) sample pretreated with ceric ion.

Figure 4.

Molecular weight of the grafted sidechain was estimated by digestion of extracted reaction product with acid and measurement of the
limiting viscosity number of a 2-propanone solution of III. Number of
grafts per gram of pulp is estimated by dividing sidechain molecular
weight into mass of sidechain retained per g of pulp. These results are
approximate because no tests were made for the end group of the sidechain. Previous studies have shown that retained polymer may be both
grafted or entangled.

Naveau has developed a novel way of chemically initiating grafting
on lignin extracted by 1,4-dioxacyclohexane extraction of hydrochloric-
acid-digested, milled wood[49]. He adds reactable ethenyl units to lignin
by reacting it with 2-methylprop-3-enoyl chloride or 2,6-dimethyl-4-oxy-
3,5-dioxahept-1,6-diene(methacrylic anhydride = VI). The 3-methyl-2-
oxabut-1-oxoyl (VII) groups added to lignin should then be reactable
with free radical initiators to produce graft poplymer. Additions of VI
to lignin dissolved in 1,4-dioxacyclohexane and in the presence of
pyridine placed sufficient VII groups on lignin to give VII groups on 8
to 72 percent of all lignin repeat units. Retention of VII groups was
measured by saponification of reacted lignin and titration of the
produced salt.

Copolymerization was attempted by thermal decomposition of benzoyl
peroxide in the presence of VI-derivitized lignin and IV. A retention
of III equal to 88 weight percent of original lignin mass during solvent
extration is used as proof of graft copolymerization of lignin and III.
This grafting method is a very innovative and effective one. All it
lacks is some way to guarantee an attack of the initiating radicals on
the ethene units on lignin only, rather than depending on statistical
factors to initiate some grafting in the presence of homopolymerization
of IV.

Poly(1-amidoethylene)

Several efforts have been made to attach poly(1-amidoethylene)
(VIII) chains to lignin. These efforts were spirred by the fact that
VIII is known to improve sheet strength in paper[50]. The presence of
lignin terminated grafting by radiation-initiation in pulp[50], however,
so most work on this sidechain has been oriented toward attaching it to
cellulose by cerium (+IV) reactions.

Recently, a method has been developed by Meister, et al. to graft
VIII onto kraft, pine lignin by free-radical polymerization in solvent[51].
The reaction is versital and can be conducted in 1-methyl-2-
pyrrolidinone, dimethylsulfoxide, dimethylacetamide, dimethylformamide,
(dimethylsulfoxide and 1,4-dioxacyclohexane) or (dimethylsulfoxide and
water) mixtures, and pyridine[52].

The reaction can be initiated by hydrogen peroxide; 3,3-dimethyl-1-
2,dioxybutane; and 2-hydroperoxy-1,4-dioxacyclohexane[53]. For polymerization at room temperature, lignin, a hydroperoxide, and chloride ion
must be present in the reaction mixture[54].

The grafting reaction is run by adding lignin and calcium chloride
to a nitrogen-flushed, 125 mL flask containing 20 mL of solvent. The
mixture is bubbled with nitrogen for 3 minutes, oxidation products of
1,4-dioxacyclohexane or another hydroperoxide are added, and the flask
contents are bubbled with nitrogen for 2 more minutes. Flask contents
were sitrred for 20 minutes and bubbled with nitrogen for 4 minutes. The
2-propenamide is then added, flask contents are bubbled with nitrogen

for 10 minutes while being stirred, 0.15 mL of 0.05M cerium (+IV)
sulfate in 1M aqueous H_2SO_4 is added, the sample is stirred and bubbled
for 10 more minutes, and is then capped with a septum stopper.

After 2 days of storage in a 30°C constant temperature bath, the
reaction is terminated with aqueous hydroquinone and product recovered
by precipitation in stirred 2-propanone.

Graft copolymmerization has been proven for these compounds by a
new method of size exclusion chromatography. The technique allows the
ultraviolet spectra of the polymer eluting from the size exclusion
column to be broken down into contributions from individual ultraviolet
absorbers in the molecule. For graft copolymers having different u.v.
spectra for backbone and sidechain, this breakdown gives a decomposition
of the absorbance signal of the eluting copolymer into signal due to
backbone and that due to sidechain[55]. This decomposition is done by a
multivariate curve resolution program (MCR-2) which processes a series
of ultraviolet spectra taken as the copolymer elutes from the chromatograph into separate absorption contributions.

For lignin-VIII graft copolymer, the ultraviolet spectra of lignin
and VIII were made distinct and readily distinguishable by n-sub-
stituting some of the 1-amidoethylene repeat units with methoxyphenyl
groups[56]. This treatment produces a graft copolymer with a lignin absorption shoulder at 286nm and a sidechain absorption peak at 246 nm. A
chromatograph of N-substituted, graft copolymer (unlabled curve)
resolved into sidechain, curve 1, and backbone, curve 2, absorbances is
given in Figure 5. These results show that lignin and sidechain are pro-
ortionally distributed throughout the graft copolymer molecular size
distribution. This type of analysis allows molecular composition as
well as molecular size to be measured for a graft copolymer sample.

These copolymers are water soluble and possess properties needed in
industrial processes. The more useful properties of these compounds are
1. The capacity to flocculate or deflocculate suspended matter such as
clay, 2. the capacity to act as a dispersing agent, 3. the capacity to
complex metal ions, and 4. the capacity to increase the viscosity of
water[54]. Aqueous drilling fluids are complex suspensions of clay in

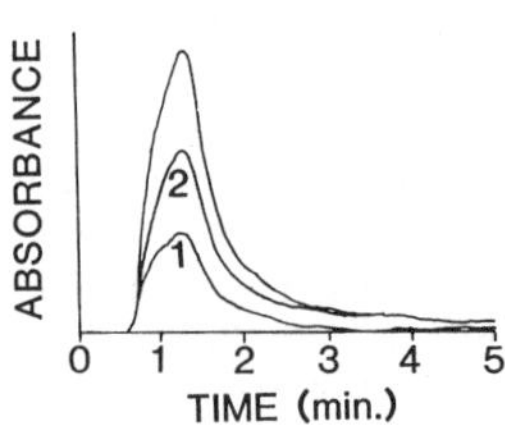

Figure 5. Decomposition of lignin
graft terpolymer chromatograph peak into
signal due to sidechain (1) and signal due to lignin (2)

salt solution which require all of the above properties. These fluids
are used to cool bits, suspend drilled solids, and seal off the wall of
a well during drilling of water, oil, or brine wells. These copolymers
have been tested and found effective as drilling mud additives.

Lignin-VIII, graft copolymers were tested for effectiveness in a
water-base mud containing 28 lb/barrel (ppb) of Wyoming bentonite and 40
ppb of Revdust, a commercial clay mixture used to represent solids
generated by drilling. The drilling muds containing graft copolymer
were tested against a base mud containing chrome lignosulfonate, a
commercial drilling mud additive, to determine if the copolymer improved
mud properties. The tests of mud rheology are standard test developed
by The American Petroleum Institute. Hot rolling is mixing the mud for
16 hours at 121°C and models exposure of the mud to well conditions.

A 25g sample of graft copolymer was base dialyzed to produce separate
fractions of graft copolymer and homopolymer. Data for the separated
products are contained in Table 5. Data from tests on mud samples
formulated with the separated products are given in Table 6 along with
data from tests of base mud and base mud formulated with 0.5 ppb of
chrome lignosulfonate.

The mud tests show that chrome lignosulfonate is a better low-
temperature thinner than either of the reaction produced fractions. The
lignosulfonate also lowers low temperature gel strength and yield point,
indicating it is acting as a deflocculant and dispersant. The poorer
performance of the reaction product fractions before hot rolling may
have been caused by incomplete solutions of the separated solids. The
lumpy lab products were not ground before being added to the mud.

The two reaction products thicken the base mud at low temperature,
possibly indicating the products have too high a molecular weight to act
as a low temperature thinner. After hot rolling, both synthesis samples
out-perform lignosulfonate as viscosity reduction agents. The synthetic
additives also reduce gel strength. The synthetic additives raise yield
point and apparent and plastic viscosity in the cold mud, but decrease
all three variables and out-perform the lignosulfonate after hot rolling.
The synthetic compounds decrease API filtrate before and after hot
rolling and also lower high-temperature-high-pressure filtrate. Chrome
lignosulfonate does not control filtrate loss.

The tests show that poly(lignin-g-(1-amideothylene)) acts as a high
temperature thinner and as a filtrate control agent. The tests also
show that reaction products prepared from reactions containing less
2-propenamide act as thinners for the base mud.

Poly(1-(1-oxoprop-1-oxyyl)ethylene)

Koshijima et al.[46] attempted to radiation graft this sidechain onto
hydrochloric acid, hardwood lignin but found that no grafting took
place. Since softwood lignin has a higher tendency to allow free radical
polymerizations, use of this form of lignin might produce a graft
copolymer with poly(1-(1-oxoprop-1-oxyyl)ethylene) (IX). However, a
2-oxo-3-oxypent-4-ene (X) is much more prone to chain transfer reactions[57] than II or IV so graft copolymeization with X will be less
effective than with other monomers.

Poly(1-cyanoethylene)

Simionescu et al.[47] have attempted to add poly(1-cyanoethylene) (XI)
sidechains to methylated, hydrochloric acid polar lignin (P. Celei) by
radiation initiation. Homopolymer was extracted from the product with

dimethylformamide. Proof of grafting was based on infrared spectra of
the extrated reaction product and inability to remove XI through extraction of the lignin. Up to 26 weight percent XI could be added to lignin
by use of up to 5Mrads of radiation. The graft copolymer began to
thermally decompose at 250°C and had less thermal stability than pure XI.

TABLE 5.   Separated Products From Lignin-(2-Propenamide) Reactions

Crude Reaction Product
    Weight (g)   18.0
    Content (wt.%)
        Lignin   7.27
        1-Amidoethylene units   73.8
        [η], water, 30°C   0.51

After Separation
Graft Polymer Fraction
    Weight (g)   9.09
    Wt.% of original sample   50.5
    Content (wt.%)
        Lignin   9.52
        1-Amidoethylene units   36.43
    Degree of Hydrolysis   49.4
    [η], in water, 30°C   6.77
Homopolymer Fraction
    Weight (g)   3.20
    Wt. % of original sample   17.8
    Content (wt.%)
        Lignin   1.53
        1-Amidoethylene units   36.94
    Degree of Hydrolysis   50.1
    [η], in 0.50 M   0.83
        NaOH, 30°C

Poly(1-oxy-3-methylprop-1,3-ylene)

Glasser has shown[58,59] that kraft, pine lignin reacts with 1-oxa-2-
methylcyclopropane (XII) at 180°C to produce a liquid polymer containing
large numbers of secondary hydroxyl groups. This reaction appears to
add poly(1-oxy-3-methylprop-1,3-ylene) (XIV) repeat units to phenolic
hydroxyl and carboxylic acid sites on lignin. The chains are
oligomeric, usually being monomer or dimer additions to each site.

Reactions containing an excess of XII produce homopolymer because of
the low reactivity of the secondary hydroxyl groups on the lignin side
groups. Less than 10 mole percent of XII copolymerizes when the mole
ratio of monomer to lignin repeat units is 20. Higher fractions of
copolymer can be achieved by lowering the monomer to lignin-repeat-unit
mole ratio. Size exclusion chromatography has been used to prove that
lignin has derivztized by these reactions[60]. The role of phenolic and
carboxylic acid groups in these reactions was verified by titration and
methylation experiments.

The polyol derivative of lignin is useful as a coating or adhesive
additive when combined with diisocyanates to form polyurethanes. The
resulting network polymer is strong, rigid, water and solvent resistant,
and provides excellent bond strength. Typical formulations for this
lignin-based material are given in Table 7.

TABLE 6. Properties of Test Muds Before and After Hot Rolling

Property	Base Mud Before	After	Graft Copolymer Fraction Before	After	Homo polymer Fraction Before	After	Chrome Ligno- sulfonate Before	After
Viscosity in Centipoise at a shear rate of:								
1020 s^{-1}	52	69	74	42	74	43	29	50
510 s^{-1}	36	47	49	24	49	24	16	33
340 s^{-1}	30	29	40	17	40	17	12	26
170 s^{-1}	21	28	27	10	27	10	7	18
Gel strength in lb/100 ft^2 mud has set for:								
10 sec	6	12	5	3	5	2	2	11
10 min	25	35	20	3	20	3	9	24
Apparent Viscosity (cp)	26	35	37	21	37	22	15	25
Plastic Viscosity (cp)	16	22	25	18	25	19	13	17
Yield Point (lb/100 ft^2)	20	25	24	6	24	5	3	16
API Filtrate Volume (mL)	12.0	13.8	7.8	8.0	7.8	8.4	11.6	14.0
pH	9.1	8.6	9.0	8.0	9.0	8.0	9.5	8.2
High Pressure, High Temperature Filtrate (mL)	–	62.8	–	52.4	–	46.0	–	64.0

Table 7. Formulation of polyurethane resins.

Components	Parts by Weight (g) Adhesvies	Coatings
Part I		
Lignin polyesterether polyol	70	80
Ethyl acetate		60
Toluene		40
DMF, benzene, or ethyl acetate	150	
Part II		
TDI		53 (index)*
MDI, HDI or TDI	100(index)*	
Ethyl acetate		80
Toluene		30
DMF, benzene, or ethyl acetate	150	
Part III		
T-9 catalyst (Union Carbide)	0.24	0.5

*index number indicates the amount of equivalent weight of isocyanate needed to be used in the system, based on the 100 equivalent weight of polyol. HDI = hexamethylenediisocyanate, MDI = 4,4'-diphenylmethane-diisocyanante, TDI = 2,4-diisocyanatotolulene.

Future

Further studies of lignin grafting are needed to define the functional groups active in derivatization reactions, find means to minimize termination and transfer reactions in reaction mixtures containing lignin, identify optimum copolymerization reactions, and, most importantly, develop methods to analyze and characterize these complex products.

Table 8

Trivial Names for Compounds and Copolymers

Name	Number	Trivial Name
poly(1-phenylethylene)	I	polystyrene
1-phenylethene	II	styrene
poly(1-(1-oxo-2-oxy-n-propyl)-1-methylethylene)	III	poly(methylmethacrylate)
2-oxy-3-oxo-4-methylpent-4-ene	IV	methylmethacrylate
2-oxy-3-oxo-4-pent-4-ene	V	methylacrylate
2,6-dimethyl-4-oxy-3,5-dioxohept-1,6-diene	VI	mathacrylic anhydride
3-methyl-2-oxobut-1-oxyyl	VII	methacrylic group
poly(1-amidoethylene)	VIII	polyacrylamide
poly(1-(2-oxoprop-1-oxyyl)ethylene	IX	polyvinyl acetate
2-oxo-3-oxypent-4-ene	X	vinyl acetate
poly(1-cyanoethylene)	XI	polyacrylonitrile
1-oxa-2-methylcyclopropane	XII	propylene oxide
poly(1-oxy-2-methylprop-1,3-ylene)	XIV	polypropylene oxide

References

1. Henry I. Bolker, "Natural and Synthetic Polymers, An Introduction," p. 580, Marcel Dekker, New Yorrk, (1974), ISBN 0-8247-1060-6.
2. Eero Sjostrom, "Wood Chemistry, Fundamentals and Applications," p.69, Academic Press, (1981), ISBN 0-12-647480-X.
3. K. V. Sarkanen, C.H. Ludwig, "Lignins; Occurrence, Formation, Structure, and Reactions", p. 1, J. Wiley, (1971), ISBN 0-471-75422-6.
4. T. Kent Kirk, T. Higuchi, H. Chang, Lignin Biodegradation: Microbiology, Chemistry, and Potential Applications, Vol. 1, p. 5, CRC Press, (1980), ISBN 0-8493-5459-5.
5. A Bjorkman, Svensk Papperstidn., 59, 477 (1956).
6. J. C. Pew, Tappi, 40, 553 (1957).
7. F. F. Nord, W. J. Schubert, Holz Forschung, 5, 1, (1951).
8. F. F. Nord, W. J. Schubert, Tappi, 40, 285, (1957).
9. G. de Stevens, F. F. Nord, Fortschr. Chem. Forsch., 3, 70 (1954).
10. G. de Stevens, F. F. Nord, J. Am. Chem. Soc., 73, 4622, (1951).
11. S. F. Kudzin, F. F. Nord, J. Am. Chem. Soc., 73, 690, 4619, (1951).
12. F. F. Nord, G. de Stevens, Naturwissenschaften, 39,479, (1952).
13. J. C. Pew, J. Am. Chem. Soc., 74, 2850, (1952).
14. E. Hagglund, Cellulosechemic, 4, 84, (1923).
15. A Sakakibara, N. Nakayama, J. Japan. Wood Res. Soc. 8, 153, (1962).
16. David N. S. Hon, Ed., "Graft Copolymerization of Lignocellulosic Fibers, Acs. Symposium Series #187, Am. Chem. Soc., (1982) ISSN 0097-1656; 187.
17. Chem. and Eng. News, 62 (#39), 19-20, (1984).
18. T. Koshijima, E. Muraki, J. Japan. Wood Res. Soc., 10, 110, 116, (1964).
19. Robert W. Lenz, "Organic Chemistry of Synthetic High Polymers", pp. 161-172, 718, Interscience, (1967), ISBN 470-52630-0.
20. J. Zoldners, A. Cinite J. Surna, R. Rasina, Khim. Drev. 9, 39-52, (1971), CA 76: 87363m and CA76: 87364n.
21. J. Zoldners, J. Surna, I. Vandana, Khim. Drev., (#12), 125-9, (1972), CA79: 20471 p.
22. J. Zoldners, J. Surna, M. Indane, Khim. Drev. 15, 153-8, 1971, CA81: 171573r.
23. J. Zoldners, J. Surna, L. Deme, Khim. Drev. (#4), 11-21, (1975).
24. J. Zoldners, Kh.D. Krivisha, J. Surna, J. Tirzina, Khim. Drev., (#5), 109-115, (1975), CA84:6714s.
25. J. Zoldners, J. Surna, J. Tirzina, Khim. Drev., (#5), 116-21, (1975), CA83:195480a.
26. T. N. Kleinert, Tappi, 50, 120, (1967).
27. A. Kobayashi, R. B. Phillips, W. Brown, V. T. Stannett, Tappi, 54 (#2), 215-221, (1971).
28. Tetsuo Koshijima, E. Muraki, J. Poly. Sci., Part A1, 6, (#6), 1431-1440, (1968).
29. T. Koshijima, Nihon Mokuzai Gakkai, 12 (#3), 144-150, (1966).
30. Tetsuo Koshijima, Einosuke Muraki, Nihon Mokuzai Gakkai, 12, (#3), 139-144, (1966).
31. T. Koshijima, E. Muraki, Zairy O., 16, #169, 834-838, (1967).
32. R. B. Phillips, W. Brown, V. T. Stannett, Jo. Appl. Poly. Sci., 15, 2929-2940, (1971).
33. M. P. Godsay, G. A. Harpell, K. E. Russell, J. Poly. Sci., 57, 641, (1962).
34. R. B. Phillips, W. Brown, V. T. Stannett, J. Appl. Poly. Sci., 16, 1-14, (1972).
35. R. B. Phillips, W. Brown, V. Stannett, J. Appl. Poly. Sci., 17, 443-451, (1973).
36. J. Marton, T. Marton, Tappi, 47, 471 (1964).

37. S. Katuscak, M. Mahdalik, A. Hrivik, V. Minarik, J. Appl. Poly.
    Sci., 17, (#6), 1919-1928, (1973).
38. J. Zoldners, J. Surna, J. Tirzina, Khim. Drev. (#5), 116-21, (1975)
    CA83:195480a.
39. J. Zoldners, J. Tirzina, J. Surna, Khim Drev. (#6), 98-102, (1975),
    CA84:6153j.
40. Adolphe Chapiro, "Radiation Chemistry of Polymeric Systems," pp.
    173, 183, 196, Interscience, New York, (19620.
41. M. S. Matheson, E. E. Auer, E. B. Bevilacqua, E. J. Hart, J. Amer.
    Chem. Soc., 73, 835, (1951).
42. Sharda Dasgupta, Canadian Spectros. 12 (#1), 16-19, 25, (1967).
43. A. A. Berlin, S. B. Chernyavaskaya, Khim. Drev, (#1), 70-73,
    (1977), CA86:191505b.
44. A. A. Berlin, S. B. Chernyavskaya, Kolloidn. Zh. 42, (#4), 731-735,
    (1980) CA93:134053y.
45. Tetsuo Koshijima, Einosuke Muraki, Nihon Mokuzai Gakkaishi, 10,
    (#3), 110-115, (1964).
46. Tetsuo Koshijima, Einosuke Muraki, Nihon Mokuzai Gakkaishi, 10,
    (#3), 116-119, (1964).
47. Cr. Simionescu, A. Cernatescu-Asandei, A. Stoleru, Cellulose.
    Chem. Tech., 9, #4, 363-380, (1975).
48. Hitoshi Kubota, Yoshitaka Ogiwara, J. Appl. Poly. Sci., 13,
    1569-1575, (1969).
49. Henry P. Naveau, Cellulose Chem. Tech., 9, 71-77, (1975).
50. A. Kobayashi, R. B. Phillips, W. Brown, V. T. Stannett, Tappi,
    54, (#2), 215-221, (1971).
51. J. J. Meister, D. R. Patil, L. R. Field, J. C. Nicholson, J.
    Poly. Sci., Poly. Chem. Ed., 22, 1963-1980, (1984).
52. J. J. Meister, D. R. Patil, "Solvent Effects and initiation
    mechanisms for Graft Polymerization on Pine Lignin",
    Accepted by Macromolecules, 2/1/85. Publication expected
    8/85.
53. J. J. Meister, D. R. Patil, H. Channell, J. Appl. Poly. Sci.,
    29, 3457-3477, (1984).
54. John J. Mesiter, Damodar R. Patil, Harvey Channell, "Synthesis
    of Graft Copolymers from Lignin and 2-propenamide and
    Applications of the products to Drilling Muds," Accepted by
    Industrial and Engineering Chemistry, Prod. Res. Dev.
    Publication expected 8/85.
55. J. C. Nicholson, J. J. Mesiter, D. R. Patil, L. R. Field, Anal.
    Chem., 56, 2447-2451, (1984).
56. John J. Meister, Damadar R. Patil, Larry R. Field, John C.
    Nicholson, "Methods for Measurement of Molecular Composition
    as a Function of Molecular Size For Random, Block, and Graft
    Copolymers", In preparation.
57. George Odian, Principles of Polymerization, 2nd Ed., p. 230-231,
    Wiley, New York, (1981). ISBN 0-471-05146-2.
58. Leo C. F. Wu, Wolfgang G. Glasser, J. Appl. Poly. Sci., 29, 1111-
    1123, (1984).
59. Oscar H. H. Hsu, Wolfgang Glasser, Wood Science, 9 (#2), 97-103,
    (1976).
60. W. G. Glasser, V. P. Sara, W. H. Newman, J. Adhesion, 14 (#3/4),
    233-255, (1982).
61. Joseph Haggen, Chem. Eng. News, 63 (#18), p. 33-34, (May 6, 1985).

NATURAL TANNINS FOR COLD-SETTING WOOD ADHESIVES

Antonio Pizzi

National Timber Research Institute
CSIR
Pretoria, South Africa

INTRODUCTION

Flavonoid tannins, from the bark and wood of trees such as *Acacia mearnsii* and *Pinus radiata*, being phenolic in nature, undergo the same well-known reaction of phenols with formaldehyde either base- or acid-catalysed, weakly basic base-catalysed reactions being predominantly used in industrial applications. Increasingly alkaline conditions lead to progressive activation of the phenol as nucleophile, especially above pH 8 where phenate ions are formed.

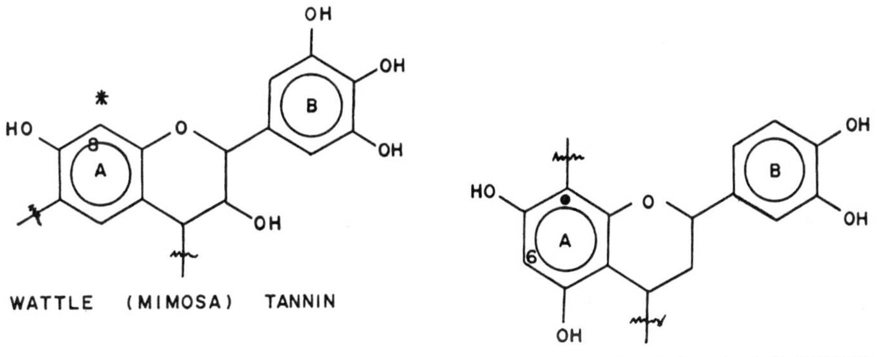

WATTLE (MIMOSA) TANNIN

4 - 6 FLAVONOID

PINE TANNIN, 4-8 FLAVONOID

* REACTIVE POSITION TOWARDS FORMALDEHYDE

Figure 1

The nucleophilic centers on the A-rings of any flavonoid unit tend to be more reactive than those found on the B-rings. This is due to the vicinal hydroxyl substituents which merely cause general activativation in the B-ring without any localized effects as on those found in the A-ring. Formaldehyde reacts with tannins to produce polymerization through the formation of methylene bridge linkages between reactive positions of the flavonoid molecules, mainly the A-rings. In condensed tannins the A-rings of the constituent flavonoid units retain only one highly reactive nucleophilic center, the remainder accommodating the interflavonoid bonds. Resorcinolic A-rings (wattle) show reactivity toward formaldehyde comparable, though slightly lower, to that of resorcinol[1]. Phloroglucinolic A-rings (pine) behave instead as phloroglucinol[2]. Pyrogallol or catechol B-rings are by comparison unreactive, and may only be activated by anion formation at relatively high pH[3]. Hence, the B-rings do not participate in the reaction except at high pH values (pH≥10) where the reactivity toward formaldehyde of the A-rings is so high that the tannin-formaldehyde adhesives prepared have unacceptably short pot-lives[1].

In general tannin adhesives practice, only the A-rings are used to cross-link the network. However, because of their size and shape, the tannin molecules become immobile at a low level of condensation with formaldehyde, so that the available reactive sites are too far apart for further methylene bridge formation[1]. The result is incomplete polymerization that leads to the weakness and brittleness which are characteristic of many tannin-formaldehyde adhesives. Bridging agents with longer molecules[1,4,5] like phenolic and aminoplastic resins have been used to solve this problem by helping to bridge distances too large for interflavonoids methylene bridges. In the case of cold-setting adhesives for timber lamination and for fingerjointing resorcinol must be added to the polyflavonoids or to the cross-linking polymers to obtain at least ambient-temperature-setting capabilities.

TANNIN-BASED COLD-SETTING ADHESIVES FOR TIMBER LAMINATING

A series of different Novolak-like materials has been prepared by copolymerization of UF resins or PF resols with resorcinol and/or resorcinolic A-rings of polyflavonoids such as condensed tannins[4,6,7]. The copolymers formed have been used as cold-setting exterior-grade wood adhesives complying to the relevant international specifications[7]. The formulations already used industrially as well as a few promising "novelties" are schematically shown below.

Resorcinolic tannins

System 1. Grafting of resorcinol on a tannin-formaldehyde resol or on a tannin/UF copolymer having free hydroxybenzyl groups. The final mixture of products of this system is an adhesive that can be set and cured at ambient temperature by addition of paraformaldehyde.

System 2. Simultaneous synthesis of resorcinol/formaldehyde and flavonoid/formaldehyde condensates[4]. Simultaneous synthesis of resorcinol/UF and flavonoid/UF copolymers. The final mixture of reaction products is an adhesive that can be set and cured at ambient temperature by addition of paraformaldehyde. The percentage of tannins on total solid resin is, in this and the previous case, of about 65 %. Resorcinol of 18 % on liquid resin at ± 53 % solids.

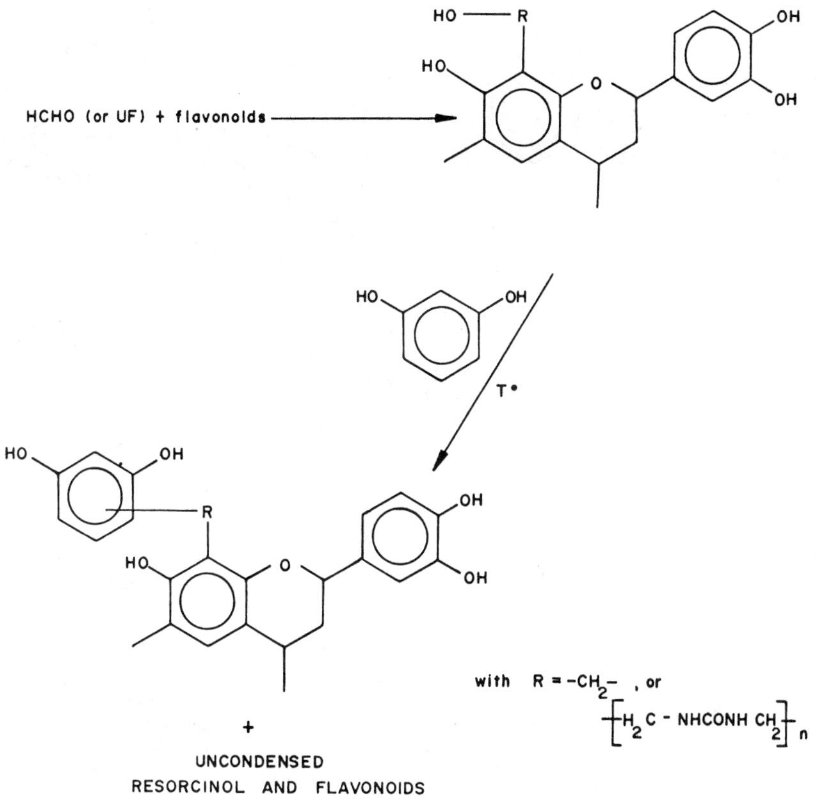

Figure 2

System 3. Synthesis of stable resorcinol-formaldehyde (RF), phenol-resorcinol-formaldehyde (PRF), or resorcinol-urea-formaldehyde (RUF) resins and subsequent addition to flavonoid polymers[4,7,8]. RF, resorcinol-terminated PF, and resorcinol-terminated UF resins are prepared and subsequently added to flavonoid polymers. The mixture of synthetic resins and tannins is used as an adhesive that can be set and cured at ambient temperature by addition of paraformaldehyde.

These three systems are the most commonly used cold-setting tannin-based resins. System 2 is the one most commonly used for its ease and safety of preparation. However, three other unusual systems also promise easy industrial application with several cost advantages (derived by the decrease in the amount or resorcinol tannins (such as mimosa and quebracho). The other is the only one suitable for the more reactive and difficult to handle phloroglucinolic tannins (such as pine tannin).

System 4. Acid hydrolysis of flavonoid etherocyclic ring and simultaneous grafting of resorcinol without formaldehyde[6]. The resorcinol must be present at the moment of the formation of the carbocation as otherwise the carbocation would attack the next best nucleophile, the flavonoid A-ring, with the formation of phlobaphenes, insoluble polymers unusable for adhesives.

325

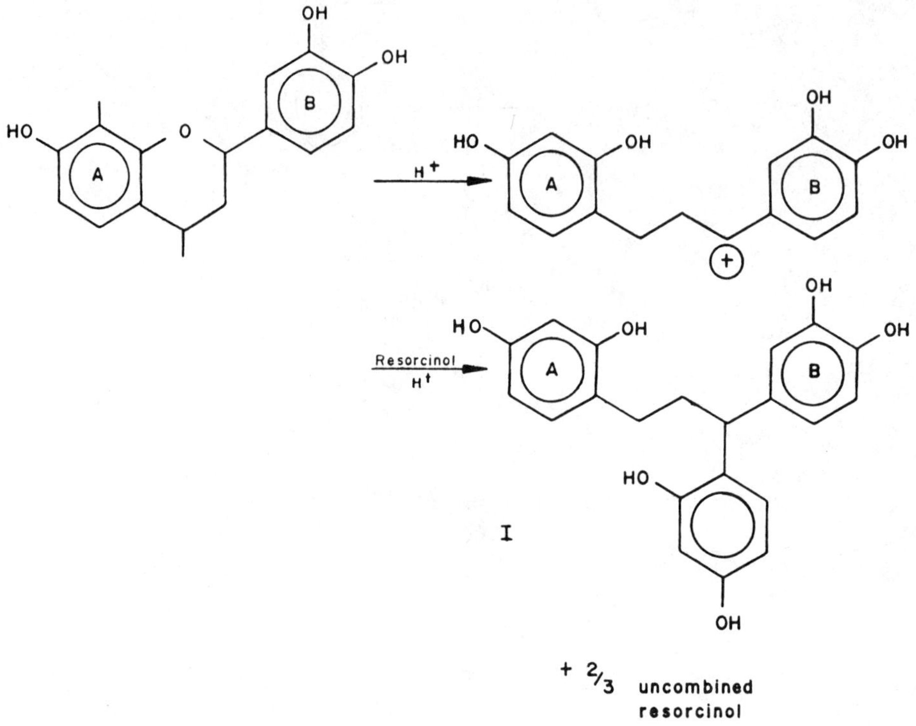

+ 2/3 uncombined resorcinol

Figure 3

Only about one-third of the resorcinol necessary to produce a good tannin-based cold-setting adhesive can be grafted in this manner. To improve the adhesive's cohesive strength, the reaction of mixture I with formaldehyde is used as in the following reaction.

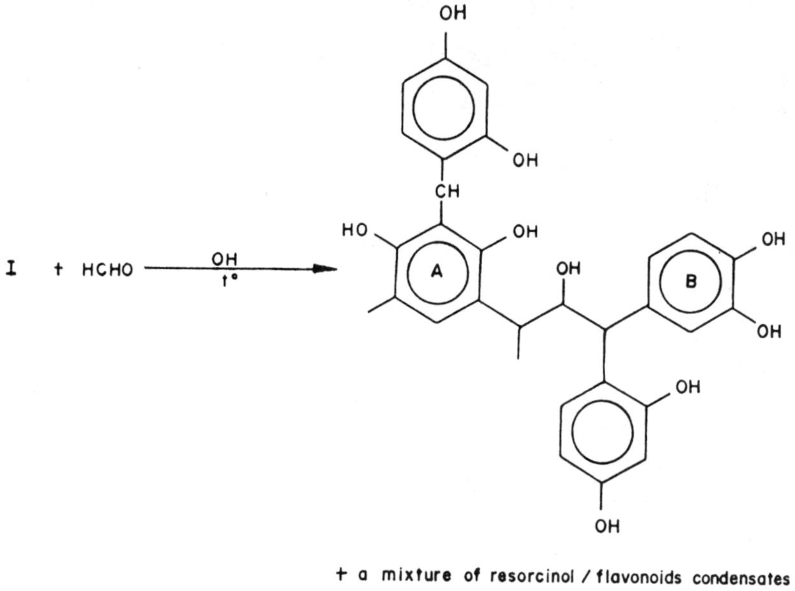

+ a mixture of resorcinol / flavonoids condensates

Figure 4

The final mixture can be hardened at ambient temperature by addition
of paraformaldehyde.

System 5[9,10]. This is identical to System 2, but half of the resorcinol chemical needed is generated directly from the flavonoid molecule
itself by direct previous sulphitation of the tannin extract. This reduces
the amount of external resorcinol added to only 9 % on liquid resin with a
considerable saving in cost.

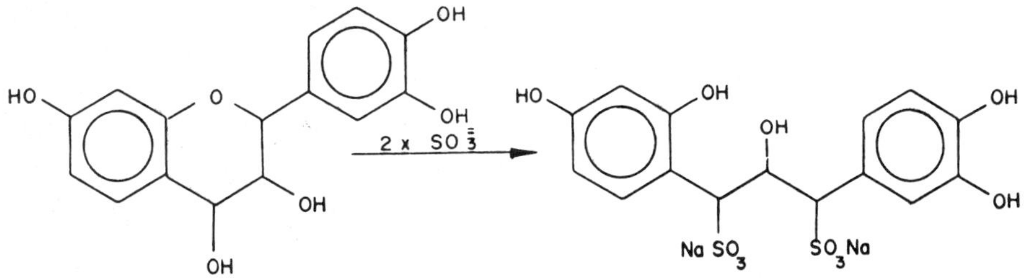

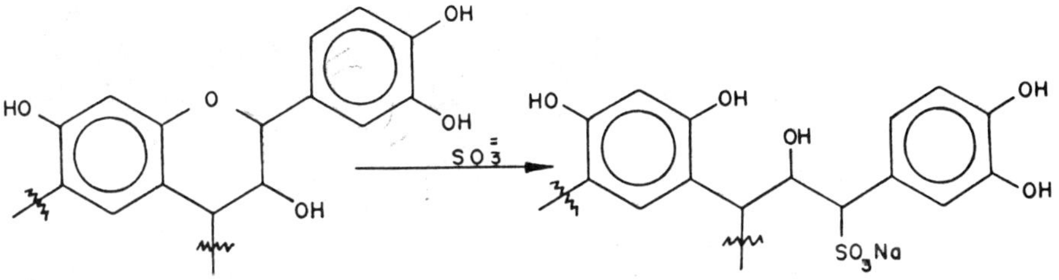

Figure 5

Phloroglucinolic tannins

System 6[11]. The etherocyclic ring is again opened by peracetic
acid-induced hydrolysis as in System 4 and the resorcinol grafted onto the
carbocation. The product obtained, similar in structure to I, is added,
in place of resorcinol, to a normal PF resin to form a product I-terminated
RPF resin. The amount of resorcinol is only of $\pm$ 7 % on liquid resin.
The preparation of this resin is however more complex than the previous one[11].

System 2 has been in extensive industrial use in a few countries in
the southern emisphere since 1973.

REFERENCES

1. A. Pizzi and H.O. Scharfetter, *J. Appl. Polym. Sci.*, **22**, 1745 (1978).
2. H.O. Scharfetter, A. Pizzi and D. du T. Rossouw, IUFRO Conference
   on Wood Gluing, Merida, Venezuela, Oct. 1977.
3. D.G. Roux, D. Ferreira, H.K.L. Hundt and E. Malan, *Appl. Polym. Symp.*
   No. 28, 335 (1975).
4. A. Pizzi and D.G. Roux, *J. Appl. Polym. Sci.*, **22**, 145 (1978).
5. A. Pizzi, *Adhesives Age*, **20**, 12, 27 (1977).
6. A. Pizzi and D.G. Roux, *J. Appl. Polym. Sci.*, **22**, 2717 (1978).

7. A. Pizzi, J. Appl. Polym. Sci., 23, 2999 (1979).
8. F.W. Herrick and R.J. Conca, For. Prod. J., 10, 7, 361 (1960).
9. K. Freudenberg and J.M.A. DeLama, Ann. Chem., 612, 78 (1958).
10. A. Pizzi and G.M.E. Daling, J. Appl. Polym. Sci., 25, 1039 (1980).